# Natural Resources and the Environment

*Natural Resources and the Environment: Economics, Law, Politics, and Institutions* provides a new approach to the study of environmental and natural resource economics.

It augments current contributions from the fields of public choice, law, and economics, and the burgeoning field of what used to be called the "New Institutional Economics," to describe, explain, and interpret how these new developments have been applied to better understand the economics of natural resources and the environment. This textbook takes a multi-disciplinary approach, which is essential for understanding complex environmental problems, and examines the issue from not only an economic perspective, but also taking into account law, politics, and institutions. In doing so, it provides students with a realistic understanding of how environmental policy is created and presents a comprehensive examination of real-world environmental policy. The book provides a comprehensive coverage of key issues, including renewable energy, climate change, agriculture, water resources, land conservation, and fisheries, with each chapter accompanied by learning resources, such as recommended further reading, discussion questions, and exercises.

This textbook is essential reading for students and scholars seeking to build an interdisciplinary understanding of natural resources and the environment.

**Mark Kanazawa** is Wadsworth A. Williams Professor of Economics, Carleton College, USA. He is author of *Research Methods for Environmental Studies: A Social Science Approach* (Routledge 2018) and *Golden Rules: The Origins of California Water Law in the Gold Rush* (University of Chicago Press 2015).

"Climate Change, Biodiversity, Population Growth, Endangered Species, Depleted Fisheries, Overharvest of Tropical Forests. These are the environmental and resource issues of the day, but confronting them typically is presented in a far too simplistic and often moralistic manner. The problems are far more complex. Mark Kanazawa provides an important and critically-needed overview of key issues that must be understood, Trade Offs. All policies, even good ones, involve trade offs and something must be given up to accomplish environmental objectives. Incentives. Unless there are proper incentives to reduce pollution or protect biodiversity, it will not happen. How do we get such incentives? Institutions, such as property rights – private, communal, or the opposite, open access, determine the long-term incentives people face in resource use. Public Choice. Government policies are designed and implemented by lobby groups, politicians, bureaucrats. When do they act to provide environmental public goods in a least cost manner? There is really no other volume that addresses these points and ties them together effectively for a more positive understanding of the choices faced and the responses provided. Mark Kanazawa has provided a valuable service in *Natural Resources and the Environment: Economics, Law, Politics, and Institutions*."

**– Gary Libecap, Distinguished Professor Emeritus, Corporate Environmental Management, UC Santa Barbara**

"Kanazawa masterfully uses the concepts from the social sciences and law to tackle the problems that we face in managing natural resources and the environment. Both students and scholars will benefit from the analysis demonstrating that as a society we face economic and political trade-offs which is why solving our environmental problems is so difficult."

**– Lee J. Alston, Professor of Economics, Indiana University, Research Associate, National Bureau of Economic Research**

"I've been searching for a textbook that studies environmental issues with an approach that blends neoclassical economics with institutional analysis and public choice. This is it! Dr. Kanazawa's book is an excellent and accessible guide for understanding the pressing management and conservation challenges of today."

**– Dominic Parker, Associate Professor, Department of Agricultural and Applied Economics, University of Wisconsin-Madison**

"This text develops a synthetic conceptual perspective from microeconomics, institutional analysis, and public choice, and then uses this paradigm to help readers understand topical environmental and natural resource policy issues. The text is well organized; the presentational style, lucid and engaging. The pedagogical craftsmanship, unique perspective, and topical range make this work an outstanding contribution."

**– Kerry Krutilla, Professor of Public Policy, O'Neill School of Public and Environmental Affairs**

"The title *Natural Resources and the Environment: Economics, Law, Politics, and Institutions* accurately encompasses the ambitious goal of this book, namely integrating sound economic analysis with a wide range of approaches and concerns. Kanazawa pulls it off in a superb fashion, explaining well the analytical tools of economics and then using those tools to shed important insights on a variety of environmental issues. The result is a textbook that, because of its range of coverage and analytical rigor, can be used for a wide variety of classes."

**– P. J. Hill, Professor of Economics Emeritus, Wheaton College, Wheaton, Illinois**

# Natural Resources and the Environment

Economics, Law, Politics, and Institutions

**Mark Kanazawa**

LONDON AND NEW YORK

First published 2021
by Routledge
2 Park Square, Milton Park, Abingdon, Oxon OX14 4RN

and by Routledge
52 Vanderbilt Avenue, New York, NY 10017

*Routledge is an imprint of the Taylor & Francis Group, an informa business*

*British Library Cataloguing-in-Publication Data*
A catalogue record for this book is available from the British Library

*Library of Congress Cataloging-in-Publication Data*
Names: Kanazawa, Mark, author.
Title: Natural resources and the environment: economics, law, politics, and institutions/Mark Kanazawa.
Description: Abingdon, Oxon; New York, NY: Routledge, 2021. |
Includes bibliographical references and index.
Identifiers: LCCN 2020048636 (print) | LCCN 2020048637 (ebook)
Subjects: LCSH: Environmental economics. | Natural resources.
Classification: LCC HD75.6 .K357 2021 (print) | LCC HD75.6 (ebook) |
DDC 333.7–dc23
LC record available at https://lccn.loc.gov/2020048636
LC ebook record available at https://lccn.loc.gov/2020048637

ISBN: 978-0-367-07760-0 (hbk)
ISBN: 978-0-367-07761-7 (pbk)
ISBN: 978-0-429-02265-4 (ebk)

Typeset in Bembo
by Deanta Global Publishing Services, Chennai, India

This book is dedicated to my wife Kathy, who has patiently endured many evenings and weekends when I was holed up in my study. I really could not have completed it without her support, love, and good humor.

# Contents

# Figures

# Boxes

# 1 Introduction

## An overview of environmental and natural resource issues

Ever since 1997, the United Nations has been publishing the *Global Environmental Outlook*, a series of reports, published every few years, on the environmental state of the world. These reports, produced as a collaborative effort by experts around the globe, represent a truly international effort and provide a periodic integrated assessment of the environmental challenges facing the world. In 2019, the United Nations published the sixth and most recent *Global Environmental Outlook* (GEO6), and its bottom line assessment gives us pause:

> Providing a decent life and well-being for nearly 10 billion people by 2050, without further compromising the ecological limits of our plane and its benefits, is one of the most serious challenges and responsibilities humanity has ever faced.
>
> [GEO6, Chapter 1, p. 4]

The report paints a sobering picture of:

- Likely continued population growth for the rest of the 21$^{st}$ century, especially in currently less-developed areas of the world;
- People increasingly crowding into growing urban areas;
- Air and water pollution contributing to risks to ecosystems and human health;
- Ongoing species extinction and continued loss of biodiversity;
- Growing demand for increasingly scarce energy and water resources;
- Continued increases in emissions of *greenhouse gases* (GHGs); and
- Steady ongoing climate change, along with predictions of warming oceans, shrinking Arctic ice cover, existential threats to island nations, changes in weather patterns, and greater incidence of extreme weather events such as flooding, droughts, heatwaves, hurricanes, and cyclones.

It is not all doom and gloom. In recent years, humankind has made much progress on various important fronts, including reducing global hunger and improving access to clean water, sanitation, and clean energy. However, it is unquestionable that much work remains to be done. The GEO6 report is a kind of clarion call to action that points to a number of tangible things that can, and in its opinion should, be done. These include: reducing greenhouse gas emissions, improving energy efficiency, increasing reliance on renewable energy sources, technological innovation, improving urban governance, adopting water-saving irrigation technologies and smart technologies in industrial production, reducing food waste, and making various lifestyle changes.

These recommendations of the GEO6 report are not new, and many of them are not even unusual. And they all speak to real environmental problems and prescribe policies based on informed scientific and social research. Indeed, to many people, the recommendations may seem like no-brainers. Yet many of them have sparked spirited public debate among people of good will – that is, not dogmatic climate deniers but careful academic scholars whose only real interest is to get to the bottom of things. For these folks, the key questions are: Do these policies work? Are they worth it?

To take only one example but an extremely prominent one, let us consider reductions in greenhouse gas (GHG) emissions. If you are taking a course in environmental economics, I am sure you have heard of climate change, and you probably know that emissions of GHGs from human-made sources are causing it. This is, of course, the rationale for reducing GHGs. Seems logical: If climate change is a serious problem and it is caused by GHG emissions, reducing emissions should solve the problem. However, as with every other big environmental problem, the situation is nowhere near this simple.

For one thing, for scientific reasons, GHGs, once emitted, can remain in the atmosphere for long periods of time. So even if we were to immediately halt all production of GHGs, this would not quickly restore GHG concentrations in the atmosphere to earlier, lower levels. In the meantime, the effects of climate change would continue. Second, there are different types of GHGs, produced through different processes and contributing in various degrees to ongoing climate change. Third, reducing GHG emissions may mean cutting back on activities that produce them, including the use of so-called fossil fuels to produce electricity, heat our homes, and drive our cars. And, finally, there may be other ways to address climate change, such as technological solutions like geoengineering, and adaptation to its impacts [Nordhaus (2013), pp. 149–56]. None of this is to argue that we should *not* reduce GHG emissions but, rather, merely to emphasize that there are many things to think about.

## The economic approach to the environment: Of bar stools and woodpeckers

In contrast to other disciplines, economics places issues of scarcity front and center in study of the environment. First, and foremost, this means that there are inevitable tradeoffs that must be acknowledged and evaluated. So when we burn fossil fuels, we contribute to global climate change. And so the tradeoff there *may* be: reducing GHG emissions means less driving or less power production.

Let me take another example, in order to elaborate on the economic approach. When we harvest timber for wood products, we may sacrifice habitat for endangered species. The tradeoffs here involve the annoying fact that we often cannot have everything we want, and sometimes more of something we want – habitat for endangered populations of red-cockaded woodpeckers – means less of something else like, say, nice wooden bar stools. Much of economics is about carefully identifying exactly what we will have to give up if we take a particular action. Like harvesting timber from a forest area that serves as critical woodpecker habitat.

Sometimes identifying the tradeoffs is tricky because the world is a complex place. If we want more wooden bar stools, do we necessarily endanger red-cockaded woodpeckers? Maybe we can harvest forest stands elsewhere so that we do not have to destroy woodpecker habitat. Or maybe, instead of clear-cutting the forests, we can selectively cut, which reduces the impact on resident woodpecker populations. Or maybe we can invent a way to produce classy-looking bar stools using other materials so that we do not have to cut down forest stands at all.

In each of these examples, tradeoffs have not disappeared: they have merely been transformed. If there are forest stands elsewhere, do they house some other endangered species that we are concerned about protecting? Do they produce different kinds of wood that are not as suited to making nice bar stools? If you selectively cut rather than clear-cut, does this reduce the amount of wood you can harvest, thus reducing the number of bar stools you can produce? If you use other materials to produce bar stools, are they more expensive, making bar stools more costly and reducing their appeal to consumers? Does using other materials require less environmentally friendly methods, potentially damaging the environment? See, no matter what you do, at some point you are confronted with a tradeoff and therefore, having to give up something you value.

But as challenging as it is, merely identifying the tradeoffs is obviously not enough. Once you have done that, the question then becomes: What do we do? Give up the bar stools, or give up the owl habitat? Or perhaps preserve the habitat while tolerating more expensive bar stools made of ugly synthetic materials? This brings me to a second key feature of the economic approach: the weighing of costs and benefits. You see, it is not merely a question of identifying *what* will be lost if we do something, like preserve woodpecker habitat. It is also a question of *how much we value* what would be lost. How much would we miss those wooden bar stools? And how does that compare to how much we value the alternative; in this case, protecting habitat so woodpeckers can thrive? The entire rest of this textbook is about understanding and evaluating inevitable tradeoffs that need to be confronted head-on when we study the environment.

Lest the economic approach be misunderstood, I need to emphasize a really important feature of the economic approach: the importance of *incentives*. Even if you have never studied economics before, you have probably heard the terms "costs" and "benefits." To many people, these terms conjure up visions of an accounting exercise where you assign numbers to these things, and then evaluating tradeoffs entails seeing which one is greater. But that is not really the essence of economics. Rather, economics is about recognizing human motivations and the incentives that people have to behave in various ways. This is what makes economics a social science.

As a simple example to illustrate what I mean: when the price of a good goes up, people generally respond by buying less of that good. Researchers who prescribe policies that drive up the price of a good need to take that effect into account. Otherwise, their policies may have unintended consequences, like producing a good that nobody wants to buy because it is too expensive. I will bet that you can see how that human response might affect our calculations regarding how much people benefit from that policy.

As another example of more direct environmental relevance, let us return to the red-cockaded woodpecker. Here is a true story. Red-cockaded woodpeckers inhabit forest stands in certain states along the eastern seaboard and prefer to nest in more mature trees. With their numbers diminishing in the 1980s, the U.S. Fish and Wildlife Service required private landowners to set aside dozens of acres of valuable timberland. The idea, of course, was to maintain habitat so the woodpeckers could thrive. But what was the result? Landowners went out and preemptively cut down whole stands of old-growth southern pine if they thought there was a danger they would become habitat for nesting woodpeckers. The very policy designed to protect red-cockaded woodpeckers actually ended up destroying woodpecker habitat [Lueck and Michael (2003)]! We will encounter this study again when we have a more detailed discussion of biodiversity and endangered species in chapter seventeen.

## Economics and institutions

When we think about the importance of incentives in influencing human behavior, it will help to think broadly. To see what I mean, let us go back to the two examples we just saw when we were thinking about incentives and their effect on human behavior. If you have taken any economics before, you probably recognized the first example as referring to what happens in a market. When the market price of a good changes, this affects people's incentives to purchase the good, given all of the other things they could spend their money on instead. Given their prevalence in most economies of the world, markets have occupied the attention of many economists, who have explored their incentive effects in great detail.

However, the second example had nothing to do with markets, which just goes to show that the issue of incentives is much broader than merely market interactions. Indeed, we human beings experience incentives in many different ways – everything from speed limit signs to job interviews to Fish and Wildlife Service policies to grades in your classes to the pat on the back you receive from your dad when you earn an A to ... the list goes on and on. So thinking about it in this way, many things can affect our behavior, and this simple fact turns out to have a number of important implications when we start to think carefully about the environment.

As you go about your business in the real world, one important set of factors that probably affects your behavior are your legal rights. You vote, exercise your right to free speech, put your car up for sale on CarSoup, do not burglarize other people's houses, and probably do not drive too many miles over the speed limit. You do all these things because there are laws that specify what you can and cannot do. In the woodpecker example, private landowners, faced with the regulations of the Fish and Wildlife Service program, decided to cut down trees while legally they still could. One way to think about all this is that there are certain legal rules that we all have to live by, as a condition for living in our society. Economists call these rules *institutions* and say that the set of institutions we live by provides the *incentive structure* for our behavior. As the Nobel prize-winning economist Douglass North has put it:

> Institutions are the rules of the game in a society or, more formally, are the humanly devised constraints that shape human interaction. In consequence they structure incentives in human exchange, whether political, social, or economic.
>
> [North (1990), p. 3]

In recent years, economists have taken this idea – that institutions are crucial in understanding economic behavior – and explored its implications in a number of different fields of economics, including industrial organization, labor economics, international trade, economic development, and, most relevant for our purposes, environmental economics. We now recognize, for example, that institutional rules derive from a number of different sources, including laws passed by legislatures, court rulings, state and national constitutions, international treaties, local resource governance arrangements, and even customs, traditions, and norms of behavior.

We also recognize that institutional rules are crucial determinants of transactions in goods and services. These include rules that determine:

- Who can buy and sell what, and under what terms and conditions;
- The extent to which individuals can profit from exchanges;
- How well individuals can maintain their rights to things they buy and sell; and
- Whether they are free to resell a good once they have ownership of it.

All of this has important implications for the operation of markets, as you might expect. Each of these factors listed here alter the incentives of people who participate in markets, and may affect their market behavior. For example, potential buyers (demanders) of a good may be discouraged by legal restrictions on their right to resell the good, or difficulties in enforcing their right to it, once they purchase it. And this might affect the price that potential sellers (suppliers) will be able to charge for the good.

However, plenty of other types of transactions are possible. You, for example, engage in transactions with other members of your study group: intramural ice hockey club or Model United Nations team. In each of these cases, nothing is being bought or sold. Rather, you are working together toward a common goal, which requires collaboration and give-and-take. Like how to split up tasks for that group paper assignment you have just been handed. Or who gets to take the crucial penalty kick when the game is on the line.

Similarly, workers in automobile factories engage in transactions with each other on the factory floor and with their supervisors and company managers. Members of your local city council engage in transactions with each other and with local community members in order to decide what local policies to enact. Villagers in low-income countries engage in transactions with other villagers in order to build and operate a local irrigation system.

The common thread in these examples is human interactions as people work to satisfy their own needs and preferences. And, in each case, there are rules that structure people's interactions with each other, whether they be study group interactions, collective bargaining agreements, or procedures for local elections and holding public hearings. So, for example, various rules and procedures govern how community public hearings and local elections are supposed to be run. And various norms of good behavior – drilled into you by your mom, I am sure – keep you from acting like a jerk and demanding that you be the one to take the crucial penalty kick or you will just take the ball and go home.

With all of this in mind, economists have devised a general way to think about all of these transactional situations and others: the notion of so-called *transaction costs*. Transaction costs refer to factors that make it difficult for transactions to occur. These factors could be legal restrictions on the ability to buy and sell goods. They could be difficulties in enforcing rights to use goods. They could even be characteristics of the goods themselves that raise doubts in people's minds that the people with whom they are transacting can be trusted. And there are various other sources of transaction costs as well. As we shall see, transaction costs play a key role in the institutional approach to the environment. They will appear everywhere in the discussion in the rest of this book.

## Institutions and public choice

In most countries of the world, legislatures comprise one key source of institutions. The U.S. Congress, German Bundestag, Japanese National Diet, and the various Parliaments of the United Kingdom, Malaysia, Sri Lanka, and Zimbabwe are all duly elected national legislative bodies that regularly convene to enact laws for their countries. These laws govern many aspects of public and private life including, of course, environmental policy. In the United States, for example, over the years Congress has passed laws like the *Clean Air Act*, the *Clean Water Act*, and the *Endangered Species Act* (as well as many others!), in order to set national policy to address a wide variety of environmental issues.

Economists have had a lot to say about the effects of these laws: how well they perform and how successful they are at achieving their programmatic goals. The verdict is somewhat mixed: some have performed much better than others, and some that seem to perform well have been subsequently repealed or emasculated. We will be talking about all of this at great length, but for now trust me that it is true. There are then two important related questions: What explains why environmental laws are such a mixed bag? And how and why do we get these laws in the first place?

When you are talking about elected representative bodies, of course one would think that politics would be an important factor. After all, politicians respond to the demands of their constituents, the people who voted them into office, and these people have all sorts of opinions about protecting the environment. At the same time, some politicians are probably genuinely interested in protecting the environment. However, there are large numbers of them: for example, in the U.S. House of Representatives, there are 435 representatives. In the U.S. Senate, there are 100 senators. And you will not be surprised to hear that they are not always on the same page. So how do 535 duly elected politicians from blue states and red states with all sorts of policy positions on different issues come together to produce environmental policy?

Traditional economic analysis of the environment has been ill-equipped to answer this question, which reflects the fact that for many years economics and political science have operated largely on parallel tracks, separated by disciplinary boundaries and very different methodological approaches. The *public choice* approach, however, represents an attempt to bring politics and economics together. And, in doing so, it provides much insight into the important question: Why and how do we actually get the environmental policies that we do?

The late economist Gordon Tullock defined public choice as "the use of economic tools to deal with traditional problems of political science" [Tullock (2008)]. Under the public choice approach, legislators are motivated by the same things in the political sphere as private individuals (like suppliers and demanders) are in the private sphere. They respond to incentives, engage in political transactions such as vote trading, and their transactions are subject to potential transaction costs. Furthermore, under the public choice approach, private individuals are not confined to operating in private settings like markets. They can try to secure political favors through activities like lobbying their representatives and joining special interest groups. Economists call these activities *rent-seeking* behavior. In engaging in these activities, they may try to shape the very institutional rules under which they operate.

In bringing all these political factors together into a coherent framework, public choice provides an important complement to traditional economic analysis of the environment. Traditional economics has done a good job of helping us understand people's private behavior: in their households, in their jobs, in markets, and in cities and local communities. However, it has been less helpful in understanding their public behavior: in the legislative arena, in public agencies, and in the rent-seeking behavior of individuals and interest groups. By expanding our analytical focus, public choice provides a broader, more complete picture of both the causes and consequences of environmental policy.

## A roadmap for the book

### I. Concepts and analytical framework

Chapters two through six provide the basic concepts and tools for your analytical toolbox. Chapters two and three begin with a summary discussion of economic concepts that are central to the traditional economic approach and that will used throughout the remainder of the book. Chapter two is basically a short course in general theoretical principles that are especially relevant for economic study of the environment. Chapter three then turns to some practical economic methods that are commonly used in applied studies of the environment: especially cost–benefit analysis, sustainability, and environmental valuation.

For courses that have a prerequisite of microeconomic principles, chapter two can be skimmed, as all of the concepts in this chapter are treated in standard microeconomic principles courses. For courses lacking such a prerequisite, chapter two can be done as a self-guided course, complete with structured problem-solving exercises designed to supplement the conceptual material.

Chapters four through six then turn to a discussion of the institutional approach. Chapter four provides a detailed introduction to the institutional approach, centering on three key foundational concepts: *institutions, transaction costs,* and *property rights.* Chapter five then focuses on a particular class of natural resources that have been the subject of much study: so-called *common-pool resources.* Finally, chapter six presents the *public choice* approach, introducing a number of concepts related to legislative lawmaking and the importance of interest groups. By the end of chapter six, students should be well versed in the institutional approach and thus feel well equipped to tackle the later topics of their choice.

### II. Topics and topic areas

The remainder of the book consists of a series of chapter-level treatments of various environmental topics, in which students can apply the concepts they have just learned. The topics are arranged in two broad topic areas, each consisting of five related chapters. The first topic area is *Energy, climate change, and pollution,* which consists of five related chapters on fossil fuel energy; climate change; air and water pollution; energy conservation and energy efficiency; and renewable energy. The second topic area is *Land, ocean, and water resources,* which consists of five related chapters on agriculture; water resources; fisheries; forests and deforestation; and wildlife, endangered species, and biodiversity.

A semester-long survey course can be organized as three consecutive modules consisting of the opening theory module (chapters two to six), followed by each of these topic area modules in turn. Shorter courses can omit one of the topic modules, depending on desired focus and course objectives, without compromising the coherence of the material covered (see Box 1.1). I should mention that discussion of common–pool resources essentially only occurs in the second short course. So the first short course can omit reading chapter five without loss of continuity.

---

**Box 1.1: Two short courses**

---

| *Short course I: Energy, climate change, and pollution* | *Short course II: Land, ocean, and water resources* |
|---|---|
| • Chapters 1–4, 6<br>• Population<br>• Fossil fuel energy<br>• Climate change<br>• Air and water pollution<br>• Energy conservation and energy efficiency<br>• Renewable energy | • Chapters 1–6<br>• Population<br>• Agriculture<br>• Water resources<br>• Fisheries<br>• Forests and deforestation<br>• Wildlife, endangered species, and biodiversity |

---

Finally, note that the issue of population occupies a chapter of its own, at the beginning of the topic chapters. This is to set the stage for everything to come, as population growth is arguably the ultimate driving force behind every single other environmental problem. In this capacity, it is included as a recommended topic in both short courses.

## Exercises/discussion questions

1.1.   A local municipality is considering setting aside land for a ten-acre municipal park in a currently unoccupied area next to an expanding residential neighborhood. Identify the tradeoffs that the municipality should consider in choosing whether or not to set aside land for this park.

1.2.   How do each of the following affect your incentives to expend effort in your spring semester environmental economics course?

  •   You have just been awarded a summer internship at an environmental consulting firm for the coming summer.

  •   You have just been selected as captain of your varsity softball team.

  •   You have just broken up with your significant other, who happens to be a fellow economics major.

## References

Lueck, Dean and Jeffrey A. Michael. "Preemptive habitat destruction under the Endangered Species Act," *Journal of Law and Economics* 46(April 2003): 27–60.

Nordhaus, William. *The Climate Casino*. New Haven: Yale University Press, 2013.

North, Douglass C. *Institutions, institutional change, and economic performance*. Cambridge: Cambridge University Press, 1990.

Tullock, Gordon. "Public choice," *The New Palgrave Dictionary of Economics Online*. 2008.

# 2 A mini-course in basic economic theory

Economists bring a certain way of thinking about environmental problems. In a world of scarcity, we constantly find ourselves asking: What are the tradeoffs, and how do we weigh those tradeoffs? You may have heard the old saying: "There is no such thing as a free lunch." In other words, to get something, we need to give up something else. If we are talking about the environment, this means that environmental problems may have solutions, but these solutions come at a cost. Mitigating greenhouse gases in order to combat climate change, for example, may have negative impacts on the world economy if it means reducing production of things we value or having to use other, non-fossil fuel, energy sources that are more costly than fossil fuels.

In this chapter, we develop the basic economist's approach. Here we will define and describe a number of important concepts and analytical tools that we will be using repeatedly in later chapters. The idea is to begin to arm you with concepts, tools, and techniques that you can use to gain insights into a variety of environmental problems. In order to help you master the material, you will be provided a series of exercises that are designed to test and reinforce your understanding of the concepts.

## Scarcity, tradeoffs, and opportunity costs

The foundational concept in economics is scarcity, which is captured nicely by that old line from the Rolling Stones: "You can't always get what you want." You have just so much money to spend, so you have to forego going on vacation to Florida during spring break. Or you have to settle for buying that clunker instead of the Beamer that you really want. From the viewpoint of an entire economy, there are just so many workers, and just so much iron ore, natural gas, arable land, and so forth to produce enough goods and services to satisfy all of its citizens. To economists, scarcity is a fact of life. And to non-economists, economics is the dismal science, largely because economists doggedly point out that we cannot have everything, which annoys everyone else no end.

Given that scarcity is a fact of life, at some point there will be tradeoffs. Taking that Florida vacation means that you have to forego – something. Maybe you cut back on eating out, or going out to shows or concerts, or by taking a vacation somewhere else, like Liverpool or Hoboken. There are many things you could spend your money on, and you only have so much to spend. Maybe you could lobby your parents to increase your allowance, but I guarantee you that, if they do, they will find other ways to take it out of your hide. Or if you borrow money from the bank, eventually you will have to pay it back. At some point, there is a day of reckoning.

The general idea of scarcity implying tradeoffs is captured in the notion of a *production possibility frontier* (PPF), which you may have encountered elsewhere, because it is a regular staple of many economics courses. If you have not, a PPF shows the greatest quantity of outputs that can be produced from a fixed amount of resources. Figure 2.1 shows one possible PPF for an entire economy with a certain amount of resources (land, water, energy, etc.) and able to produce two things for its citizens: computers and soy burgers. Every point on this graph represents some combination of computers and soy burgers produced by this economy. For example, at point A the economy is producing 50,000 computers and 1,000,000 soy burgers.

The position of this PPF is determined both by the total amount of resources available and the technologies available to turn those resources into soy burgers and computers. Here, the most computers this economy can produce is $C^{Max}$, where the PPF touches the vertical axis. To produce this many computers, all of the economy's resources must be devoted to computer production, meaning that there are none left to produce soy burgers. Similarly, the most soy burgers that can be produced is $S^{Max}$, where all resources are being devoted to soy burger production. Intermediate points on the PPF represent combinations of computers and soy burgers that are possible if all of a society's resources are devoted to producing some of each, using the best technologies currently available, loosely speaking.

It may be apparent that there is a big difference between points on the PPF and points in the interior, like point A. If you are at point A, it is possible to have more of both computers and soy burgers, for example by moving along the arrow from A in a northeasterly direction. However, once you are on the PPF – say point B – there is no way to increase the number of soy burgers you produce without decreasing the number of computers, and vice versa. You just do not have enough resources to produce more of both. So there are two aspects of scarcity. If you are not doing as much with your resources as you could be, it may be possible to stave off tradeoffs for a while. But ultimately there comes a point when you have no choice: having more of one good will require having less of another. At this point, you have a real either/or choice to make.

When there is a real choice to make, we say that there is a cost to taking any action, no matter what we choose to do. If we are on the PPF and choose to produce more computers, we incur a cost in terms of foregone soy burgers. If we choose to produce more

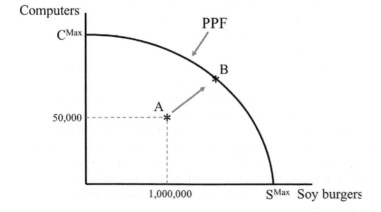

*Figure 2.1* Production possibility frontier

soy burgers, there is a cost in foregone computers. In this simple example, these costs are known as *opportunity costs*. That is, the opportunity cost of producing extra computers is the value to us of the soy burgers we have to forego.

Now let me expand this example slightly in order to get at a general definition of opportunity costs. Suppose there is a third option: to produce bowling balls. You can probably imagine the same principles applying. If the economy uses all of its resources to produce computers, it will not be able to produce any soy burgers or bowling balls. Similarly if it chooses to use all its resources to produce soy burgers, and similarly for bowling balls. In between these extremes, there will be various combinations of the three goods that can be produced given available resources. In short, you can imagine a three-dimensional PPF.

I am introducing this extra good in order to get at a more general definition of opportunity costs; namely, the value of the *best* foregone alternative. Here is the difference from the previous example. *Now* if you decide to produce more computers, what you forego is more complicated. It is not simply foregoing so many soy burgers. It is not simply foregoing so many bowling balls. It could be either. It could be a combination of the two. The idea of opportunity costs is to imagine NOT having used those resources to produce those extra computers. If you did not, what combination of soy burgers and bowling balls would you produce? The value of that combination is what we would call the opportunity costs; that is, the value of the best action you could take instead of producing those computers.

In the economic approach to the environment, opportunity costs are crucial. Since scarcity is a fact of life, making informed choices depends on acknowledging the tradeoffs, identifying the alternatives, and having a reasonable basis for comparing them. Opportunity cost provides just such a basis for evaluating choices we make, by providing a sense for exactly what we would be giving up, and how much we would lose.

**Question 2.1**: Suppose there are two countries, England and France, and suppose that there are two goods that can be produced: wine and cheese. Suppose each country has 32 units of labor in total that can be used to produce wine and/or cheese.

Consider the following table of production of two countries. *The numbers represent the number of labor units* that it takes to produce either a bottle of wine or a pound of cheese.

|  | One bottle of wine | One pound of cheese |
|---|---|---|
| England | 4 | 4 |
| France | 2 | 8 |

In the table below, calculate the opportunity costs (OC) of producing the goods in either country.

|  | OC of a bottle of wine | OC of a pound of cheese |
|---|---|---|
| England |  |  |
| France |  |  |

## Thinking in terms of costs and benefits

Once we acknowledge scarcity and the annoying fact that there will be tradeoffs, the question then becomes: What do we do now? The economist's short answer is: The best we can. The more complete answer is: Let's make our choice based on the costs and benefits associated with that choice. In the simple computers–soy burgers example, producing more computers provides a benefit to the economy. At the same time, it has an opportunity cost in terms of foregone soy burgers. For the economy as a whole, the question is whether the benefits of those extra computers outweigh the cost of the foregone soy burgers.

In order to make all of this operational, let us introduce another important weapon in the economist's analytical arsenal, which I will call *marginal thinking*. Here is the basic idea. We are often interested in knowing what will happen if we do things a bit differently. A Fortune 500 company might wonder whether to offer a job to a qualified applicant. A convenience store might wonder whether to stay open an hour later on Saturday nights. A student with a term paper due tomorrow might wonder how it will affect her grade if she works on it for an extra hour this evening. Your overweight uncle might wonder if he should take a third helping of mashed potatoes at Thanksgiving dinner.

In the larger scheme of things, in none of these examples is there an earthshaking decision to be made. However, when you add them all up, they can make a big difference. Marginal thinking is a general way of addressing these sorts of questions, which turns out to have many useful applications when studying the environment.

Under marginal thinking, you first consider everything *you are already doing*: how many workers you already employ, how many hours you have already spent on that term paper, how many helpings of mashed potatoes you have already eaten. And then you ask the question: What happens if I do a little bit more – if I hire that extra worker, work that extra hour, help myself to that third serving? Am I better off or worse off?

Let us illustrate what is going on here for that student trying to figure out whether to spend that extra hour polishing that term paper. For the sake of this exercise, I have created a little graph, which is shown in Figure 2.2. On this graph, I have drawn two functions – the *marginal benefit* (MB) function and the *marginal cost* (MC) function. Notice I am calling them *functions*, as they both vary with (*are a function of*) the number of hours spent working on the paper. These functions represent the *additional* benefit and *additional* cost of working an extra hour on the term paper. That is, given how many hours she has already worked, how much extra benefit or cost does she derive from working one more hour?

In Figure 2.2, for example, given that she has already worked two hours, the additional benefit she derives from working the third hour is nine, while the additional cost is three.

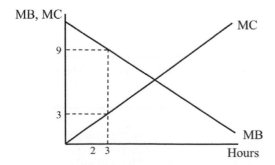

*Figure 2.2* Thinking marginally

For now, let me skirt the issue of exactly what these numbers represent. If you allow me to do that, however, it certainly looks like the third hour is worth doing, as she derives more additional benefit than she incurs in additional cost.

You may be wondering why I have drawn the MB and MC in this particular way, with MC increasing, and MB decreasing, in hours worked. Here is the basic idea. Generally speaking (and I think this applies to a lot of things), *the more we do something, the harder it becomes to do more, while at the same time, the less we get out of the extra effort.*

Think of the opportunity cost of writing a term paper. If you are juggling your schedule around, those first hours spent on it probably come at the expense of things that you probably do not care much about. Maybe the first hour comes at the expense of that fifth extracurricular activity meeting that you did not really want to go to anyway. So that first hour has a really low opportunity cost. However, as you spend more and more time on that paper, it becomes harder to find insipid, boring things to skip. Extra time starts cutting into things you really value, like going out with your friends to that rock concert featuring your favorite band. Or time spent on that organic chemistry assignment, which you really need to do well on because your parents are making you go to med school. Or dining out with your significant other on the three-and-a-half-month anniversary of your first date. For these things, which you saved until last to sacrifice, the opportunity cost is probably pretty high.

At the same time, extra hours spent writing that term paper are probably yielding less and less additional benefit. The first hour, where you outline the paper and give it a coherent structure, probably does a lot to boost your grade. Additional hours spent locating extra important sources, or writing rough drafts on key sections, also help but probably not as much as the first hour spent writing the original outline. However, by the time you get to the 25$^{th}$ hour, you are probably doing things like moving a few words around, or trying to decide whether the correct word is "affect" or "effect." These final polishing efforts probably do not add much to your grade (though as a teacher, I cannot in good conscience encourage you to skip them!).

This framework of marginal benefits and marginal costs is extremely important, and it has many applications in environmental economics, as we shall see in later chapters of this book. However, sometimes it will be important to have information on the *total* costs and *total* benefits of actions. For example, what did all your herculean efforts on that paper get you in terms of an overall letter grade? Did all those helpings of mashed potatoes make your uncle happy and good company for the rest of the evening? In addition, it sometimes helps to try to answer the question: Was it worth it? Did the total benefits exceed the total costs?

Fortunately, another nice feature of the marginal framework is that it permits easy calculation of the total costs and total benefits if you know the marginal costs and marginal benefits. To see this, consider Figure 2.3, where I have taken Figure 2.2 and filled in a few more numbers. Looking at the MB schedule, we see that she derives 11 units of marginal benefit from the first hour of study and ten from the second. When you add these to the nine units she derives from the third hour, we see that she is deriving 30 units of total benefit from working a total of three hours.

Notice that this total benefit is equal to the summed area of the tall, skinny rectangles. Why? Because the area of each of these shaded-in columns represents the marginal benefit of each additional hour. That is, the area of each column equals the height of the column (MB) times the width, which equals one. Adding the areas of all the columns together – the entire shaded area – gives the total benefit from working three hours. More generally,

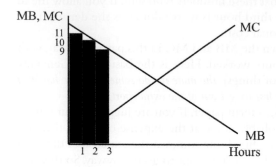

*Figure 2.3* Total cost as area under MC

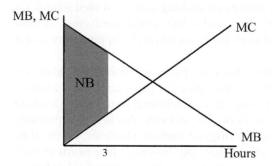

*Figure 2.4* Net benefits of working 3 hours

except for the tiny, tiny, tiny little triangles above the columns, the area under the marginal benefit schedule equals total benefit. In actual applications, it is generally safe to ignore the triangles, giving us the following general relationship between MB and TB: *TB equals the area under MB.* This argument illustrates a general principle: if you want to calculate total benefits, simply add all of the individual increments to benefits (which is what we mean by marginal benefits).

By the same argument, you can probably also see that in working these three hours, she is incurring six units of total costs (Right?). So similarly, we can calculate total costs by simply summing up all of the individual increments to costs. Equivalently, TC will simply equal the area under MC.

With this interpretation, it is now easily possible to use the MB and MC schedules to calculate the net benefits (NB) associated with any outcome. With NB defined as simply the difference between TB and TC, or (TB − TC), we could calculate NB as the difference in the areas under MB and MC. Figure 2.4 illustrates this, where the shaded area represents the net benefits associated with working a total of three hours on that paper.

**Question 2.2**: To see this relationship between marginal and total, let us assign some numbers to the costs and benefits of working on a term paper. Suppose you can algebraically express the marginal benefits as: MB = 8 − H, where MB equals marginal benefits and H represents the number of hours. What are the total benefits of working on the paper for four hours?

## Three important notions

One reason all of this will be extremely useful is that economists often apply certain criteria based on total costs and total benefits when evaluating environmental outcomes or policy. These criteria are the notions of *efficient*, *cost-beneficial*, and *cost-efficient* outcomes.

### I. Efficient outcomes

Efficient outcomes are defined as ones that maximize net benefits. This case is shown in Figure 2.5, where the number of hours is H★ and the associated net benefits are the entire shaded area, or NB$^{\text{Max}}$. Why are NB maximized at H★? Well first, it is easy to see that the NB at H★ have to exceed the NB at any smaller number of hours, like H$_0$. This is because by producing H$_0$, we would be foregoing all of the NB represented by the shaded triangle between H$_0$ and H★. And it is easy to see that the NB at H★ have to exceed the NB of any larger number of hours, like H$_1$. This is because every hour that you work past H★ adds more to costs than it does to benefits, which means that NB would have to go down.

### II. Cost-beneficial outcomes

Cost-beneficial outcomes are ones where total benefits exceed total costs (alternatively, that net benefits are positive). This concept is often applied using a criterion called the *benefit-cost ratio* (BC), which is defined simply as:

$$BC = \text{Benefits/Costs} \tag{2.1}$$

So, if the benefit–cost ratio of an outcome is greater than one, it is said to be cost-beneficial. For example, H★ is certainly cost-beneficial. However, it should be obvious that this is true of many other numbers of hours worked. For example, it was true of three hours in

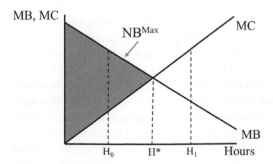

*Figure 2.5* Efficiency

Figure 2.4. This argument tells us that the cost-benefit criterion is less stringent than the efficiency criterion: many outcomes could be cost-beneficial but not necessarily efficient.

Despite the fact that it is less stringent than the efficiency criterion, the cost-benefit criterion is commonly used because it is generally much easier to implement. It does not require us to compare the TB and TC for all outcomes, which would require knowledge of the entire MC and MB schedules. All it requires is knowledge of TC and TB for the one outcome being evaluated. The downside is that by choosing an outcome based on a positive benefit–cost ratio, we may be ignoring other more efficient ones waiting in the wings. Still, though it may not allow us to *achieve* efficiency, it certainly moves us *toward* efficiency by increasing total net benefits.

**Question 2.3**: Consider that term paper again. Suppose you can algebraically express the marginal benefits as: MB = 8 − H, where MB equals marginal benefits and H represents the number of hours. Now suppose in addition that the marginal costs (MC) can be expressed as: MC = 2 + H.

(a)  What is the efficient amount of time to work on this term paper?
(b)  What are the net benefits of working this amount of time?
(c)  Verify that doing this is cost-beneficial.

### III. Cost-efficient outcomes

The third criterion is cost-efficiency. Cost-efficient outcomes are ones that are achieved at the lowest cost. The idea here is that you may have a specific target outcome and multiple ways to achieve it. Applying the cost-efficiency criterion simply means you choose the lowest-cost way. For example, you may be a homeowner wanting to decrease your energy consumption by ten percent. There may be several ways to do this, including adding more insulating material to your exterior walls, installing new energy-efficient windows, or investing in a new passive solar heating system. A cost-efficient solution would be to choose that option, or combination of options, that requires you to spend the least amount of money.

Notice that the cost-efficiency criterion says nothing about benefits. By choosing a target outcome up front, you are implicitly assuming that it is worth doing. However, without knowing the benefits of achieving your target, there is no guarantee that it is. You can probably see that if those benefits are sufficiently low, a cost-efficient action could be both inefficient and non-cost-beneficial. Nevertheless, the cost-efficiency criterion is often applied because it does not require information on benefits, which is really hard to get for many environmental goods (see chapter three). What are, for example, the benefits of mitigating so many tons of greenhouse gases, or creating a new wildlife preserve to protect endangered species? So you can think of cost-efficiency as a less-than-ideal tool, which is nevertheless useful because it provides actionable guidance on what strategy to take.

The efficiency, cost-benefit, and cost-efficiency criteria are widely used in environmental economics. However, it is important to keep in mind that they are not the only criteria used by environmental economists. Another important criterion is *equity*: how the net benefits are divided up among the rich and poor, for example. Another criterion we will encounter later in this textbook is the notion of *sustainability*, which will be defined in due course. We do not discuss it here because it is not easily described using the simple benefit and cost framework that we are presently concerned with.

## Resource allocation mechanisms: Markets

Many goods and services are exchanged in markets, which bring together buyers and sellers. Because markets are used for so many different goods and services – including many natural resources – it will be useful to explore how, and how well, they function. When considering markets, the key concepts are demand and supply, so let us begin by considering these one at a time.

### I. Demand

The market demand for a good can be thought of in two complementary ways. First, it shows the amount of the good that is demanded in the market at all possible prices. Figure 2.6, for example, shows the demand for Jimmy John submarine sandwiches from consumers. According to this figure, there are 1,200 Jimmy John subs demanded when the price of a sub is $6.50. This demand curve slopes downwards, which makes intuitive sense: we would expect that when a sub becomes less expensive, people will demand more subs. So here, when the price of subs has fallen to $4 per sub, the quantity demanded has increased to 1,500 subs.

The second interpretation of a demand curve is that it reflects how much consumers are willing to pay for the last unit sold of a good, when different quantities are being sold. So, in Figure 2.6, the willingness to pay for the 1200th sub is $6.50. Since the demand curve slopes downward, this implies that when more subs are being sold, consumers are willing to pay less for the last sub sold. Notice how this is being phrased: as the willingness-to-pay *on the margin*, for the last sub.

This second interpretation is important because to economists, the amount people are willing to pay for a good should be an accurate indicator of how much benefit they derive from it. All of this is to argue that demand curves can be interpreted as representing the marginal benefit consumers derive from consuming a good. This is why we are also labeling demand as MB.

### II Supply

Now let us turn to the other side of the market: Jimmy John submarine shops, the producers of Jimmy John subs. Figure 2.6 also shows a market supply curve for a bunch of these shops. As with demand, there are two interpretations of the supply curve. The first is that it shows, for all possible prices, how many units of a good are supplied at each price. So

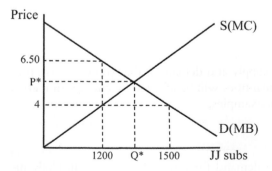

*Figure 2.6* Supply, demand, and market equilibrium

according to Figure 2.6, when the price of a sub is $4, sub shops are willing to produce 1,200 subs. In contrast to demand, we generally draw supply as sloping upwards because at higher prices you would expect shops to be willing to supply more subs.

The second interpretation of the supply curve is that it shows how much it costs to produce the last unit of a good, for all possible quantities. So Figure 2.6 shows that when 1,200 subs are being supplied to the market, the last sub costs $4 to produce. This interpretation makes sense, right? If it cost more than $4 to produce the 1,200[th] sub and the market price was $4, the shops would be unlikely to want to supply the 1,200[th] sub. I know I wouldn't want to. On the other hand, if it cost less than $4 to produce the 1,200[th] sub, the shops would probably not need to be enticed to supply that last sub by a price of $4, but should be willing to take less.

Again, notice the marginal language: "the last sub." So: given that 1,199 are already being produced, the question is how much it costs to produce one more. So with supply, we seem to be referring to a notion of marginal cost. This is why in Figure 2.6, we are also labeling the supply curve as MC.

### III. Market equilibrium

So: demand and supply show how much of a good is demanded and supplied at different prices. In order to determine how much of the good will actually be sold in the market, and at what price, we need to bring demand and supply together. So what is happening in Figure 2.6 is that consumers and sub shops are coming together and trying to strike deals with each other, and prices are adjusting to make sure that all supplied goods are being purchased and all demands are satisfied. In Figure 2.6, this occurs where supply is equal to demand, where the price of a sub is P★ and Q★ subs are sold. If the price is below P★, there are too many subs being demanded by consumers relative to the amount shops are willing to supply, and consumers will bid prices upward. If the price is above P★, there are too many subs being supplied relative to what consumers want, and shops anxious to unload their subs will bid prices downward. P★ and Q★ are called the *equilibrium price* and *equilibrium quantity*. The term "equilibrium" is used because, at these levels, there is no tendency for the price and quantity to change, as long as market conditions do not change.

**Question 2.4**:The annual market demand for bananas in New York City can be expressed algebraically as: P = 120 − 0.01Q, where P = price of bananas, in dollars per ton; and Q = quantity of bananas, in tons. Market supply, on the other hand, is: P = 20 + 0.01Q. Calculate equilibrium price and quantity, and total expenditures on bananas in equilibrium.

### IV. Comparative statics analysis

One nice feature of market analysis using supply and demand is that you can make predictions for how equilibrium prices and quantities will be affected by changes in market conditions. Let us see this with a couple of examples.

*A. Increases in consumer income*

If consumer income were to increase, the demand for Jimmy John subs will likely increase. Right? When people feel richer, they tend to buy more. Thinking of our two interpretations, we expect both that, at given prices, more subs will be demanded, and that the (marginal) willingness-to-pay for subs will increase at all quantities. The result, then, will be an upward/outward shift in demand, as shown in Figure 2.7. As a result of this shift, we would predict both equilibrium price and equilibrium quantity to increase, as they do in Figure 2.7 to P' and Q'.

## B. Increases in the cost of bread

Now, instead of assuming that consumer income increases, suppose that the price of bread increases. In this case, the cost of producing subs will likely increase. This means an upward/leftward shift in the supply curve, as shown in Figure 2.8. Here, we would expect the equilibrium price to increase as before, but now the equilibrium quantity should decrease. In Figure 2.8, the new equilibrium price and quantity are P' and Q'.

There are, of course, many factors that could cause either the demand and/or supply curve to shift up or down. This makes supply and demand analysis very useful, as we can use it to predict the effects of all sorts of factors that affect markets. As a general rule, when doing this kind of analysis it is helpful to use the following rules of thumb. Anything that affects the cost of production will tend to shift the supply curve. Anything that affects consumer willingness-to-pay will tend to shift the demand curve. Other things can matter

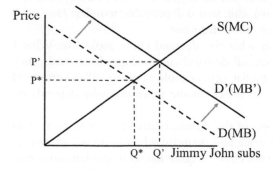

*Figure 2.7* Effect of an increase in consumer income

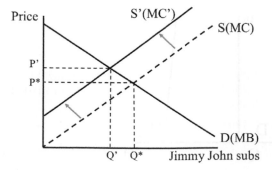

*Figure 2.8* Effect of an increase in the price of bread

as well – for example, population growth resulting in more consumers – but you will get a lot of mileage out of thinking in terms of these two rules of thumb.

**Question 2.5**: Consider the market for air travel. Suppose that a robust economy increases consumer incomes, while **simultaneously** jet fuel prices decrease. What would you predict will happen to the price of air flights?

(a)  Increase;
(b)  Decrease;
(c)  Stay the same;
(d)  Cannot tell.

## An extremely important concept: Elasticity

The concepts of demand and supply are extremely useful tools of analysis for making predictions regarding the effects of changes in market conditions. Now let us make them even more useful by introducing the notion of an elasticity.

In many situations, it will be useful to know not just *whether* prices and quantities will change given a change in market conditions, but *how much* they will change. For example, suppose you are trying to decide whether to buy a new computer. You know that the technology for making computers is constantly improving, which means that the price of a computer is likely to go down if you just wait (the same goes for a lot of things, like televisions, smart phones, and microwave ovens). But: you will probably wonder: *How much* is the price going to go down? I need a new computer now!

It turns out that the answer depends on what the demand curve looks like. What I mean by this is that even though (just about) all demand curves slope downwards, they could still look quite different. Some could be flat, others could be steep, still others could be extremely steep (see Figure 2.9). And the effect on the price crucially depends on which one it is.

As you can tell from Figure 2.9, the slope of the demand curve is very important in determining what happens to equilibrium price and quantity when there is a change in supply conditions. For example, if the demand curve is quite flat, as in the left-hand-side graph of Figure 2.9, the effect of a supply increase is to lower the price hardly at all, while

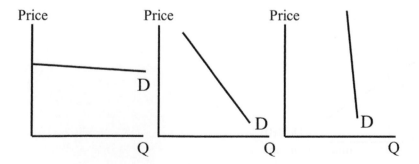

*Figure 2.9* Three possible demand curves

dramatically increasing the quantity sold. In this case, you would be ill-advised to wait, because computers are not going to be a much better deal next year. This is the case of so-called *price-elastic* demand, where small reductions in price result in large increases in the quantity demanded.

On the other hand, consider the right-hand-side graph in Figure 2.9 where the demand curve is quite steep. In this case, the exact same shift in the supply curve would cause the price to decrease dramatically, while the quantity sold would increase hardly at all. In this case, it makes much more sense to wait to purchase your new computer. This is the case of so-called *price-inelastic* demand, where large reductions in price result in relatively small increases in the quantity demanded.

The question of the elasticity of demand for various goods will appear over and over again in the topic chapters. To prepare for those later discussions, you might ask yourself the following question: What sorts of goods would tend to have price-elastic, vs. price-inelastic, demand? Here, there are two important related factors: how necessary is the good, and are there other goods that would serve just about as well? If the answers are *very* and *no*, demand will tend to be quite price-inelastic. For example, consider the good "food." We literally cannot live without it, and there are no good substitutes for it. So we would expect the demand for food to be quite price-inelastic (and economics studies have shown this to be the case). The same goes for "housing," "clothing," and, for people suffering from diabetes, insulin. All of these are things that people "have to" have, and no other things exist that would serve equally well instead.

On the other hand, we would probably expect the demand for "Visits to Disneyland" to be relatively price-elastic. Even if we love to go to Disneyland (and I personally do), most of us cannot really say that it is necessary. And there are plenty of other vacation destination spots where we could go and still have a really good time.

## Economic efficiency and equity

In addition to being able to make predictions regarding equilibrium prices and quantities, demand and supply analysis is also useful in permitting us to evaluate those market outcomes in terms of how well-off they make society as a whole. Before proceeding further, however, I need to explain an important distinction made by economists: the difference between a *positive* statement and a *normative* statement. In essence, the difference is that positive statements are about what causes outcomes to occur, and normative statements are about how we feel about those outcomes.

To understand the difference, consider this poor guy, who accidentally hit his finger with a hammer (Figure 2.10). There are actually two separate things going on here. One is that when you hit your finger with a hammer, you cause pain. This is a positive statement: A causes B. The second is that this guy is obviously not very happy. This is a normative statement: Pain is bad.

OK, now that we have made that important distinction, let us go back and consider the operation of markets again. The statement that prices will go down when consumer incomes decrease is a positive statement. In fact, all of the comparative statics that you can imagine doing with demand and supply analysis is about making positive statements about what affects what, and why, and how much. By contrast, an example of a normative statement would be: When prices go up, is society better-off or worse-off? In this case, we are taking an outcome and applying a criterion to evaluate that outcome. This is what we are about to do right now.

*Figure 2.10* Positive and normative. *Source*: Shutterstock

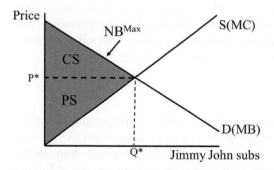

*Figure 2.11* Efficiency of market outcomes

**Question 2.6**: Which of the following are positive statements and which are normative statements?

(a) When the price of a good increases, the quantity demanded will tend to decrease.
(b) Consumers tend to benefit from competition among suppliers.
(c) When production costs increase, prices tend to rise.

Remember the notion of efficiency? It turns out that we can use the framework we have just developed to evaluate the efficiency of market outcomes. So not only can we *predict* outcomes using the framework, we can also *evaluate* whether they are good or bad outcomes. To see this, consider Figure 2.11, which brings together two results that we have already seen (and for now, ignore the mysterious looking "CS" and "PS"). The first result is that Q★ is the equilibrium quantity, the value of sub sales that you would observe in this market with freely adjusting prices. The second is that Q★ is efficient: it maximizes total net benefits to society (the entire shaded area NB^{Max}). In a nutshell, this is the argument that economists often make for favoring the use of markets.

In order to develop one more important concept, let us consider what happens to net benefits if a different quantity is bought and sold. In Figure 2.12, this quantity is Q', an amount less than the efficient market equilibrium quantity Q★. If only Q' is produced and sold, it looks like some net benefits will be foregone because we lose all of the net benefit represented by the shaded triangle between Q' to Q★ (Sound familiar?). These foregone net benefits relative to the efficient quantity is known as *social deadweight loss* (SDWL). It represents what we lose in net benefits if we produce anything but the efficient quantity.

**Question 2.7**: In the market for wheat, demand may be expressed as: $P = 140 - 0.1Q$, where P is the price of wheat, in dollars per ton, and Q is the quantity sold, in tons. Supply may be expressed as: $P = 20 + 0.2Q$.

(a) Calculate the total net benefits to society in market equilibrium.
(b) Now calculate the social deadweight loss if, for some reason, only 200 tons were allowed to be sold in this market.

All of this analysis tells us how well off society as a whole is, as measured by total net benefits. However, it does not tell us anything about how those net benefits are divided up among consumers and the owners of sub shops. In other words, so far it has not told us anything about how equitable these market outcomes are. In order to rectify this omission, we need to first talk about how we might measure net benefits to consumers and to sub shops.

To begin with, let us go back to Figure 2.11. Now ask yourself: *How well off are these consumers?* Adding up all the marginal benefits, consumers are enjoying benefits equal to the entire area under the demand curve, all the way out to Q★, the quantity of subs they are buying. But notice how much they are paying for all these subs. They are paying a price of P★ for each sub, and they are buying Q★ subs. This means they must be paying a total of (P★ X Q★), the area of the rectangle with sides of length P★ and Q★. All of this means that consumers are enjoying net benefits of the triangle above P★ and below the

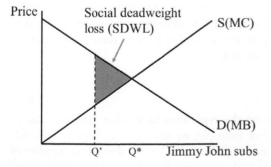

*Figure 2.12* Social deadweight loss

demand curve, the difference between their total benefits and the amount they are spending. This measure of net benefits to consumers is known as *consumer surplus*, and is labeled CS accordingly.

Similarly, in supplying Q* subs to the market, sub shops are incurring total costs equal to the area under the supply curve, which is interpreted as the marginal cost of production. On the other hand, they are earning total revenues of (P* X Q*), the price per sub times the total number of subs they sell. In other words, expenditures by consumers translate directly into revenues for sub shops. The net benefits of the sub shops are then the difference between their revenues and their total costs, or the lower triangle labeled PS. This measure of net benefits to the suppliers is known as *producer surplus*. So total net benefits equal the sum of consumer surplus and producer surplus. This figure thus shows a pretty equitable division of total net benefits between consumers and sub shops, because the areas of CS and PS look pretty comparable. I should add, however, that this question of the division of net benefits between consumers and producers in one single market is only a very small element of the entire issue of equity, which we will explore more fully in subsequent chapters.

**Question 2.8**: In the market for Jimmy John subs, demand may be expressed as: P = 15 − 0.01Q, where P is the price of a sub, and Q is the number of JJ subs sold. Supply may be expressed as: P = 5.

   (a)  Calculate the equilibrium price and quantity of subs sold.
   (b)  Calculate consumer surplus and producer surplus in equilibrium.

## Two important qualifications

The market model is an extremely useful and powerful model that is used widely to (a) make predictions regarding market outcomes and (b) evaluate those outcomes. However, you should recognize that that model is based on certain assumptions that may or may not be satisfied in the real world. One key assumption is that there are many consumers and suppliers on each side of the market, engaging in vigorous competition. Another is that all firms are selling an identical product. A third is that each firm supplies a tiny fraction of overall market supply, so that no one firm can influence the price of the product. Finally, it assumes that all suppliers are bearing all the costs of their production activities.

Though these assumptions are reasonable under many real-world conditions, they obviously do not come close to capturing the reality of many other situations. They are made for simplicity, to help us predict and evaluate market outcomes under conditions that come close to being described by these assumptions. Sometimes, however, we need to use different models to capture other, different situations. Let us now turn to two important alternative models: *market power*, and *externalities*.

### I. Market power

One situation that often arises when studying the environment is where there is one or a few large suppliers of a natural resource. For example, we shall see in chapter eight that the *Organization of Petroleum Exporting Countries* (OPEC) has been a dominant supplier of petroleum in world oil markets for decades. Other entities have been dominant in various other world resource markets, including markets for bauxite, zinc, copper, gold, and diamonds. Still others have dominated local and regional markets for timber, iron, and steel. The question is: How do we understand the production and pricing decisions in these cases?

In the simplest economic model of market dominance, there is one supplier producing for the entire market. This situation is captured in the *monopoly* model. Implicit in this model is the existence of so-called *barriers to entry* keeping other firms from entering the market. Examples of entry barriers include high startup costs for new firms, brand-name recognition for existing firms, or government-sanctioned limits on entry. These entry barriers permit the supplier to operate completely free from competition. In this case, the monopolist is free to service all of market demand. The questions are: What does all this mean for how much output it will produce and sell and at what price? And from the viewpoint of society, is it good or bad for a market to be serviced by a monopoly?

To see how a monopolist would behave under these conditions, we begin by assuming that he is interested in maximizing his profits, defined as the excess of his revenues over his costs:

$$\text{Profit} = \text{Total revenue} - \text{Total cost} \tag{2.2}$$

We also assume he has an upward-sloping marginal cost (MC) curve that resembles the market supply curve in our earlier supply and demand analysis. We can thus calculate his total costs for all possible output levels as the area under his MC curve, as we did before. This means that if we can calculate his total revenues at all output levels, we should be able to calculate his profits as well, which allows us to predict what output he will want to produce: that is, whichever quantity maximizes his profits.

To calculate his total revenue, it will be useful to define another marginal concept: *marginal revenue* (MR). Similar to other marginal concepts we have seen, we define MR as the addition to total revenue when you produce and sell another unit of output. In the earlier supply and demand analysis, it was easy to calculate MR: it was simply the price of the good. There, when the market price of a Jimmy John sub was $7, a sub shop gained an extra $7 in revenue when it sold one more. The key to understanding what happens under monopoly conditions is that *it is no longer true that price equals MR.*

The reason has to do with the fact that the monopolist has the market all to himself. Being in the enviable position of servicing the entire market, he experiences something that the small competitive firms did not: *as he sells more, the price he is able to charge goes down.* This occurs as he moves along the downward-sloping demand curve, as in Figure 2.13. Suppose he is originally selling the quantity $Q_0$ at the price $P_0$. If he sells one more unit, the price will fall to $P_1$. Notice what happens to total revenue: it was originally $(P_0 \times Q_0)$, but it is now $(Q_0 + 1)$, the new quantity sold, times the new lower price $P_1$. It is *not* $(Q_0 + 1)$ times the *original* $P_0$, which it would have been if the price had remained where it was, at $P_0$.

So what is the MR, the additional revenue to the monopolist of selling the extra unit? Yes, he is gaining extra revenue on the sale of the extra unit, represented in Figure 2.13 as the shaded area A, which happens to be equal to the price $P_1$. However, this extra revenue is offset by the *lost* revenue from having to sell all the units he had previously sold at the price $P_0$ at the new lower price $P_1$. This lost revenue is the shaded area B. This means that his marginal revenue MR is equal to $(A - B)$. So: we have just established that the marginal revenue in moving from $Q_0$ to $(Q_0 + 1)$, is definitely less than the price.

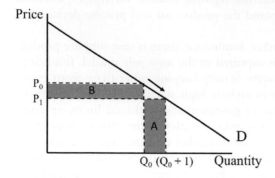

*Figure 2.13* Marginal revenue for a monopoly

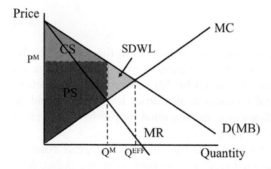

*Figure 2.14* Monopoly equilibrium

You can probably see that this will be true of all quantities that this monopolist might want to sell. In fact, associated with every demand curve faced by a monopolist is a lower MR curve, like in Figure 2.14. This MR curve shows the addition to revenue from sell-ing an extra unit, for all possible quantities sold. You do not need to know why, but in the special case where the demand curve is linear, the MR curve will also be linear, and it will be exactly twice as steep.[1]

To find the monopolist's profit-maximizing output, Figure 2.14 also shows the demand curve, interpreted as always as the marginal benefits to consumers. The question is: How much output will this monopolist produce? The answer is: If it wants to maximize prof-its, it will produce $Q^M$, the quantity of output where MC = MR. Why? Because for any other quantity, either profits are being left on the table or some units are costing more to produce than they are worth.

Consider, for example, quantities less than $Q^M$. For every one of these quantities, MR is greater than MC. This means that expanding output will add more to revenues than it will to costs and hence, profits will go up. So quantities less than $Q^M$ cannot be profit-maximizing. This is what I mean by profits being left on the table.

Now consider quantities greater than $Q^M$. For these quantities, MC is greater than MR. This means every one of those additional units is costing the firm more to produce than it is providing in extra revenues. So those quantities cannot be profit maximizing either: those last units are costing more to produce than they are worth.

So this profit–maximizing monopolist has incentive to produce exactly $Q^M$ units of the good. If he does, he will want to set the highest price that will enable him to sell this quantity. This price is $P^M$, the most that consumers are willing to pay for the good given that $Q^M$ is being sold. $P^M$ is called the *monopoly price*. You should be able to verify that the triangle labeled CS represents the consumer surplus at this monopoly price and quantity, while the area labeled PS represents producer surplus.

Now: it is important to see that the monopoly output $Q^M$ is not efficient; that is, it does not maximize net benefits to society. To see this, ask yourself the counterfactual: What quantity of output *would* be efficient? The answer is $Q^{EFF}$, where MC = D. At this quantity, the marginal cost of production equals the marginal benefit to consumers, recalling the interpretation of demand.

Another way to see that $Q^M$ is not efficient is to notice that at $Q^M$, MB to consumers exceeds MC of production to the monopolist. This means that starting at $Q^M$, producing extra units of output would add more to societal benefits than to societal costs, meaning that total net benefits would go up. All of those marginal increments of benefits over costs for the entire range of output from $Q^M$ to $Q^{EFF}$ are net benefits foregone by producing only $Q^M$ rather than $Q^{EFF}$. The area of this triangle is thus social deadweight loss, which is why we have labeled it SDWL.

This discussion makes two crucial points about monopoly. First, monopolists have incentive to restrict output in order to maximize profit. Second, by restricting output, they impose social deadweight losses on society. These results will be important in later chapters when we examine certain resource markets with suppliers that dominate the market.

**Question 2.9**: Consider the market for iron ore and assume the demand for iron ore can be written as: $P = 1000 - 0.1Q$, where P is the price of iron ore, in dollars per ton, and Q is the quantity of iron ore, in tons. Assume also that iron ore can be produced at a constant marginal cost of $200 per ton.

(a) Calculate equilibrium monopoly price and quantity.
(b) Calculate how much better-off/worse-off society is under monopoly conditions than under competition.

## II. Externalities

A key assumption of the market model was that producers of goods bear all the costs of that production process. They must hire labor, buy capital equipment, build physical plant, and so forth. If they bear all these costs, they have incentive to keep costs down and not incur any unnecessary expenses. At the same time, they will only produce up to the point where it is worth it to do so; namely, where the marginal cost is no greater than the marginal benefit to them. If producing additional output costs more than they can recoup in additional revenue, they will have no incentive to produce that extra output.

An important missing ingredient in that analysis was the fact that sometimes – often, in fact – producers of goods do *not* bear all the costs of production. For example, in producing electricity, a coal-fired power plant spews smoke laden with sulfur dioxide into the air. When this smoke is carried downwind, it can aggravate heart conditions and cause respiratory ailments for residents of downwind communities. It can also cause damage to vegetation, lakes, and ecosystems. All of these impacts impose real costs, both on the downwind communities but also on society in general due to a degraded environment. Since these costs are generally not borne by the owners of the power plant, this smoke is an example of what economists call an *externality*.

There are many real-world examples of externalities. When fossil fuels are burned to heat buildings and to power factories and automobiles, carbon dioxide is produced as a by-product, which contributes to climate change. When farmers fertilize their fields, run-off into nearby rivers and streams can cause algal blooms that decimate fish populations. Chemical factories produce chemical products that can leach into local groundwater supplies. Coastal municipalities divert storm water and runoff from septic tanks into nearby lakes and estuaries. In all these cases, the party that generates the externality is not bearing the cost to others.

In order to understand how externalities can lead to inefficient outcomes, let us consider a particular market, the market for steel. And to keep things simple, assume that it is a competitive industry with many steel producing firms, but that steel production generates polluting smoke. In buying iron and coal, hiring workers, and so forth, the steel firms incur production costs. These decisions collectively generate an industry supply curve for steel, which you can think of in the standard way as reflecting marginal costs.

If these firms were not producing polluting smoke, we could stop here, in thinking about the cost side of things. However, the fact that they are producing this smoke means that there is another component of costs: namely, the costs imposed on everyone else, such as people living in downwind communities. The problem is that firms have no incentive to consider these costs when deciding how much steel to produce. But from the viewpoint of society, these costs are real: just as real as the costs the firms themselves have to incur. So let us add them to the analysis. And since they are costs to everyone else, let us call these *external marginal costs*. In Figure 2.15, these are labeled $MC^{EXT}$.

Now that we have these two different types of costs – ones considered by the polluting firms and others not – let us define another concept I will call *social marginal costs* ($MC^{SOC}$). As the name implies, social marginal costs are the total marginal costs borne by all of society. So here, where there are only two sources of marginal costs – private and external – social marginal costs are the sum of the two. Figure 2.15 shows private, external, and social marginal costs and the relationship among them. So, for example, if the *private marginal cost* of producing the 100th ton of steel is b and the *external marginal cost* of producing the 100th

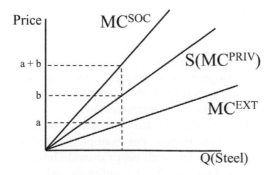

*Figure 2.15* Deriving social marginal costs

ton is a, the social marginal cost is (a + b). And the same relationship, of course, holds for all possible quantities of steel produced.

To see the efficiency implications of the existence of negative externalities, let us now bring benefits into the picture. In line with the earlier discussion, let us assume that the demand for steel captures the marginal benefit to consumers of steel. And let me make it explicit that we are assuming that there are no other benefits to society aside from the benefits to these steel consumers. In this case, we can add the demand curve to the analysis and interpret it as social marginal benefits. To be clear, this is because by definition, private marginal benefits (here, to consumers of steel) equal social marginal benefits if there are no other benefits.

Now that we have a way to represent all of the relevant benefits and costs, let us now consider the efficiency of the market outcome. The first important point is that the quantity that will be produced in this market is found at the intersection of demand and private marginal cost, that is, supply. In Figure 2.16, this quantity is labeled as Q⋆. Firms have incentive to produce this quantity because they only care about the costs they incur themselves. Though you may think they *should* care about the costs imposed on others, in general we think that unless given a reason to do so, they probably will not. In real-world debates, you may hear something like the following argument: firms mainly have a duty to their stockholders, who care about the value of their shares and, thus, the profitability of the firms. Thus, the firms have no incentive to incur any extra costs that hurt their bottom lines.

The problem is that Q⋆ is not efficient. That is, Q⋆ is not the quantity that maximizes net benefits to society. The efficient quantity occurs where social marginal costs are equal to social marginal benefits. In Figure 2.16, this quantity is denoted Q$^{EFF}$. By not considering the costs they impose on the rest of society, including the downwind communities, the firms end up producing too much steel and inflict social deadweight loss on society. This social deadweight loss is shown in Figure 2.16 as the triangle bounded by the social marginal cost and social marginal benefit curves between Q$^{EFF}$ and Q⋆.

**Question 2.10**: Suppose that the supply curve for steel can be written as: P = 20 + 0.1Q, where P is in dollars per ton and Q is in tons. Now suppose that steel production is accompanied by a negative externality, where external marginal costs may be written as: MC$^{Ext}$ = 0.2Q, where MC$^{Ext}$ is also in dollars per ton. Finally, assume that the steel

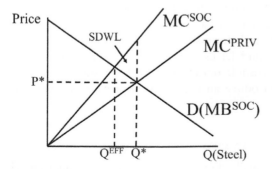

*Figure 2.16* Market inefficiency with negative externalities

market is competitive and the demand for steel may be written as: $P = 140 - 0.1Q$. From the viewpoint of economic efficiency:

(a) Will steel be overproduced and, if so, by how much?
(b) What will be the magnitude of social deadweight loss in dollars?

## Intertemporal allocation

Many economic decisions involve more than one time period. For example, suppose you have an oil field containing so many thousands of barrels of oil. You may not want to pump all the oil and sell it today. Rather, you may prefer to hold off on pumping some of it so that you can sell it later. After all, you figure, there is still going to be demand for my oil in the future: people are still going to want to drive their cars and heat their homes. Maybe, you figure, I will be able to sell it at a higher price next year, or maybe my pumping costs will go down. To wait or not to wait: that is the question.

Trying to figure out how to much to produce now and how much to produce later, if I may make a strange-sounding analogy, is kind of like eating a birthday cake. Every slice that you eat today means one less slice for you to eat tomorrow. If you think about this, the question of tapping the oil field becomes a question of timing. When should you extract oil? If you extract it today to meet a present need, you leave less available for tomorrow, when you will almost certainly have need for it as well. If you decide to leave it in the ground today, you will have more available tomorrow but today you will have to do without.

The key, then, to sensibly exploiting an oil field is to consider its value both now and in the future and allocate it accordingly; that is, develop and use it when it is more valuable to you. To continue the analogy, if you are really hungry and are also concerned your cake will go bad, you would be ill-advised not to just eat it right now. If, however, you have just come back from a pastry-eating contest where you won first prize and it is hard to envision ever eating another thing, you might want to hold off.

This careful weighing of relative value now and in the future is central to the question of efficiently exploiting any resource that you have in fixed amounts. This includes not just oil fields, but also deposits of natural gas, shale oil, bauxite, iron, copper, and a whole slew of other natural resources. All of the resources I just mentioned have one important feature in common. They have all, through natural processes, been deposited in the earth in geologic time, with no tendency to replenish themselves. This means that when we extract them now, we reduce the amount available to extract in the future. We call these types of resources *exhaustible resources*. This is to distinguish them from other resources that can replenish themselves over time, like fisheries and forests, which we call *renewable resources*. For now, let us hold off on thinking about this latter case.

In any case, keeping this distinction in mind, let us now turn to a framework that we can use to analyze situations involving exhaustible resources and other resources fixed in amount. But before we do, we have to introduce an extremely important concept: the notion of *discounting*.

### I. Discounting

The basic idea of discounting is that we generally do not view costs and benefits realized at different points in time as equivalent. Rather, we generally think of dollar amounts that we receive in the future as worth less to us, just because they happen later. This will have important implications for many decisions involving tradeoffs that occur over time.

One way to see the argument here is to ask yourself the following: If I were to offer you $1 million, which you could receive either today or one year in the future, which would you prefer? Most folks would say today. Certainly this would be the case if people had access to investment markets. On such markets, there are investment instruments such as stocks, bonds, and mutual funds, each of which provide some rate of return. This means that if I gave you $1 million dollars today, you could invest it and realize a larger dollar amount one year from now. Certainly you would prefer "more than $1 million" to $1 million. This is why we generally view future dollar amounts as worth less than nominally equivalent amounts received today.

To formalize all this, suppose that the rate of return on an investment is some percentage r. For example, if an investment has a return of five percent, then r = 0.05. What this means is that if you invest $1 million now, then exactly one year from now you will receive back your $1 million plus the return on your investment, ($1 million X 0.05), or $50,000. So at a rate of return of five percent, $1 million now is equivalent to $1,050,000 one year from now. More generally, $X now equals $X(1 + r) one year from now, for any dollar amount X and any rate of return r.

Now let us compare amounts two years apart. The question is: How much is $X now worth in *two* years? Well, in this case, you get to realize your return next year and then reinvest it to realize another, larger return two years from now. So if you invest $X now, and your return next year is $X(1 + r) and if you reinvest *that*, then two years from now you should have $X(1 + r)(1 + r), or $X(1 + r)^2$. If you keep rolling over these investments – that is, just take your money and keep on reinvesting it – you will have $X(1 + r)^3$ after three years, $X(1 + r)^4$ after four years, etc. This means that after N years, where N is any number of years in the future, you would have $X(1 + r)^N$.

Now: in order to define a very important concept, let us reverse the question and ask ourselves how much we would need to be given *now* in order to feel just as well off as if we were given $1 million one year from now. This should be ($1 million)/(1 + r), right? This is because ($1 million)/(1 + r) invested in an instrument yielding a rate of return of r equals [($1 million)/(1 + r)] X (1 + r). Voila, the (1 + r)'s cancel and we are left with $1 million next year. Economists refer to ($1 million)/(1 + r) now as the *present discounted value* of $1 million realized one year from now. And, generalizing this argument, $X/(1 + r)^N$ is the present discounted value of $X realized N years from now, again where X is any dollar amount and N is any number of years.

## II. Efficiency in dynamic settings

It turns out that discounting has a number of applications in environmental economics, which we will see in later chapters. For now, let us return to the idea of an exhaustible resource. To be concrete, let us consider our particular example of an oil field. We saw earlier that tapping an oil field is a question of timing: how much we extract now vs. how much we leave in the ground to extract in the future. Let us now develop a simple intertemporal model that allows us to draw conclusions regarding how markets would allocate the oil over time, and how efficient this allocation would be.

To keep things simple, let us assume only two time periods, the current year (t = 0) and next year (t = 1). You can think of next year as representing "the future." To make the analysis a bit more rigorous, let us introduce some mathematical notation to represent the value of the resource both now and in the future. Let $MB_t$ denote the marginal benefit of the resource in time period t. So in the current year the marginal benefit would be $MB_0$ and next year it would be $MB_1$. As always, we assume that both $MB_0$ and $MB_1$ are declining functions of the amount of resource used in each time period.

As we have seen, in general the market demand for a resource will accurately reflect its marginal benefit to society. To keep things simple, assume that there are zero marginal costs of producing the resource. In Figure 2.17, the current value of the resource is measured on the left-hand-side vertical axis and the future value is measured on the right-hand-side. The marginal benefit functions are written as $D_0$ and $D_1$. The $D_1$ function is drawn "backwards" because the quantity of oil used in the future year is measured from right to left. In addition, the length of the horizontal axis measures the total stock of oil. This means that any point along the horizontal axis represents a division of extraction of oil between the current year and the future year.

In addition to the demand function $D_1$, Figure 2.17 shows the discounted demand function $D_1/(1 + r)$. This discounted function is shown because, as we have seen, buyers and sellers of resources tend to discount the future. Because they do, market allocations will be governed by the discounted function, which shows the discounted value of the resource in the future period. The equilibrium allocation of the resource $Q^\star$ is given by the intersection of *present* demand $D_0$ and *discounted* future demand $D_1/(1 + r)$. At this allocation, the current marginal value of the last unit of resource is equal to the discounted marginal value of the last unit in the future year. Any other allocation will cause production of resource to shift from the low marginal-value period to the higher marginal-value period, until the marginal values (in discounted terms) are equal again.

I should add that this market equilibrium will be efficient. You can see this if you consider the total net benefits across both years, keeping in mind that you need to appropriately discount future net benefits in order to make them comparable to current net benefits. So in Figure 2.17, the entire shaded area under the $D_0$ curve out to the quantity $Q^\star$ represents total net benefits in the current year. And the entire shaded area under $D_1/(1 + r)$ represents total *discounted* net benefits in the future year. Adding up these areas gives total discounted net benefits combined over both years. This area represents the maximum discounted net benefits possible, as it is not possible to increase total net benefits by dividing up production of the total stock in any other way. Hence, $Q^\star$ is called the *dynamically efficient* market equilibrium.

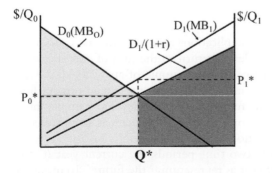

*Figure 2.17* Dynamic market equilibrium and efficiency

**Question 2.11**: Oil is produced under conditions of perfect competition in each of two years, after which the world blows up. In each year, the market demand for oil is given algebraically as: $P = 120 - 0.12Q$, where $P$ = price of oil, in dollars per barrel; and $Q$ = quantity of oil, in millions of barrels. Suppose that oil stocks total 1,450 million barrels, oil production is costless, and the market interest rate is 20%. How much oil is produced in each period, and at what prices?

## Key takeaways

- Under conditions of scarcity, all actions have an opportunity cost.
- With many firms and consumers, vigorous competition, and no negative externalities, markets tend to perform well in terms of producing efficient outcomes.
- It is often important to quantify demand and supply responses to changes in price changes and changes in other factors. This is typically represented with the notion of an *elasticity*.
- Monopoly conditions tend to result in underproduction and social deadweight loss.
- Uninternalized negative externalities tend to result in overproduction and social deadweight loss.
- When outcomes occur at different points in time, future outcomes must be discounted to make them comparable to earlier ones.
- Dynamic efficiency is achieved by markets in multiple-period settings, under the same competitive assumptions as in single-period settings.

## Key concepts

- Opportunity cost
- Marginal thinking
- Comparative statics
- Elasticity
- Consumer surplus
- Producer surplus
- Social deadweight loss
- Market power
- Discounting
- Intertemporal allocation
- Dynamic efficiency

## Answers to "Test your understanding" questions

**Question 2.1** (a) *England*: (i) OC of 1 bottle of wine: 1 lb of cheese; (ii) OC of 1 lb of cheese: 1 bottle of wine. (b) *France*: (i) OC of 1 bottle of wine: ¼ lb of cheese; (ii) OC of 1 pound of cheese: 4 bottles of wine.

**Question 2.2** 24.

**Question 2.3** (a) 3 hours. (b) NB = 9. (c) Yes, because BC = 1.86. Equivalently, NB are positive.

**Question 2.4** Equilibrium quantity = 5,000 tons. Equilibrium price = $70 per ton. Total expenditures = $350,000.

**Question 2.5** Cannot tell. Increasing consumer incomes should shift demand upward/ outward, while decreasing fuel prices should shift supply downward/outward. The overall effect on prices depends upon the relative magnitudes of the demand and supply shifts. Without more information, we cannot tell whether prices will go up or go down.

**Question 2.6**: (a) Positive. (b) Normative. (c) Positive.

**Question 2.7** (a) Net benefits = $24,000. (b) SDWL = $6,000.

**Question 2.8** (a) Equilibrium price = $5 per sub, equilibrium quantity = 1,000 subs. (b) Consumer surplus = $5,000, producer surplus = $0.

**Question 2.9** (a) Equilibrium monopoly price = $600 per ton, equilibrium quantity = 4000 tons. (b) SDWL = $800,000

**Question 2.10** (a) Steel is overproduced by 300 tons. (b) SDWL = $18,000.

**Question 2.11** (a) 750 million barrels of oil are produced in year 0, 700 million barrels are produced in year 1. The year 0 price per barrel is $30 and the year 1 price is $36.

## Additional problems/exercises

2.1.  The market demand for bread may be written as: $P = 10 - 2Q$, where $P$ = price of bread, in dollars per loaf; and $Q$ = quantity of bread, in thousands of loaves. Market supply, on the other hand, is: $P = 2 + 2Q$. Calculate equilibrium price and quantity, and total expenditures on bread in equilibrium.

2.2.  In the Twin Cities market for housing, the annual market demand for a typical three-bedroom house is initially: $P = 200,000 - 2,000Q$, where $P$ = price of a house, in dollars; and $Q$ = thousands of houses. Market supply, on the other hand, is: $P = 2,000Q$. Then, as a result of a boom in emigration to Minnesota, market demand increases to: $P = 300,000 - 3,000Q$. What is the predicted change in the price of a house? What is the predicted change in the number of houses sold?

2.3.  Prior to 1966, the market demand for fish in Boston was: $P = 24 - 0.01Q$, where $P$ = price of fish, in dollars per pound; and $Q$ = quantity of fish, in tons. Market supply, on the other hand, is: $P = 0.02Q$. Then in 1966, the Pope issued a decree that it was no longer obligatory for Catholics to abstain from eating meat on non-Lent Fridays *[True story!]*. As a result, demand for fish decreased to: $P = 18 - 0.01Q$. What was the change in total expenditures on fish?

2.4.  The market demand may be written as: $P = 30 - 0.01Q$, where $P$ = price of gasoline, in dollars per gallon, and $Q$ = quantity of gasoline, in thousands of gallons. Market supply, on the other hand, may be written as: $P = 10 + 0.01Q$. (a) Calculate equilibrium price and quantity, and total expenditures on gasoline in equilibrium.

   After a number of oil-exporting countries decrease oil exports, the market supply of gasoline decreases to: $P = 20 + 0.01Q$. (b) Calculate the change in the equilibrium price and quantity of gasoline.

2.5.  The world market for diamonds is currently monopolized by the DeBeers company. Market demand for diamonds is: $P = 2000 - 0.1Q$, where $P$ = price of diamonds, in dollars per diamond, and $Q$ = quantity of diamonds. DeBeers has two mines, the ABC and XYZ mines, each of which can produce diamonds in unlimited quantities. Diamonds from ABC can be produced at $200 per diamond, whereas diamonds from XYZ can be produced according to the following MC schedule: $MC = 600 + 0.1Q$, where $MC$ = long-run marginal costs, in dollars per diamond. Assuming that DeBeers acts to maximize profits, calculate its total diamond output, output from each mine, associated profits, and social deadweight loss.

2.6. Assume that the market for steel is perfectly competitive. Market demand for steel may be written as: $P = 1000 - 0.1Q$, where $P$ = price of steel, in dollars per ton, and $Q$ = quantity of steel, in thousands of tons. Market supply, on the other hand, may be written as: $P = 200 + 0.1Q$. In addition, suppose that production of steel is accompanied by production of air pollutants, which impose external costs on surrounding communities of: $MEC = 0.1Q$. Calculate the amount of resulting social deadweight loss.

2.7. Consider the market for bauxite ore and consider the two-period intertemporal model. Suppose that the demand for bauxite ore is: $P_i = 242 - 0.11Q_i$ in each of two time periods 0 and 1, where $P$ is the price of bauxite ore, in dollars per ton, and $Q$ is the quantity of bauxite ore, in tons. Furthermore, assume there are 830 total tons of bauxite. Finally, assume the discount rate is 10%. Calculate the total amount of bauxite ore produced, and the market price, in each of the two periods. In addition, calculate total discounted surplus.

## Note

1  If you are mathematically inclined, it is really easy to show this using a bit of simple calculus. Ask your instructor to show you how.

# 3  Valuation

Iguazu Falls, on the border between Argentina and Brazil, are the largest, most spectacular waterfalls in the world (Figure 3.1). In the native Guarani language, Iguazu means "big water," and it really lives up to its name. When viewed from a central vantage point, the falls seem to go on and on, stretching as far as the eye can see. In all, there are 275 distinct falls spread out over an area nearly three kilometers wide on the Iguazu River, the tallest being nearly twice as tall, and nearly three times as wide, as the much more famous Niagara Falls. During the rainy season, the amount of water that flows through the falls can be enough to fill five Olympic-sized swimming pools *every second*. It is no wonder that when Eleanor Roosevelt saw the falls for the first time, her reaction was reportedly: "Poor Niagara."

Got all that? Now imagine there was a threat to its existence. Perhaps plans to divert water for other uses or a hydropower dam that would impound the water upstream. Given its remote location and iconic status (it was designated a UNESCO World Heritage site in 1984), none of this is likely to ever happen, but humor me. How would we weigh the tradeoffs between preserving Iguazu Falls and developing these other uses of the water? Is it even possible to apply the economic approach of weighing costs and benefits in this case?

I ask these questions because they, or questions very much like them, commonly arise when we apply the economic approach to the environment. And to be honest, they are tough questions for economists to answer because they require us to confront the fiendishly difficult issue of how to place a value on the environment. Other examples include: developing a wilderness area for recreational activities, preserving critical wildlife habitat for endangered species, curbing air pollution to preserve scenic views, and protecting the Amazon rainforest. I am sure you can think of many others. In each of these examples, there is an environmental amenity of obvious social value – pristine wilderness, biodiversity, beautiful scenic views, and rainforest – that must be weighed against another activity with social value: developing a down-hill ski resort, hydropowered or coal-fired electricity generation, or logging and agriculture. If assessing these trade-offs seems like an incredibly daunting task, that is only because it is. In this chapter, we will explore how economists have thought about these tradeoffs, and the specific approaches and methods they have proposed to weigh these tradeoffs as best they can.

*Figure 3.1* Iguazu Falls. *Source*: Shutterstock

## An important distinction: Marketed vs. non-marketed goods

It may be obvious that one key distinction between many environmental amenities and activities we weigh them against is that the other activities are often exchanged on markets, while environmental amenities typically are not. When they are sold on competitive markets, timber, food products, and electricity have market prices that provide a reasonable measure of their marginal social value. In Figure 3.2, for example, the equilibrium price P★ reflects both consumers' (marginal) willingness to pay for the product and the (marginal) cost of producing it. This implies that we could in principle evaluate the social costs of preserving an environmental amenity like a stunning view by quantifying the amount of foregone power production needed to preserve that view and ascribing a market value to that foregone power. It is not quite this simple of course, but the main point here is that ascribing social value to marketed goods is relatively straightforward, at least conceptually.

A considerably greater challenge is to try to place a value on many environmental amenities. Here, there is usually no market that we can rely upon to provide measures of social value. Indeed, for many amenities like Iguazu Falls, it may be difficult for you to imagine how we could possibly ever get a handle on their social value. What, for example, would be the value of a scenic view or preserving the habitat of an obscure species of fish?

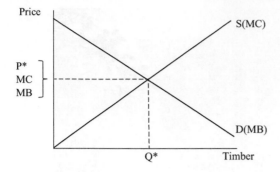

*Figure 3.2* Market prices reflecting societal value

## Use of valuation numbers: Cost-benefit analysis again

Before going into specific approaches to valuing the environment, let us first develop in more detail how these values would be used. Here, we will start with the economic idea that society is best served by making our scarce resources go as far as possible in serving social goals. In practical terms, this means a general presumption only to take actions where the social benefits exceed the social costs. This discussion is returning, of course, to the concept of cost-benefit analysis introduced in the mini-course in chapter two.

To be concrete, let us assume that we are evaluating a dam construction project that provides electricity but floods a low-lying area, which inflicts costs on local inhabitants and causes damages to habitat for waterfowl. Assuming for the moment that we have reasonable numbers to plug into a benefit-cost formula, we will have measures of benefits B and costs C of dam construction, which we can evaluate in one of two equivalent ways. The first is simply to take the difference between benefits and costs, what we will call net benefits NB:

$$NB = B - C \qquad (3.1)$$

If benefits exceed costs, net benefits are positive. In this case, we would conclude that the dam is worth building. The other method is to calculate a benefit-cost ratio:

$$\text{Benefit-cost ratio} = B/C \qquad (3.2)$$

Here, if the benefit-cost ratio exceeds one, then the dam is worth building. It should be apparent that applying (3.1) and (3.2) in this way will yield the same conclusion regarding the desirability of building the dam.

In performing cost-benefit analysis, it is often assumed that the *distribution* of the benefits and costs across different members of society does not matter for measuring overall social welfare. As you might expect, this is a controversial assumption, which flows directly from application of a principle known as the *Kaldor-Hicks criterion*. The Kaldor-Hicks criterion, named after two famous economists Nicholas Kaldor and John Hicks, says that an action is worth doing if beneficiaries could hypothetically compensate losers and still make both sets of parties better off. Applying this criterion does not require that compensation actually be made, however, and thus it reduces to simply whether the overall benefits exceed the overall costs.

Since many people would argue that distributional consequences do matter, many cost-benefit analyses assume, at least implicitly, that the distributional consequences of any proposed action are small. Such an assumption, of course, would need to be evaluated on a case-by-case basis. And in cases where the distributional consequences are not small, the final prescription may need to be modified when the distributional consequences are seen as undesirable. However, generally speaking, economists have less to say regarding precisely how to take distributional considerations into account.

## Important special case: Cost-benefit analyses with long time horizons

An important complication enters into cost-benefit analysis when the consequences of an action are long-lived, resulting in an entire stream of benefits and costs over time. Dams, for example, can last for decades and so, in the case of the hydropower dam, there will be an entire stream of benefits over the lifetime of the dam, deriving from the value of the produced power. On the other hand, there will be upfront construction costs and perhaps a stream of costs in operation and maintenance of the dam as long as it continues to operate. Furthermore, there will be a stream of environmental costs to consider as well.

When we consider the longer term, we have to modify our cost-benefit analysis accordingly. This means that we need to consider an entire stream of net benefits over time and, recalling our earlier discussion in chapter two, we need to discount future values appropriately. So our benefits B and costs C become:

$$B = \sum B_t / (1+r)^t \tag{3.3}$$

$$C = \sum C_t / (1+r)^t \tag{3.4}$$

Here $B_t$ and $C_t$ are the benefits and costs incurred in time period t. Then, the net benefits formula becomes:

$$NB = \sum (B_t - C_t) / (1+r)^t \tag{3.5}$$

Once we start thinking about costs and benefits in this way, the choice of discount rate becomes an important issue. Recall that larger discount rates imply smaller discounted values of future dollar amounts. This fact means that the choice of discount rate can dramatically affect bottom line prescriptions of cost-benefit analyses. This is because costs and benefits are typically not realized at the same time.

In some cases, costs are incurred now and benefits are realized in the future. For example, building a hydropower dam entails a sizable upfront expenditure for the prospect of years of electricity production. If future benefits are discounted heavily, dams tend not to be worth it because those upfront costs will dominate the net benefit calculation. On the other hand, if future benefits are discounted little, they are more likely to outweigh upfront costs, making those dams more attractive.

Let us use some numbers to make this example more concrete. Suppose we are considering building a dam, which would require an upfront expenditure of $50 million. Suppose that if we build it, we will enjoy a stream of benefits of $2 million per year for the next 50 years. If the discount rate is 10%, it is easy to apply the formula in equation (3.5) to calculate that the PDV of NB of this dam is −$30.18 million, definitely failing the cost benefit test. At a discount rate of 5%, the PDV of NB is −$13.66 million, so it still fails the test. However, if the discount rate is only 2%, the PDV of NB is +$12.10 million. So the choice of discount rate makes a huge difference to the bottom line question of whether or not to build the dam.

In other cases, benefits are largely enjoyed now and costs are largely borne in the future. Climate change provides a good example of this, as we obtain benefits now from burning fossil fuels in exchange for future costs, such as from rising sea levels and more extreme climate events. In this case, alternative assumptions about the discount rate have the opposite effect. High discount rates will lead us to devalue future costs more heavily, which will make it more likely for current actions to pass cost-benefit tests. So, from this discussion, you can see that the choice of discount rate is a matter of huge practical importance.

It turns out that economists hold varying opinions regarding the appropriate discount rate to use, especially for actions that may have large consequences years off into the future. In chapter two, we developed the concept of discounting in the context of an investment opportunity. An approach like this tends to be the position of economists who believe that real-world markets provide accurate information regarding social preferences about tradeoffs between the present and the future, and that those preferences should be honored in assigning discount rates.

However, issues of ethics and fairness also enter into discussions among economists over the appropriate choice of discount rate. And sometimes fairness concerns are at odds with this market-based approach. The basic argument is that discounting the future means that the welfare of future generations is weighted less heavily than the welfare of the current generation, which is unfair to those future generations. Since the current generation makes policies that often affect future generations who have no say in the creation of these policies, some have called this the "tyranny of the present" [Pearce et al. (2003), p. 134].

To help you understand this fairness argument, let me define a concept that is related to the discount rate, the so-called *discount factor*. The discount factor (DF) is simply equal to the inverse of $(1 + r)$ taken to the power of t. This is shown in equation (3.6). So you can probably see that when you apply the net benefit formula shown in equation (3.5), the discount factor is the value that is multiplied by every realization of costs or benefits in any time period. To make this explicit, equation (3.7) rewrites the net benefit formula.

$$DF_t = 1 / (1 + r)^t \qquad (3.6)$$

$$NB = \sum (B_t - C_t) \times DF_t \qquad (3.7)$$

Equation (3.7) tells us that the total net benefit formula is simply a weighted sum of the net benefits in each individual time period, where the weights are given by the discount factors. As we know, in the net benefit formula $(1 + r)$ is taken to higher and higher powers as t increases. This implies that the discount factors, and hence the weights applied to the net benefits enjoyed in every time period, are getting smaller and smaller as you move further into the future. So, for example, if the discount rate is 5%, the discount factor 50 years out is 0.087: that is, a dollar realized 50 years in the future is only valued at 8.7 cents today. One hundred years in the future, it would be worth less than one penny today. Thus, distant future outcomes are being weighted very little when we apply the net benefit formula. The interests of future generations are barely being factored in at all.

This means, of course, that when we apply *lower* discount rates, we are in a real sense being fairer to future generations, because we are weighting their interests – for example, the costs they incur – more heavily when deciding what actions to take today. The question then becomes: How far do we wish to take this argument? It seems to imply that it would be fairest to apply a discount rate of zero – that is, not discount at all – in which case the interests of all generations would be weighted equally. Unfortunately, not discounting at all has a couple of implications that we may not feel entirely comfortable with.

First, it means that every single generation is weighted equally, no matter how far into the future they live. Now, many of us may feel comfortable with the idea of weighting the interests of the next generation – our children – equally to ours. Perhaps even our grandchildren. But what about ten generations out, or a hundred, or even a thousand? Should the interests of people living thousands of years in the future receive the exact same consideration as the people living right now? I think many would say no. But this is one logical implication of arguing that the future should not be discounted.

Second, in a world of scarce resources – that is, the world we live in – not discounting the future implies that the current generation may need to reduce its resource usage to extremely low levels, in order to leave sufficient resources to service all of those future generations. The basic problem is that there is only one current generation, and potentially thousands and thousands of future generations. Weighting all of those future generations equally in our current decisions may mean leaving those resources in the ground so they are available in the future. In principle, the current generation could be reduced to abject poverty.[1]

So what is to be done? Well, this is a tough question, for which economists have no easy answers. But there is a compromise position that has been around for a while that has been slowly gaining traction among many economists: that it might be best to have discount rates that are not constant but, rather, that decline over time. The idea is that we would apply relatively high discount rates to near-term outcomes while applying lower discount rates to outcomes further off in the future. This idea is known by the fancy name *hyperbolic discounting*.

One attractive feature of discounting in this way is that it seems to accord with the way we all behave. You behave this way, for example, when you procrastinate by going to the movies when you should really be studying for your economics final exam. Or your overweight uncle behaves this way when he takes that extra helping of mashed potatoes rather than sticking to his diet. In each of these cases, a high discount rate is being applied in the short term (you want to see that hit movie now!), despite the fact that deep down inside, you really do value the longer-term objective. So because hyperbolic discounting reflects the actual choices we make, it appeals to many economists who believe that discounting should be based on our preferences. Even when those preferences keep us from getting what we want, like an A in economics or being able to wear a swimsuit on the beach without embarrassment.

But, even more than this, it has the appeal of embodying a more equal weighting of the interests of current and future generations. To see this, consider the following thought experiment, which captures a lot of real-world issues. This thought experiment is based upon the work of the late environmental economist Martin Weitzman, who contributed in important ways to better our understanding of a wide range of issues in environmental economics, including discounting, environmental policy, climate change, and biodiversity.

Weitzman posed the following hypothetical situation. Suppose the future is uncertain. To keep things simple, suppose there are three possible future states of the world. In one of these states, the Antarctic ice sheet breaks down completely and sea levels rise dramatically and cause serious damages to low-lying coastal areas around the world. In the second state of the world, it breaks down partially, causing more limited coastal damages. In the third state of the world, it remains intact and coastal areas are fine. I will call these the *Very Bad*, *Bad*, and *Not So Bad* scenarios. Finally, suppose that these scenarios are all equally likely so that in this simple example, each will occur with a probability of one-third.

It makes sense that you might want to apply different discount rates depending on the scenario. For example, you might want to apply lower discount rates for the Very Bad scenario, which would weight those large future costs more heavily. So let us suppose that each of these scenarios corresponds to a different discount rate: assuming the Very Bad scenario, we would apply a discount rate of 1%; for the Bad scenario, we would apply a discount rate of 5%; and, lastly, for the Not So Bad scenario, we would apply a discount rate of 10%.

Now suppose we adopt the following policy. Since we do not know which state of the world will occur, let us hedge our bets. We will do this by taking the discount factors associated with each scenario and weighting them by the probabilities that each scenario will occur. Notice that, from our earlier discussion, this is like taking the weights we assign to future outcomes and in turn weighting them by the probability of those future outcomes. It turns out that doing this translates into discount rates that decline for outcomes further into the future.

A simple example will hopefully make this argument clear. Suppose each of these outcomes were to occur next year. A discount rate of 10% implies a one-year discount factor of 0.91. Similarly, discount rates of 5% and 1% imply one-year discount factors of 0.952 and 0.99. Weighting each of these discount factors by 0.33 yields a discount factor of 0.94. That is:

$$(0.91 \times 0.33) + (0.952 \times 0.33) + (0.99 \times 0.33) = 0.94 \tag{3.8}$$

This discount factor translates into an average discount rate of 0.063 for outcomes that occur next year. Now apply this argument to outcomes two years out. In this case, these 10%, 5%, and 1% discount rates translate into discount factors of 0.826, 0.907, and 0.98. Equally weighting these discount factors yields an average discount factor of 0.896, implying a discount rate of 0.057. Notice that the discount rate has gone down.

Figure 3.3 shows what happens to the discount rate under these conditions for future outcomes up to 20 years. By the time we get to 20 years out, the discount rate is barely over 0.04, or 4%. Extending this example further, for outcomes 50 years out, the discount rate falls under 3%. Weitzman showed that in the limit, as time goes to infinity, the average discount rate approaches the lowest individual scenario discount rate: in our example, 1%.

I mentioned earlier that declining discount rates represent a compromise position between those who believe that discounting should be based on people's preferences and those who are concerned that discounting is unfair to future generations. I hope this example illustrates why. Declining discount rates are consistent with people's preferences in general in two ways. They take into account the fact that people do in fact discount the future. And they are consistent with the fact that many people seem to engage in hyperbolic discounting, which is reflected in their actual behavior. In addition, they allow for outcomes that are fairer to future generations, while helping us avoid potentially undesirable implications of not discounting at all.

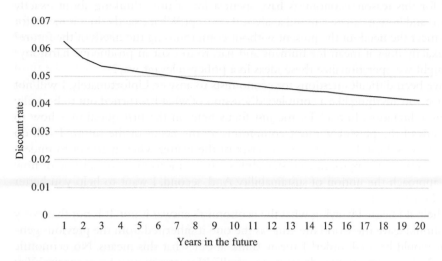

*Figure 3.3* Implied discount rate, hyperbolic discounting example

## Sustainability

The issue of fairness to future generations – what many economists call *intergenerational equity* – is tied up in the broader notion of *sustainability*. Sustainability is an idea that has received a lot of play in recent days in public debates about appropriate environmental policies. Even though I have not defined it yet, you probably have some idea of what I am talking about. Loosely speaking, you probably think of sustainable policies as ones that protect the environment. You may also have the sense that they refer to policies that do not "use up" or degrade resources so that they will continue to be available for us and for people in the future.

Even though you already have a pretty good sense of what I am talking about, I should probably give you a more formal definition of sustainability. It turns out that sustainability means slightly different things to different people. This can be seen in the following various ways that sustainability has been defined by different entities such as the United Nations and the U.S. Environmental Protection Agency.

- "To pursue sustainability is to create and maintain the conditions under which humans and nature can exist in productive harmony to support present and future generations." [U.S. EPA]
- Sustainable development "is development that meets the needs of the present without compromising the ability of future generations to meet their own needs" [UN World Commission on Environment and Development].
- "In a sustainable world, society's demand on nature is in balance with nature's capacity to meet that demand" [Global Footprint Network].

But in most cases, there is a clear implication that the needs of future generations are important, which requires both a healthy environment and the conservation of resources so that they will be available in the future.

The notion of sustainable environmental outcomes has intuitive appeal, which helps explain why it has become such a common staple in public discussions of the environment.

And, partly for this reason, economists have spent a lot of time thinking about exactly what it means and how it can be a useful analytical concept. What exactly does it mean, for example, to meet the needs of the present without compromising the needs of the future? And what exactly does it mean for humans and nature to exist in productive harmony? And how might we operationalize these ideas in a policy-relevant way?

These have been difficult questions for economists to answer. Unfortunately, I will not be able to give you anything like a complete discussion of what has turned out to be really large economic literature. Instead, let me just focus here on the first question – how to meet the needs of the present without compromising the needs of the future. I will do this for two reasons. First, I think this is the crux of the matter when it comes to understanding the economist's approach to sustainability. So I want to give you a sense of how economists approach the notion of sustainability. And, second, I want to help you better understand the fairness issue.

It might help if I begin by saying what the economist's answer is not. It is not that every generation should leave the world the same as the one it inherited from the previous generation. This would be a tall order! I mean, think about what this means. No economic development? No environmental degradation at all? No extinctions of any species? You may think I am setting up a straw man here, but something very close to these sentiments have been stated in the past as official positions of agencies of the United Nations [Solow (1993), p. 180].

Instead, the economic approach to sustainability focuses more on *capabilities*: the ability to enjoy things we value, like adequate consumption, vital services, and a healthy environment. So a sustainable outcome is one that provides all generations present and future the capacity to attain a good standard of living, construed broadly to include environmental quality. And the basic idea is that there are a lot of ways to accomplish this. These include: using resources more efficiently, inventing new production processes and products, and switching to new factor inputs. So, for example, it is not necessarily a tragedy to run out of rubber to produce automobile tires, if there are other products that can be used to produce perfectly good tires, or if driving becomes obsolete because we all fly around using jetpacks or futuristic mini-helicopters. In this example, the thing of value is not driving in cars per se but, rather, the ability to get where we want to go.

If you buy the argument that what matters is overall standard of living rather than specific consumption goods or environmental objectives, this allows for flexibility in how we maintain quality of life. But the amount of flexibility we have depends upon our ability to produce things in different ways. So if we run out of rubber, or aluminum, or petroleum, we are still able to produce tires and lightweight airplane fuselages, and we can still operate our automobiles or we have resources to produce jetpacks instead. But this is the $64 question: How possible is it to do without these things and keep the things we value?

The general idea is that substitutability is key: if we run out of something, are there good substitutes that permit us to continue to meet our needs at comparable cost? In some cases, there will be: synthetic composites instead of aluminum to produce lightweight fuselages, or rechargeable hybrids instead of gasoline-fueled cars. In these cases, we are not as concerned about running out of particular resources, because we can simply use other resources for production instead and still obtain what we want.

However, in many other cases, there will not be good substitutes. Everyone would probably agree, for example, that if you want a spectacular scenic vista, there is simply no good substitute for the Grand Canyon, or Iguazu Falls. And there is no good substitute for a large Arctic polar icecap that reflects a great deal of sunlight and helps keep the earth from warming. In these cases, however, some economists would argue that we just might want to preserve them for their own sake. And they would say that there is no need to invoke any notion of sustainability to argue that we should protect the Grand Canyon, or Iguazu Falls, or the polar icecap from destruction [Solow (1993), p. 181].

## How does one put a number on the value of the environment?

I did not want to leave the discussion of weighing costs and benefits without giving the notion of intergenerational fairness more consideration. But, now that we have a sense for the basic economic approach of weighing costs and benefits both now and in the future, let us turn to the thorny question of how precisely one ascribes a value to the environment. This value will be interpreted as the benefit of preserving an environmental amenity, like a salmon run, or a rainforest, or Iguazu Falls. We would then use this value number in an analysis where there is a clear tradeoff against some other use of the same environmental amenity, like constructing a ski resort or building a hydropower dam. A cost-benefit analysis would weigh the alternative uses and then, in its simplest form, prescribe preserving the amenity if the benefits exceed the costs, in terms of the foregone value of the alternative use.

Economists have put a lot of thought into how we might go about quantifying the benefits of environmental amenities. To be completely honest about it, all of the methods we are about to discuss have the common feature that they rely on the value ascribed by humans, in one form or another, as opposed to any sort of intrinsic value. Economists may disagree on whether a critical habitat for an endangered species should be protected. However, they generally agree that the decision should be based on how *humans* feel about providing protection. They generally do not assume that things have value in and of themselves. This puts economists at odds with moral philosophers who have argued that various environmental entities have inherent worth that exists independent of human values. Consider, for example, these statements about endangered species:

- "Species count, whether or not there is anybody to do the counting" [Rolston (1989)].
- "Species have value in themselves, a value neither conferred nor revocable, but springing from a species' long evolutionary heritage and potential" [Soule (1985)].

These statements reflect the view that endangered species have what is sometimes called *inherent intrinsic value*. Though this is a vigorous ongoing debate, we shall not pursue it here.

The economic approach to valuation can be divided into two basic types of approaches: *revealed preference* and *stated preference*. Let us consider these in turn.

### I. Revealed preference

Revealed preference refers to approaches that are based on people's actual behavior; that is, the actions they actually take. The term revealed preference refers to the economist's general assumption that people's actions are based on what they prefer to do. If they are at the store and trying to decide whether to buy apples or oranges (that happen to cost the same), they will buy whichever they like more. If they end up buying apples, then economists say they have revealed a preference for apples. That is, preferences are revealed in people's behavior.

At this point, let me introduce some terminology that will come in handy not just for this discussion but for the entire rest of the textbook. Economists put great stock in the idea that people are motivated by self-interest, which leads them to take actions that make them happiest or most satisfied. The way that economists often put this is to say that people behave in order to maximize their *utility*. So the way we explain why they might choose apples instead of oranges is that they derive more utility from apples, which is the same thing as revealing a preference for apples.

Okay, now that we have that idea under our belts, let us turn to the two main approaches to valuation that are based on people's actual behavior. These are the *hedonic pricing*, and the *travel cost* approaches.

### Hedonic pricing approach

Before I actually define hedonic pricing, let us first consider the following real-life situation that I would wager you have found yourself in. Imagine going to the supermarket, grabbing a shopping cart, and walking up and down the aisles filling your cart with grocery items. When you are done, you wander over to the checkout counter and the checkout person tallies up your bill. As usual, you gulp at the expense but insert your debit card into the billing machine anyway.

I start with this shopping example because I think it provides good intuition for the hedonic pricing approach. Under this approach, you take the value of a piece of property, like a house, and you use it to infer how much people value an associated environmental good, like local air quality. The idea is that when you buy a house, you are not so much buying the house itself as much as you are buying a number of things you like about the house. People tend to prefer (read: *derive more utility from*) houses that are larger, have more bathrooms and bigger lots, and which have central air conditioning. This means that people are willing to pay more for houses that have these features.

Similarly, your grocery bill reflects the characteristics of your shopping cart full of goods. Each of the different items in your cart contributes to the final grocery bill. Those two boxes of Raisin Bran add so much to the final amount you have to pay, as does that bag of apples and that case of Mountain Dew. Similarly, each of the characteristics of your house – the number of bedrooms, the number of bathrooms, and so forth – contribute to the final value of the house. In the terminology of hedonic pricing, each of these features is known as an *attribute* of the house.

The idea behind the hedonic pricing approach to valuing the environment is that under certain circumstances, we can treat an environmental amenity, or disamenity (like dirty air), as an attribute of a house. So say that a house with so many bedrooms, bathrooms, square footage, and so forth also happens to be located in a neighborhood suffering from air pollution. That house would probably be valued less than an otherwise equivalent house located in a neighborhood with clean air. Controlling for all other attributes of the houses, the difference in the house prices would reflect the amount that people value clean air. Just like your grocery bill, controlling for everything else in your shopping cart, reflects the case of Mountain Dew you just stuck in there.

Without going into too much gory detail, hedonic pricing studies collect data on the prices of a number of houses and various attributes of those houses, including a measure of some environmental good like local air quality levels. The attributes could include features of the houses themselves as well as characteristics of the neighborhood, like local income levels, local crime rates, local school quality, and distance from the city center.

Then something like the following (obviously incomplete) model is estimated using statistical techniques that economists call *econometrics*:

$$\text{House Price} = a + b \left( \text{\# of bedrooms} \right) + c \left( \text{\# of bathrooms} \right)$$

$$+ d \left( \text{house square footage} \right)$$

$$+ e \left( \text{median income in neighborhood} \right) \tag{3.9}$$

$$+ f \left( \text{Air quality measure} \right)$$

Where the coefficient on each attribute variable represents the contribution of that attribute to the value of a house. So, for example, the estimate of f – the coefficient on a measure of air quality – would represent our best guess as to the value of a unit of air quality, as measured by people's *willingness to pay* (WTP) for it when they buy a house. These estimated WTPs are then used to estimate demand curves for air quality, from which are derived measures of the benefits of air quality protection. For a summary of a well-known hedonic pricing study, see Box 3.1. If you are interested in exploring further, there are a few additional representative studies using hedonic pricing in *Further readings* at the end of the chapter.

---

**Box 3.1: Hedonic pricing air quality study, Zabel and Kiel (2000)**

The economists Jeffrey Zabel and Katherine Kiel did a hedonic pricing study of air quality in 2000 [Zabel and Kiel (2000)]. For this study, they collected data on housing prices in four U.S. cities: Chicago, Denver, Philadelphia, and Washington, D.C. over a roughly fifteen year period. During this period, there was generally improving air quality as well as significant differences in air quality within each city and across the cities. They were able to collect prices for individual houses, along with a wide variety of attributes of each house, and then associate these with air pollutant levels from local monitoring stations. They calculated total benefits from $171 million in Denver all the way up to $953 million in Philadelphia from improving air quality levels sufficiently to meet national air quality standards.

---

*B. Travel cost approach*

The second revealed preference approach is the travel cost approach. To understand this approach, recall the last time you visited a national park, like Yellowstone, Yosemite, or the Cairngorms. You probably had to spend a lot of time and money to get there. Maybe you flew to the nearest airport and then rented a car. Or maybe you took a long road trip. For the experience of visiting Yellowstone, you had to pay a price: the cost of traveling to get there. And that was probably a pretty big cost: I know it is for me because I live in Minnesota, and that is a *long*, mostly boring, drive. So, why did you go? Presumably because it was worth it to you. That is, your willingness to pay for a trip to Yellowstone exceeded the price you had to pay in terms of the cost of traveling to get there. Or else you would not have done it, right?

The idea behind the travel cost approach is that for some environmental goods like trips to national parks, the cost of traveling to get there can be treated as the price of the good. So, even though, like many environmental goods, these goods are not being transacted on markets, there is still a "price": namely, the cost of the trip. This travel cost can be used to estimate WTP for the good and, therefore, how much benefit is derived from consuming it. For a thumbnail summary of an actual travel cost study, see Box 3.2. For some additional examples of travel cost studies, see the studies at the end of the chapter under *Further readings*. These are attempts to value fishing trips, hunting trips, visits to marine parks in low-income countries, Civil War battlefields, and cultural heritage sites.

---

**Box 3.2: Travel cost study, Offenbach and Goodwin (1994)**

The economists Lisa Offenbach and Barry Goodwin did a study where they applied the travel cost methodology to determine the demand for hunting trips in Kansas. For data on hunting trip outings and travel costs, they asked the following questions:

(1) Approximately how many times last season did you hunt your favorite windbreak hunting site? _____

(2) Thinking back to the last time you hunted on your favorite windbreak site, please estimate the total amount you spent on this trip for:

Gas _____ Dollars
Food _____ Dollars
Lodging _____ Dollars

Controlling for a number of factors, they concluded that hunters derived between $161 and $177 dollars per trip.

---

As you can probably tell from this discussion, economists have mostly applied the travel cost method to recreational activities, where there is an important travel component. It obviously could not be used to value environmental goods like climate change mitigation, or national water quality standards. Furthermore, there are numerous complications that economists must address in practical applications. For one, how do you address the likely fact that a given trip may have multiple destinations? If so, how does one ascribe a specific price to any one of those destinations?

Another issue is that travel cost generally does not merely consist of physical inputs into travel like gasoline costs and meals on the road but, also, the opportunity cost of the traveler's time.[2] The question then becomes: what value do you ascribe to the time cost involved in traveling to the site? And even if you can settle on an hourly rate, what do you assume about how long it takes to drive there? These and an entire slew of other challenging questions must be addressed whenever you perform a travel cost study.

## II. Stated preference

The second general type of valuation method is the stated preference approach. There are two approaches you should be aware of: *contingent valuation* (CV), and the *choice experiment*. Contingent valuation is the more well-known of the two, but the choice experiment has been attracting increasing attention in recent years.

### A. Contingent valuation

The idea behind contingent valuation is pretty intuitive. How do you find out how much value people derive from the environment? Answer: *You ask them*. That is, rather than inferring people's preferences based on what they do, you simply ask them to tell you what their preferences are. Sounds easy, right? Plus, it would seem to have much wider applicability than the revealed preference approaches we just discussed, which seem to be confined to environmental amenities that you can travel to and ones that are associated with your property.

Doing a CV study involves creating a survey containing questions designed to elicit the WTP of respondents for a particular environmental good. But, before diving into the details, let me expand on the idea that it has wide applicability. When your strategy is simply to ask people how much they value an environmental good, this opens up a world of possibilities. For example, you could ask them how much they value:

- Restoring eroded ocean beaches in New Jersey [Silberman et al. (1992)];
- Medicinal herbs and other non-forest products in India [Chopra (1993)];
- Visibility over national parklands in the U.S. Southwest [Schulze et al. (1983)];
- Wildlife relocations in Botswana [Thomas and Mmopelwa (2012)];
- Preservation of Indigenous moss in subantarctic Chile [Cerda et al. (2012)].

As you can see, all of these have been the subject of actual studies, which are shown here and also referenced at the end of the chapter.

This list indicates a world of possibilities in the types of environmental goods that people can be asked to value. This list suggests that they could be goods that people use, do not use, and even ones they never intend to use. Economists use these possibilities to categorize types of value that could be calculated in stated preference studies. These types of value are known as: *use value, non-use value,* and *existence value.* In this last category, the value is derived merely from the knowledge that a good – such as a particular tropical rainforest – exists. So even if I know that I may never visit Iguazu Falls, I may nevertheless derive existence value from it.

To illustrate the CV approach, let us begin by considering this question from a CV study conducted in 1996 for the U.S. Forest Service (see Box 3.3). In this study, the Forest Service wanted to estimate the benefits of a fire prevention program designed to protect old-growth forests in Oregon that provided critical habitat for an endangered species, the northern spotted owl [Loomis et al. (1996), p. 20]. In this question, $X is a dollar amount that varies across respondents, so that different respondents with different preferences would reply with different amounts that they would be willing to pay.

---

**Box 3.3: Contingent valuation WPT question**

Suppose this Oregon Old Growth Fire Prevention and Control Program proposal was on the next ballot. This program would reduce by half the number of acres of old-growth forests in Critical Habitat Units that burn in Oregon each year. If it cost your household $X each year, would you vote for this program?

a. YES  b. NO

---

I chose this example because it reflects many standard components of the CV approach. This question is an example of a *dichotomous,* or *binary, choice* question. This terminology refers to the fact that there are simply two possible responses, yes or no. If a respondent answers "yes," this implies that their WTP for this program is at least as high as $X. If they answer "no," this implies that their WTP is less than $X. So you can think of X as being the "price" of the program to the respondent and, depending upon their answer, that they would either be willing to "buy" the program at that price, or not. By varying X across different respondents, we get "demand" responses to different prices, which enable us to construct what is essentially a demand curve for the program.

You might be wondering why the question was worded in this yes/no way. I say this because I can think of other formats. For example, the question could have asked a completely open-ended question: What is the most you would be willing to pay each year for this program? And then left space for the respondents to fill in a number. Or it could have provided a range of options: $0 to $5, $5 to $10, $10 to $15, and so forth, and then directed respondents to pick one. This is sometimes called the *payment card* format. Economists have considered all of these options (and more!) when performing contingent valuation.[3]

However, the dichotomous choice format is the one that is most widely used in contingent valuation studies, as it is generally recognized to have certain important advantages over other options. For example, one distinct advantage that the dichotomous choice approach has over the open-ended approach is that it is much easier to answer. When studies in the past have tried to pose open-ended WTP questions, researchers have gotten an annoyingly large number of "don't know" responses, all of which were useless to the researchers and had to be discarded [Carson and Hanemann 2005, p. 870].

The payment card format would seem to have the advantage of being easier to answer, if not quite as much as the dichotomous format. After all, it is not too hard to pick from a limited number of options. However, it poses other problems. One important one is something called *range bias*. It turns out that the very ranges of values that you choose for the response options can themselves affect the responses. For example, if you specify the ranges $1–$5, $5–$10, $10–$15, $15–$20, and $20–$25, you will tend to get lower responses than if you specify the ranges of, say, $15–$20, $20–$25, $25–$30, $30–$35, and $35–$40. This is because people are impressionable. They take cues from you, the researcher. And, if you specify the higher ranges of values, many figure that must be the "correct" range of responses.

But perhaps even more important than ease of answering is that, under certain known conditions, the dichotomous choice format is believed to encourage respondents to answer WTP questions more truthfully. To understand why, it may help to put yourself in the shoes of a respondent who happens to care deeply about the environment. Suppose you were asked to take the USFS CV survey and it led with the following statement: *"Your answers will make no difference, none, zero, nada, in influencing fire prevention policies to preserve spotted owl habitat."* It may be hard to imagine why any CV survey would ever say this (and the USFS survey did not!), but humor me. If it did, you would probably not be inclined to take the survey seriously. I mean, your first impulse might be to throw it in the trash. But even if you took the time to fill it out, my guess is that you would not take two seconds to think seriously about whether you would vote yes or no for that program.

So the first principle of good CV method is to try to convey to respondents that their answers will matter. Or, in the terminology of the CV literature, that their answers are *consequential*. All good CV studies, and the USFS study is no exception, try to do this.

However, if they *do* make their answers consequential, it creates a potential problem. To see the problem, consider this image of a couple of pandas (see Figure 3.4). Adorable, aren't they? Now suppose the survey was about preserving panda bear habitat. And suppose some respondents believed that their answers might influence whether the habitat

*Figure 3.4* Pandas. *Source*: Shutterstock

was actually going to be preserved. I can see panda enthusiasts answering that they would be willing to pay a large sum of money – say, $1 million – to preserve their habitat, if they thought this would help save the panda bears. I mean, look at them: Wouldn't you? And they can say this with impunity because no one is going to actually make them pay. But, of course, if push came to shove, it seems highly unlikely that they would actually be willing to pay $1 million, which economists say is what really matters.

So this is another danger of the open-ended format – and here I am going to pull out another $64 term from the CV literature – that open-ended questions are not *incentive compatible*. That is, they tend not to reveal the true preferences of respondents on conse-quential questions. Researchers have found that dichotomous choice questions tend to be much better in revealing true preferences. So the bottom line is that when doing a CV study, ask consequential questions. And, if you do, your best bet of getting accurate answers is to ask them in dichotomous choice form [Carson and Groves (2007)].

For these and other reasons, dichotomous choice appears to be the best question for-mat, in terms of being able to elicit the most accurate responses. Nevertheless, you should be aware of various concerns that economists have raised about the reliability of contin-gent valuation as a method for accurately gauging the value of environmental goods. In general, it turns out that asking people questions in a way that reveals accurate information about their true preferences is much more difficult than it sounds. Ask anyone who has ever wondered whether their significant other thinks they should lose some weight.

Perhaps the biggest concern of economists is that respondents are being confronted with a hypothetical situation that is, well, hypothetical. This fact alone raises serious ques-tions regarding the credibility of their responses. For one thing, talk is cheap. Anyone can *say* they are going to do something, like vote for fire prevention programs that would cost them some money. But if you required them to pay, how do you know that they actually would?

It is not even a question of respondents being dishonest or disingenuous. It may simply be difficult for them to know how much they would be willing to pay because they may not have a clear idea of what the good actually is. If I were to ask you how much you would pay for a *schmoozel*, what would you say? My guess is that this would be hard for you to answer because you would have no idea what I was talking about (In fact, I made that word up). More generally, the less actual experience respondents have with an environmental good, the harder it will be for them to give an informed answer that tells you how much the good is actually worth to them.

As a matter of good practice, therefore, CV surveys typically include information regarding the environmental good in question. In the fire prevention example, the authors provided concrete information on the impact of fire on old-growth forests and described the fire prevention programs. Others have gone further and conducted intensive workshops that provided detailed information about the good, or even used virtual reality software to convey images of goods, like landscapes [Atkinson et al. (2012), pp. 30–1]. Pre-testing the survey beforehand with a smaller group of respondents can also alert CV practitioners to potential problems, like respondents having no idea how big a hectare actually is.

However, despite all this care given to these issues, when economists have gone out and done CV studies, they often get stated response values that are quite different from actual willingness to pay. When this happens, we say that there is *hypothetical bias*. In most cases, when studies find hypothetical bias, it is in the upward direction; that is, that stated willingness to pay exceeds actual willingness to pay [Hausman (2012), pp. 44–5; Kling et al. (2012), p. 15].

And, in some cases, the difference is really large. For example, one well-known study examined a number of different studies and found that, on average, hypothetical values exceeded actual values by 300% [List and Gallet (2001)]. As an extreme example, one CV study conducted in the early 1990s concluded that public WTP to save whooping cranes from extinction was around $32 billion per year, or nearly 12 times larger than annual giving to *all* environmental non-profit organizations in the United States at the time [Stewart (1995), p. 237].

These and other issues have led some economists to throw up their hands and conclude that there is really no point in doing a CV study. I will call this group the *pessimists* because they tend not to believe that the CV method can be improved sufficiently to yield useful value numbers. To pessimists, contingent valuation is so unreliable that it is better to go with no number at all rather than the number estimated from a CV study. One prominent economist has famously called it "hopeless" to do a CV study, arguing that there is simply no way to get reliable, usable results. Others have similarly argued that the difficulties are so serious and uncorrectable that CV studies "should simply be abandoned."[4]

Other economists are more sanguine, however. I will call this group the *optimists*. Optimists argue that we are continually learning about the pitfalls of doing contingent valuation studies and how to avoid those pitfalls [See, for example, Haab et al. (2013)]. One approach that may limit biased results is the so-called *cheap talk* approach. Under this approach, prior to taking the survey participants are warned that people tend to overstate the values they report when taking a survey [Cummings and Taylor (1999)]. Apparently, just being made aware of this danger can make people respond more accurately.

Other researchers have found that adding extra script in the survey that reminds respondents that they can use their discretionary income to support other environmental causes also tends to lower their WTP [Li et al. (2005)]. So simply being made conscious of the tradeoffs in supporting different environmental goals can also help.

When it comes down to it, the problem is that for many environmental goods, there seems to be little choice.[5] If you actually want to do a cost-benefit analysis, you require some measure of the benefit derived from the good. And, for a lot of goods, asking people to state their preferences may be the only way to get information regarding that benefit. It is hard to see how else to come up with valuation numbers, for example, for Chilean indigenous moss! Pessimists respond that rather than relying on people off the street, we should ask experts: at least their opinions are likely to be informed. Optimists are less sanguine that using experts will yield any more accurate results.[6]

In the end, even the optimists recognize that much work needs to be done to correct the problems associated with the CV approach. While recognizing merit in the positions of the pessimists, however, they argue that even if their results are not terribly precise, contingent valuation studies can still give us useful general guidance. At the very least, they argue that CV studies might permit us to rule out certain options from consideration. And they advocate ongoing research in order to continue to refine and improve the CV methodology, to allow us to obtain more reliable responses in the future.

## B. Choice experiment

The other stated preference approach we will discuss here is called the *choice experiment*. This approach has been growing in popularity among economists in recent years. It does not address the pessimist's objection that it is unreliable because it is inherently hypothetical and not based on real choices. However, it does address some of the potential sources of bias in the CV approach. On the other hand, as we shall see, it has some disadvantages of its own.

To understand the choice experiment approach, let us go back to the shopping cart metaphor that we saw earlier. As you recall, a shopping cart full of different items is similar to a house with different attributes, with each individual item contributing so much to the cost of the whole. Then the hedonic pricing approach consists of taking the price of the house and using this information to infer the value of each attribute, including an associated environmental good. Under the choice experiment approach, respondents are presented with two packages of attributes, and simply asked which one they prefer. So it is kind of like being given your choice of two shopping carts full of different grocery items and being asked to pick one. Responses can then be used to infer the value they ascribe to individual attributes, including environmental goods.

For example, consider two shopping carts, one containing lots of bottles of Coca-Cola and just a few bottles of Pepsi-Cola and the other with the proportions reversed. To keep things simple, assume bottles of each cost the same. If a shopper said he would rather have the first cart, you can be pretty sure that he prefers Coke over Pepsi. In an actual choice experiment study with many grocery items, an econometric analysis would be used to actually quantify the difference in value ascribed to different items.

A simple example from an actual study may help clarify the approach [Adamowicz et al. (1998)]. The study was concerned with whether to increase moose and caribou habitat in western Canada. Respondents were presented with various scenarios consisting of different combinations of attributes. The attributes presented to respondents were: moose/caribou population sizes, wilderness area, various restrictions on recreation, forest industry jobs, and changes in income tax. Each respondent was presented with a series of head-to-head competitions of scenarios containing different amounts of each attribute. As is usual with choice experiments, respondents were also given the option of picking no change in any of the attributes, an option that represents the status quo.

Based on an econometric analysis of the responses, they were able to quantify the effects of the individual attributes. Respondents supported greater moose and caribou herds and greater wilderness areas, opposed greater restrictions on recreational use and higher taxes, and exhibited no preferences on employment [Adamowicz et al. (1998), pp. 70–1].

Choice experiments are believed to have certain advantages over CV studies in obtaining true valuations of environmental amenities. For example, respondents to CV surveys may experience a so-called *yea-saying bias*: that it may be psychologically difficult to say no to a request to give money. Choice experiments may not suffer from this problem because respondents are just being asked which attributes they like. On the other hand, choice experiments face certain challenges as well. For one, they enable valuation of individual attributes of an environmental amenity, but do not permit valuation of the overall amenity itself. Furthermore, if not done right, choice experiments may overload respondents with too many scenario choices. Or respondents may overly focus on certain attributes, especially if presented with a large selection. But overall, choice experiments may represent a reasonable alternative to CV studies.

## Key takeaways

- The fundamental rationale for cost-benefit analyses is improved resource allocation efficiency, which may come at the expense of fair or equitable outcomes, including for future generations.
- In the case of long-lived projects or policies, the choice of discount rate may have a huge impact on the bottom line of cost-benefit analyses.
- There are two basic approaches to environmental valuation: *revealed preference*, and *stated preference*.
- All valuation approaches present methodological challenges to deriving accurate, usable valuation numbers.
- There is a sharp divide between CV practitioners and other economists who do not believe the CV method can produce useful results.
- Stated preference approaches are highly controversial but for many environmental amenities, they may be the only game in town.

## Key concepts

- Cost–benefit analysis
- Kaldor–Hicks criterion
- Hyperbolic discounting
- Sustainability
- Revealed preference
- Hedonic pricing

- Travel cost approach
- Stated preference
- Contingent valuation
- Choice experiment

## Exercises/discussion questions

3.1. Verify the PDV numbers at different discount rates in the example about the dam that requires an upfront expenditure of $50 million. This can easily be done using a spreadsheet program like Excel.

3.2. Assume that under ongoing climate change, there is a 50–50 chance that the Antarctic ice sheet could break off 10 years in the future, inflicting $100 billion of damages per year after that into the indefinite future. Call this the *high-damage scenario*. On the other hand, if it does not break off, it would inflict zero damages. Call this the *low-damage scenario*. Suppose you want to discount the high-damage scenario by 2% and the low-damage scenario by 10%. (a) Calculate the implied (declining) discount rate over the next 20 years. (b) Suppose that, in order to keep the high-damage scenario from occurring, you need to incur $10 billion of costs annually beginning right now. Using your calculated discount rates, would doing so be cost-beneficial? {You may assume that nothing that occurs after twenty years affects the costs or benefits.}

3.3. You are interested in using the travel cost approach to wilderness valuation to determine the value of protecting Mono Lake, a scenic lake with unique geological formations east of the Sierra Nevada Mountains in California. Toward this end, you go out and collect the following information. First, you find out that visitors to Mono lake come ONLY from New York City (population 10 million), Los Angeles (population 3 million), and San Francisco (population 1 million). From New York City come 10,000 tourists per year, paying on average $600 in travel costs. From Los Angeles come 9,000 tourists per year, paying on average $400 in travel costs. From San Francisco come 6,000 tourists per year, paying on average $100 in travel costs. Assuming tourists are made to pay zero entrance fees to visit the lake, estimate the implied value of preserving the lake.

3.4. Suppose you want to do a contingent valuation study to assess the value of a Taiwanese wetland (see Hammitt in Further Readings). One option you have is to pull together a team of ecologists, hydrologists, and economists with specialized expertise in the uses of wetlands. Why would you not use the opinions of these experts rather than solicit the WTP of tourists and local residents with no particular expertise in wetlands?

3.5. We saw that when running choice experiments, practitioners generally include the status quo as one option. Why do you think it is good practice to do so?

## Notes

1 See Olson and Bailey (1981) for this argument.
2 For a couple of studies that discuss this issue, see McConnell and Strand (1981); Bockstael et al. (1987).
3 See, for example, Carson and Hanemann 2005, pp. 869–73.
4 For these views, see Hausman (2012); Boudreaux et al. (1999), p. 776.
5 See, for example, Portney (1994), p. 14.
6 For the pessimist's view on using experts, see Hausman (2012), p. 44. For the optimist's view, see Haab et al. (2013), pp. 604–6.

# Further readings

## I. Hedonic pricing studies

Irwin, Elena G. "The effects of open space on residential property values," *Land Economics* 78 (November 2002): 465–80.

Ozdenerol, Esra et al. "The impact of traffic noise on housing values," *Journal of Real Estate Practice and Education* 18(2015): 35–54.

Wasson, James R. et al. "The effects of environmental amenities on agricultural land values," *Land Economics* 89(August 2013): 466–78.

## II. Travel cost studies

Alberini, Anna and Alberto Longo. "Combining the travel cost and contingent behavior methods to value cultural heritage sites: Evidence from Armenia," *Journal of Cultural Economics* 30(2006): 287–306.

Androkovich, Robert A. "Recreational visits to the Adam's River during the annual sockeye run: A travel cost analysis," *Marine Resource Economics* 30(January 2015): 35–49.

Melstrom, Richard T. "Valuing historic battlefields: An application of the travel cost method to three American Civil War battlefields," *Journal of Cultural Economics* 38(August 2014): 223–36.

Mwebaze, Paul and Alan MacLeod. "Valuing marine parks in a small island developing state: A travel cost analysis in Seychelles," *Environment and Development Economics* 18(August 2013): 405–26.

## III. Contingent valuation studies

Cicchetti, Charles J. and V. Kerry Smith. "Congestion, quality deterioration, and optimal use: Wilderness recreation in the Spanish Peaks Primitive Area," *Social Science Research* 2(1973): 15–30.

Hammitt, James K.; Jin-Tan Liu; and Jin-Long Liu. "Contingent valuation of a Taiwanese wetland," *Environment and Development Economics* 6(May 2001): 259–68.

Loomis, John B. and Douglas M. Larson. "Total economic values of increasing gray whale populations: Results from a contingent valuation survey of visitors and households," *Marine Resource Economics* 9(Fall 1994): 275–86.

## III. Choice experiment studies

Grant, Kara; R. Karina Gallardo; and Jill J. McCluskey. "Are consumers willing to pay to reduce food waste?" *Choices* 34(1$^{st}$ Quarter 2019): 1–7.

Krueger, Andrew D.; George R. Parsons; and Jeremy Firestone. "Valuing the visual disamenity of offshore wind power projects at varying distances from the shore: An application on the Delaware shoreline," *Land Economics* 87(May 2011): 268–83.

# References

Adamowicz, Wiktor; Peter Boxall; Michael Williams; and Jordan Louviere. "Stated preference approaches for measuring passive use values: Choice experiments and contingent valuation," *American Journal of Agricultural Economics* 80(February 1998): 64–75.

Atkinson, Giles; Ian Bateman; and Susana Mourato. "Recent advances in the valuation of ecosystem services and biodiversity," *Oxford Review of Economic Policy* 28(Spring 2012): 22–47.

Bockstael, Nancy E.; I.E. Strand; and W. Michael Hanemann. "Time and the recreational demand model," *American Journal of Agricultural Economics* 69(May 1987): 293–302.

Boudreaux, Donald J.; Roger E. Meiners; and Todd J. Zywicki. "Talk is cheap: The existence value fallacy," *Environmental Law* 29(Winter 1999): 765–809.

Carson, Richard T. and Theodore Groves. "Incentive and information properties of preference questions," *Environmental and Resource Economics* 37(2007): 181–210.

Cerda, Claudia; Jan Barkmann; and Rainer Marggraf. "Application of choice experiments to quantify the existence value of an endemic moss: A case study in Chile," *Environment and Development Economics* 18(2012): 207–24.

Chopra, Kanchan. "The value of non-timber forest products: An estimation for tropical deciduous forests in India," *Economic Botany* 47(July–Sept 1993): 251–7.

Cummings, Ronald G. and Laura O. Taylor. "Unbiased value estimates for environmental goods: A cheap talk design for the contingent valuation method," *American Economic Review* 89(1999): 649–65.

Haab, Timothy C.; Matthew G. Interis; Daniel R. Petrolia; and John C. Whitehead. "From hopeless to curious? Thoughts on Hausman's 'Dubious to Hopeless' critique of contingent valuation, " *Applied Economic Perspectives and Policy* 35(December 2013): 593–612.

Carson, R.T. and W.M. Hanemann. "Contingent Valuation," *Handbook of Environmental Economics* 2(2005): 821–936.

Hausman, Jerry. "Contingent valuation: From dubious to hopeless," *Journal of Economic Perspectives* 26(Fall 2012): 43–56.

Kling, Catherine L.; Daniel J. Phaneuf; and Jinhua Zhao. "From Exxon to BP: Is some number better than no number?" *Journal of Economic Perspectives* 26(Fall 2012): 3–26.

Li, Hui, et al. "Testing for budget constraint effects in a national advisory referendum survey on the Kyoto Protocol," *Journal of Agricultural and Resource Economics* 39(August 2005): 350–66.

List, John A. and Craig A. Gallet. "What experimental protocol influence disparities between actual and hypothetical stated values?" *Environmental and Resource Economics* 20(2001): 241–54.

Loomis, John B.; Armando Gonzalez-Caban; and Robin Gregory. "A contingent-valuation study of the value of reducing fire hazards to old-growth forests in the Pacific Northwest," United States Department of Agriculture. Forest Service. *Research Paper PSW-RP-229-Web.* July, 1996.

McConnell, K.E. and I.E. Strand. "Measuring the cost of time in recreation demand analysis," *American Journal of Agricultural Economics* 63(1981): 153–6.

Offenbach, Lisa A. and Barry K Goodwin. "A travel-cost analysis of the demand for hunting trips in Kansas," *Review of Agricultural Economics* 16(January 1994): 55–61.

Olson, Mancur and Martin J. Bailey. "Positive time preference," *Journal of Political Economy* 89(1981): 1–25.

Pearce, David; Ben Groom; Cameron Hepburn; and Phoebe Koundouri. "Valuing the future: Recent advances in social discounting," *World Economics* 4(April–June 2003): 121–41.

Portney, Paul R. "The contingent valuation debate: Why economists should care," *Journal of Economic Perspectives* 8(1994): 3–17.

Rolston, Holmes. *Philosophy gone wild.* Buffalo: Prometheus, 1989.

Schulze, William D. et al. "The economic benefits of preserving visibility in the national parklands of the Southwest," *Natural Resources Journal* 23(January 1983): 149–73.

Silberman, Jonathan; Daniel A. Gerlowski; and Nancy A. Williams. "Estimating existence value for users and nonusers of New Jersey Beaches," *Land Economics* 68(May 1992): 225–36.

Solow, Robert M. "Sustainability: An economist's perspective," in *Economics of the Environment* (3rd ed.), Robert Dorfman and Nancy S. Dortman (eds.). New York: Norton, 1993: pp. 179–87.

Soule, M.E. "What is conservation biology?" *Bioscience* 35(1985): 727–34.

Stewart, Richard B. "Liability for natural resource injury: Beyond tort," in *Analyzing Superfund: Economics, Science, and Law.* (Richard L Revesz and Richard B. Stewart (eds.) Washington: Resources for the Future, 1995.

Thomas, Elizabeth and G. Mmopelwa. "International tourists' willingness to pay for relocation of elephants to manage herd size in Botswana," *Botswana Notes and Records* 44(2012): 144–53.

Zabel, Jeffrey E. and Katherine A. Kiel. "Estimating the demand for air quality in four U.S. cities," *Land Economics* 76(May 2000): 174–94.

# 4 Institutions, property rights, and transaction costs

One of the key objectives of this book is to make explicit, and sometimes to challenge, some of the assumptions that are often unstated or implicit in standard economic analysis. Consider, for example, the market analysis developed in chapter two, which is the standard treatment of markets in many economics courses. In this analysis, shown in Figure 4.1, suppliers (typically firms) have something that demanders (typically households) want, and they come together in a market to engage in exchange at a price that is determined by competition among demanders and among suppliers. The result is a prediction regarding how much will be sold and at what price, what we call the equilibrium quantity $Q^\star$ and equilibrium price $P^\star$. The market model also yields predictions regarding what will happen to the equilibrium price and quantity when conditions change. Like an upward shift in demand to D' that results in the new equilibrium quantity Q' and equilibrium price P'. This is a very powerful model with real-world applications in many areas of the economy.

However, it leaves many questions unanswered. What exactly is the good that buyers are buying? If it is a parcel of land, are there legal restrictions on how it may be developed? How do we know the suppliers have the legal right to sell the good they are bringing to the market? If it is a house, is there a lien on it and will the lienholder, say a bank, object to the transaction? How easy will it be for sellers and buyers to write a contract that accomplishes the transaction? Will the potential buyer of a car suspect that the seller is trying to sell her a lemon? If so, what is the legal liability for engaging in fraudulent activity? Will the buyer be sufficiently reassured that she will go ahead with the transaction? All of these questions may come up in real life even when considering something as fundamental to economics as the operation of a market. And the answers may well determine whether a car, house, or parcel of land actually gets sold, and on what terms.

In this chapter, we develop several foundational concepts that will be central to our analysis of a wide range of environmental issues in the rest of this book. These concepts will make explicit many of the implicit assumptions of standard economic analysis. They will provide a conceptual framework for understanding environmental issues that enables additional, deeper insights into the true nature and causes of environmental problems. Finally, they will shed light on certain shortcomings of traditional environmental policies and permit us to propose solutions that go beyond these policies.

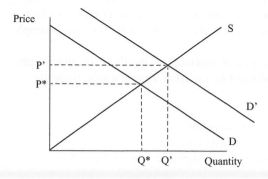

*Figure 4.1* Standard market analysis

## Institutions

Let us begin with the economist's notion of the crucial concept of an *institution*.

### I. What are institutions?

It will help to be extremely careful in defining the concept of an institution. I want to be careful here because different people use this word to mean different things. One common meaning that people ascribe to the word institutions is that it refers to entities or organizations. For example, we hear people talk about the institutions of City Hall, or Congress, or General Motors, or your local credit union. Importantly, in the sense we will be using them throughout this book, you should *not* think of institutions as entities or organizations. Rather, we will use the word to describe what the economist Douglass North has called the *rules of the game.*

To understand the distinction, consider the times you may have played the board game Monopoly. In that game, we would not use the word institutions to refer to Electric Company, or Water Works, or B&O Railroad. Rather, the institutions are the rules you play by, like if you pass *Go*, you get to collect $200. Or if you are unfortunate enough to land on Boardwalk when it has a hotel on it, you have to pay the owner a lot of money. These are the rules you are required to abide by when you play Monopoly. Douglass North would call these rules the institutions of that board game.

Similarly, when we go about our business in the real world, there are many rules we have to abide by. We cannot just walk up to a stranger and hit him with a baseball bat. When driving on the highway, we need to obey the speed limit. When we take out a loan from a bank, we have to pay it back with the agreed upon interest. If we do not do these things, there may be consequences, like going to prison, receiving a speeding ticket, or losing the car that you put down as collateral for the loan.

In the context of the earlier market example, many institutional rules govern how that market will actually work. The parcel of land may be subject to zoning restrictions, which determine how the land can be developed. When you contract to buy a house or a car, there are rules and penalties that impose costs on people who renege on those contracts;

say, for non-payment or not living up to the servicing terms of a warranty. Even simple consumption activities like buying groceries from your local supermarket may be affected by whether or not there are shoplifting laws, which make it more costly for you to simply steal the groceries instead of buying them.

To be clear, institutional rules do not just govern market activity: there are many institutional rules that govern activities not conducted in a market. For example, there are:

- Anti-littering laws;
- Laws that determine liability in automobile accidents;
- Laws against driving under the influence or while texting;
- Laws against various personal crimes like assault, burglary, arson, and murder;
- Constitutional rights to free speech;
- Prohibitions against racial segregation in public schools;
- Regulations against smoking in closed public spaces;
- Zoning restrictions on certain types of development;
- The tradition of trick-or-treating on Halloween;
- Social norms against talking with your mouth full;
- And so forth (I could go on and on).

All of these are rules governing individual behavior and, importantly, all of them may be understood in economic terms using the approach developed in this book. Of course, in this book we will be emphasizing rules relating to environmental issues.

Before proceeding, I would like to point out two things about the items on this list. First, you may notice that institutional rules come from various sources. Some of these rules are the result of legislators passing laws. Others come from the rulings of courts. Still others are based on the provisions of the Constitution. In the real world, there many sources of institutional rules, and it will be important to keep in mind where those rules come from. This is because the sources of those rules will be important in helping us interpret how and why those particular rules were adopted. Keeping the sources in mind will also help us interpret the economic content of those rules.

Second, you may notice that not all of these rules derive from official organs of government, like courts and legislatures. Some are simply based on longstanding traditions, customs, or norms of behavior. For example, it is believed that Halloween originates in centuries-old Celtic traditions. And talking with your mouth full is simply something you try to avoid doing, probably because your parents raised you well. But if you ever asked them *why* you should not talk with your mouth full, you probably got back that annoying parental rejoinder: "Because I said so!" I am pretty sure they did not cite some local ordinance or Act of Congress. All of this goes to show that institutional rules can be based on social norms, customs, or traditions. Economists call these types of institutions *informal institutions*. But just because they are informal does not mean that they do not exert a powerful influence on people's behavior.

## II. The economic function of institutions

### A. What do institutions do?

Institutions are important because they provide structure to our interactions with each other. They tell us what actions must be done (abide by the speed limit), what actions must not be done (assault a stranger with a baseball bat), and what actions may be done (exercise your right to free speech). They also provide mechanisms for resolving disputes and for enforcing our agreements with each other. In general, they perform two valuable economic functions.

First, they provide a particular set of incentives that influence our behavior in certain ways. For example, speed limit laws enforced by tickets and fines make it more costly for us to drive 90 mph in a 55 mph zone, making us less likely to do it. Second, they reduce uncertainty regarding the outcomes of our economic activity. For example, patent laws forbid others from copying our inventions. This makes it more likely that we will be able to profit from engaging in inventive activity. In serving both of these functions, institutions support our ability to live our lives in peaceful harmony with others, and to engage in orderly, productive economic activity.

To begin to appreciate the importance of institutions, let us engage in a thought experiment. Imagine an economy with few rules that govern economic interactions, what I will call an *Anything Goes* economy. In such an economy, people could not rely on others to live up to the terms of their contracts. People would not even know what they own, if there were no laws defining property or what you are entitled to do with property (in fact, I am not sure the word *property* would even exist). This would obviously make it difficult to sell goods in markets. In fact, it is hard for me to see anyone wanting to produce much except maybe for their own consumption, because it would be difficult to realize any returns on production. Furthermore, people might have little incentive to make productive investments in land or resource development or production of goods and services, because they would be unlikely to be able to enjoy the fruits of these investments. Under *Anything Goes*, the incentives for productive activity would be all wrong.

Now imagine that there are rules. Here, I can see some rules being better than others, in terms of encouraging economic activity and efficient use of resources. Rules that clearly define what people can do with their property would probably be better than rules that do not. This includes, for the most part, rules that allow people to trade, or buy and sell. However, rules that allow people to do whatever they bloody well please with their property would not be ideal, if their actions impose costs on others. Rules that protect people in the possession of their property would be better than rules that do not. Rules that allow people to realize returns on their investments would be better than rules that do not. This is by no means a comprehensive list of desirable qualities of rules. The point here is simply comparative: there can be good rules and bad rules.

### B. Institutional change and economic performance

The reason this is important is that it provides a way to think about desirable institutional rules, "good" being better than "bad." In addition, it allows us to think about how institutions might change, and why. Finally, it allows us to evaluate how well those institutions serve the economic needs of a society. A key question, of course, is: What do we *mean* by good and bad? For the reasoning behind all this, let us return to the ideas of Douglass North (Figure 4.2).

Douglass North is, it is fair to say, a giant in the field of the economics of institutions. Though he has passed away, his ideas remain and they have had a profound influence on an entire generation of scholars. Primarily for his work in this area, he was a co-recipient of the Nobel Prize in economics in 1993.[1] Let us consider what he had to say about institutions.

*Figure 4.2* Douglass North. *Source*: No attribution necessary: Wikimedia Commons

To North, institutions did not merely have economic consequences, as we have just argued. They were also themselves explainable by economic factors. So, for example, a particular set of institutions might prevail under a particular set of economic conditions. And then conditions might change to the point where those institutions become untenable, at which time new institutions emerge. In economic terms, changing economic conditions might increase the societal net benefits of a different set of institutions.

North gives the example of ancient Greece, where the city state of Athens transformed from an oligarchy to a democracy because of developments in military technology. At the time Athens was constantly under threat of external attack. The development of the phalanx as the most effective battlefield strategy necessitated giving more political power to Athenian citizens because the strategy required a citizen army [North (1981), p. 30]. So, in essence, North argues that this little technological advance shifted the benefit-cost calculus in favor of democracy, resulting in a major institutional change.

North goes on to argue, however, that this is not the end of the story. If institutions could change quickly and seamlessly in response to changes in economic conditions, they would tend to be good rules: that is, reasonably efficient. The problem is that there are all sorts of real-world institutions that persist over time that no one would consider to be efficient. The communist systems in the former Soviet Union and in 20th century China under Mao Tse-tung are two good examples of this. North himself gives the example of 16th century Spain, where severe fiscal challenges faced by Spanish kings led to crushing taxation and confiscation of property, which undoubtedly harmed economic performance [North (1981), p. 151; North(1990), p. 7].

So if bad, inefficient institutions exist, what explains their existence? The answer is a bit complicated, but here is part of North's basic reasoning developed over several books and numerous articles. Self-interested leaders of countries, like the 16th century Spanish kings, tend to try to skew institutional rules in their favor, enriching themselves at the expense of the citizens of a country. They can succeed, at least for a time, especially if they can bend legislators and important government officials to their will with bribes and other compensation, or by appealing to ideology.

However, North did not believe that this fully explains the persistence of inefficient institutions because it does not explain why those crummy institutions would not be weeded out over time. The idea here is that inefficient institutions can be quite costly for an economy to endure. Just ask the poor Chinese people who had to endure the Cultural Revolution. So if other, better, institutions are available, there should be strong pressures for them to be adopted.[2] So here is the economic idea of the virtues of competition applied to institutions: just as competitive market forces drive out weak firms, similar competitive forces should drive out weak institutions, or so the argument goes. And since North believed that institutional competition was potentially quite strong, he found it puzzling that inefficient institutions could persist over time.

## C. An important concept: Path dependence

North struggled with this puzzle for some time, and here is the answer he came up with: that institutions are subject to inertia. Once they are created, they set in motion self-reinforcing tendencies to persist over time. This is the idea, very important in institutional analysis, of *path dependence*.

To understand path dependence, consider the following simple example, made famous by the economic historian Paul David: the typewriter (now word processor) keyboard, based on the QWERTY system [David (1985)].

If you have written any term papers for your college courses, you are certainly familiar with the QWERTY system, which simply refers to the arrangement of letter keys on your keyboard. The system, of course, takes its name from the six letters on the upper left-hand-side of your keyboard. David argues, and presents supportive evidence, that the QWERTY keyboard is inefficient, based on studies that have found that typists can type faster with an alternative keyboard configuration that has been available for years, the so-called DVORAK keyboard system (see Figure 4.3). I would guess that you have never heard of the DVORAK system and, if so, you are not alone: very few people have. The puzzle is: why does (virtually) everybody continue to use the QWERTY system when it is significantly slower? Why have we not collectively switched over to DVORAK and saved ourselves some X% of our time, which could be much better spent doing other things?

David's answer is that use of the QWERTY keyboard has self-reinforcing tendencies. Just about everyone uses it. And, because everyone uses it, keyboard manufacturers have a strong incentive to produce only keyboards that use the QWERTY configuration. And, because QWERTY keyboards abound, people only learn how to type using QWERTY keyboards. Which gives keyboard producers incentive to continue to produce QWERTY keyboards. And so forth. And so inefficient QWERTY keyboards continue to be used, with little sign that this is likely to change in the foreseeable future.

David's QWERTY example clearly illustrates the notion of path dependence in the development of a technology over time. There are in fact many other real-world examples of path dependence at work. The concept has been invoked to explain, for example, how the old VHS recording technology won out over its main rival at the time, Betamax. It also may explain why economic activities cluster together in cities, why we get certain kinds

*Figure 4.3* Dvorak keyboard. *Source*: Karl432 / CC BY-SA (https://creativecommons.org/licenses/by-sa
/4.0): Wikimedia Commons

of technological change, why poor people get stuck in poverty traps, and why economies
are slow to recover from economic slumps. Later in this textbook, we will see several environmental examples. In all cases, there will be self-reinforcing tendencies to remain on the
same development path.

North argues that institutions have similarly self-reinforcing tendencies. Once a set of
institutions is established, this sets in motion a chain of events that tend to perpetuate those
institutions. These events vary depending upon which institution we are talking about.
Players of Monopoly get used to, and devise winning strategies based on, existing rules and
would squawk mightily if the rules were changed. In the real world, interest groups will
form that have an interest in the perpetuation of those statutes that benefit them. Those
interest groups will tend to exert political pressure for those laws to be kept, and they will
tend to be successful if they are unified and if they have a lot at stake (see the discussion
of public choice in chapter six). Under certain legal systems, judges often base rulings on
previous rulings; that is, they rely on *precedent*. Such laws will tend to persist over time, even
when conditions change. There are many examples, argues North, of such self-fulfilling
tendencies in real-world institutions.

Maybe now you have a sense for why economists call this process *path dependent*. The
idea is simple: once you are on a historical path, you tend to stay on that path. The very
fact that we are so heavily QWERTY-dependent now makes us more likely to continue
to be QWERTY-dependent in the future. This notion emphasizes the importance of historical process and the importance of events that happened back in time. Way back when
the QWERTY keyboard was invented, there was absolutely no guarantee that it would
grow to dominate as a keyboard configuration. But small, individual decisions by users
to adopt QWERTY slowly snowballed and eventually became a QWERTY juggernaut.
Poor DVORAK never had a chance.

As it turns out, this argument provides insight into many interesting economic questions. One question that many economists have puzzled about is relative economic performance. Why is it, for example, that some countries are wealthy and others are poor? Or
what explains why a struggling, poor country can suddenly experience rapid economic
growth?

Sometimes, differences in economic performance may be explained by happenstance, relative resource endowments, differences in technology adoption, or various cultural factors (such as different tendencies (propensities) to save and invest money). However, many economists now believe that institutional differences may also help explain why some countries are rich and others are poor. This is because institutions vary in the extent to which they support sustained economic growth, with some institutions being much better than others (see Box 4.1). And, as we have just seen, once installed, institutions, including bad ones, can last for a long time.

---

**Box 4.1: Institutions and economic growth**

In a classic article in institutional analysis, Douglass North and his co-author Barry Weingast were interested in explaining the emergence of England as a dominant world power in the 18$^{th}$ century. They identify a key factor: institutional reforms that took place in England during the Glorious Revolution of 1688. Prior to that date, the English monarch, perennially deeply in debt, was in the habit of simply seizing wealth from private citizens in various ways. The 1688 Revolution ushered in reforms – including increasing the power of Parliament and greater independence of the courts – that dramatically reduced the monarch's ability to unilaterally seize wealth. The result was dramatic growth in public finance and private capital markets, which launched England on the path to sustained economic growth [North and Weingast (1989)].

---

## Transaction costs

Let us now dig a little deeper. As we were going through the discussion of inefficient institutions, you may have been puzzled by so many potentially profitable exchanges going undone. If not, put on your economist's hat and consider the example of 16$^{th}$ century Spain, where private citizens were being hurt by taxation and confiscation of property. Assume that this was inefficient, meaning that the costs to them of the system were greater than the benefits to the monarch. Economists might ask why some alternative institutional arrangement did not emerge. For example, instead of taxes and property confiscation, wasn't there an alternative form of compensation that was not so detrimental to economic growth? Couldn't such a system be worked out, perhaps in negotiations between representatives of Spanish citizens and the monarch? If the costs to citizens of the existing system were greater than the benefits to the monarch, one might think that such a change – one that would have made parties as a whole better off – should have been possible. Here is that old gains from trade argument again that economists use a lot.

Or consider again the QWERTY keyboard. Under the story told by Paul David, you currently have millions of users incurring perhaps hundreds of millions of dollars in costs every year by being slowed down by the QWERTY keyboard. Is it not possible for some enterprising firm to come along and offer to retrain users and work with manufacturers to replace QWERTY keyboards with DVORAK keyboards, in order to eliminate these losses? If this would cost less than hundreds of millions of dollars, it seems like it should be possible, again because there would be gains from trade.

All of this brings us to another extremely important concept: the notion of *transaction costs*. It is difficult to overstate the importance of transaction costs in environmental economics. They provide a way to think about and understand why potentially welfare-improving exchanges of goods or resources may not occur, including exchanges that benefit the environment. As we shall see later on in this book, the environmental applications are many, including climate change, energy policy, clean air and water policy, water resources, deforestation, fishery management, protection of endangered species, and many other issues. Transaction costs help explain seeming anomalies in a variety of areas of environmental economics. And they help us evaluate environmental outcomes in all these areas in terms of economic efficiency.

## I. What are transaction costs?

To understand the concept of transaction costs, consider again the basic market model that started off this discussion. In that model, there is an implicit assumption that it is easy (in fact, costless) for demanders and suppliers to engage in transactions. This assumption is conveyed by the simple metaphor of a produce marketplace. In this simple market, a bunch of buyers and sellers get together in a well-lit, safe, clearly defined area. Buyers walk around squeezing melons and comparing prices, trying to figure out where they can get the best goods for the best price. On the other side, sellers set prices after carefully sizing up the competition, and they set up attractive displays of unblemished melons and apples to try to get buyers to buy from them instead of one of their competitors. The whole process is quick, easy to navigate, and painless.

Simply put, transaction costs are the costs associated with engaging in transactions with others. In this marketplace example, the buyers and sellers are all in one place. Buyers can easily observe the goods and compare different sellers. Sellers have incentive not to jack up their prices because they know their potential customers can easily go elsewhere. It is a simple matter for buyers to ascertain the quality of the produce. Exchanges can occur quickly and easily with the exchange of cash or the swipe of a debit card. In this case, we would say that transaction costs are low.

However, in many cases transaction costs are not low. Indeed, in many situations they are probably quite high, as in the examples of 16th century Spain and the QWERTY keyboard. We have argued that in principle, gains from trade from negotiations between 16th century Spanish citizens and the Crown were possible. However, in practice such negotiations would have been extremely difficult to pull off. What exactly would be the alternative form of benefits that the citizens could offer to the Crown? A portion of their harvests? Military service? Furthermore, even if there was something they could offer, there were literally thousands and thousands of citizens to all get on the same page, to agree on the form and size of the payment and who should contribute how much to it. Finally, the Crown might have simply been resistant to the idea of negotiating with peasants and probably did not even need to because it could simply take what it wanted and throw people into prison or worse if they resisted. All of these things would have probably placed insuperable obstacles in the way of such exchanges actually occurring.

To place this discussion in a general framework, let us now turn to an extremely important concept: the famous *Coase Theorem*.

## II. The Coase Theorem

The Coase Theorem may be the single most important idea in all of modern institutional economics. It is named after the economist Ronald Coase (Figure 4.4), who advanced the idea in a paper published in 1960 entitled *The Problem of Social Cost*. Chiefly for this idea, Coase was awarded the Nobel Prize in economics in 1991.

*Figure 4.4* Ronald Coase. *Source*: University of Chicago Law School: Wikimedia Commons

The Coase Theorem is about the importance of transaction costs in determining economic outcomes and the efficiency of those outcomes. To help me state the theorem correctly, let us first consider the following famous example used by Coase in his paper.

Suppose you have a farmer growing crops in a field, and a rancher raising cattle on the neighboring field. Sometimes the rancher's cattle stray onto the farmer's fields, trampling the farmer's crops. So here is what seems like a textbook example of an externality, right? The rancher's cattle are inflicting a cost on the farmer and the rancher seems to have no incentive to take those costs into account. So, what should be done?

Coase's insight was that looks are deceiving. It is not that the rancher is inflicting costs on the farmer. Presumably, the rancher is deriving some benefit from his cattle being allowed to roam freely. Maybe this spares him from having to build a fence to keep the cattle in, or from having to hire extra workers to herd the cattle. If so, *not* being able to let his cattle roam freely imposes a cost on him. So, in a real sense, *the farmer is inflicting costs on the rancher*. In Coase's view, both the rancher and the farmer are contributing to the damages that are occurring. Remove either one, and the problem goes away.

In order to better understand this argument, let us consider it in a different, modern context: a busy airport. The people who live near the airport have to contend constantly with the noise of the planes taking off and landing. So it looks like the planes are inflicting a negative externality on the people. But, again, *not* being allowed to fly noisy planes would impose costs on the airlines. So, according to this argument, the people living nearby are "causing" the problem just as much as the planes are.

To many people, this is a strange argument. Isn't it the farmer's crops that are getting trampled? Isn't it the nearby residents who are being deprived of sleep at night or who cannot hear each other in conversation around the dinner table? But Coase's interpretations follow directly from the economic notion of opportunity cost, that (virtually) every action comes at a cost: be it in trampled crops, fencing costs, lost sleep, or having to install jet engine mufflers.

When we start viewing things in this way, we are led to the following inescapable conclusion: that from the viewpoint of economic efficiency, the solution is *not* to penalize the party who is inflicting costs on the other party. It is to pursue whatever actions are less costly. So it is not that we have to deal with the cost of free-roaming cattle. Rather, we should compare the costs (to the farmer) of free-roaming cattle to the costs (to the rancher) of penned-up cattle. And whichever activity imposes greater costs is the one that we should avoid. This argument says nothing, of course, about equity, but we will get to that.

The question then becomes, practically speaking, how we weigh these costs against each other. And the following strategy may have occurred to you. Why not let the parties work it out among themselves? So, if the farmer sees the cattle trampling her crops, she could go knock on the rancher's door, point it out, and ask him to do something about it. Most people are reasonable and willing to work together to address a problem of mutual concern.

The thing is: I can see this working in some cases but not others. To see what I mean, let us return to the notion of transaction costs, which is central to the Coase Theorem. For Coase, the magnitude of the transaction costs was everything. And, in particular, it was central to the question of whether private negotiations to deal with the problem would work.

To see all this, let us go back to the rancher-farmer example and conduct a thought experiment using some simple numbers. Suppose that the cattle are causing $15 worth of damage to the farmer's crops. But suppose there is the possibility that the rancher and the farmer could try to work things out between them. In addition, suppose that one option is for the farmer to build a fence, which would cost $10 in materials and, if built, would effectively keep the cattle out.[3] What do you think would happen?

I think the answer depends upon two key issues. First, under the law, is the farmer entitled to undamaged crops or not? By this, we mean that the rancher has to pay the farmer for the damages inflicted by his cattle. Second, it would depend upon how large transaction costs are. Let us consider two scenarios.

*A. Farmer is not entitled to undamaged crops.*

In this case, the farmer has no legal leg to stand on to get the rancher to restrain his cattle. However, she does have the option of building the fence. And it looks like she would choose this option, because by spending only $10, she could avoid $15 worth of damage. So building the fence would leave her $5 better off. *Outcome: fence gets built.*

*B. Farmer is entitled to undamaged crops.*

In this case, the rancher has to pay for all damages to the farmer's crops. Here, the farmer loses all incentive to build the fence because even though she loses $15 in crop damages, she knows the law will compensate her by making the rancher pay. Certainly she would not pay $10 for a fence when the law compensates her for all her crop losses. This, however, is not quite the end of the story and here is where transaction costs enter the picture. In particular, there is the possibility of negotiations between the two parties to have the farmer build the fence. But whether or not this occurs depends upon the magnitude of the transaction costs.

## I. TRANSACTION COSTS LOW

Let us begin by assuming that it is easy for the rancher and the farmer to get together and negotiate this little deal. In this simple example, it is not hard to believe. Maybe the rancher just walks down the road and they sit down over a cup of coffee at the farmer's kitchen table, calmly discuss the situation, and agree that the extent of the damages is clear, as is the fact that it was his cattle that trampled the crops. What is going through the rancher's mind is the bill for $15 in legal damages he is facing. However, he knows that the fence only costs $10 to build. So he might offer to pay her, say, $12 to build the fence. He would happily make this offer because he would be out only $12, not $15. And she would happily accept this offer because the fence only costs $10 to build and so she could pocket the difference. *Outcome: fence gets built.*

Several things are important to note here. First, this result – that the fence gets built – is the same as in the previous case where the farmer was not entitled to damages. In other words, what the law says about legal liability does not matter in determining which outcome we get. Second, not to belabor the obvious, this outcome is the efficient one: rather than losing $15 worth of crops, society loses only $10 worth of fence-building materials.

Finally, I need to point out that the outcomes under the two legal regimes are different in an important sense: how well-off they leave the rancher and the farmer. Notice that in our example, when the rancher is liable for damages, the farmer enjoys net benefits of $2 – the $12 that the rancher pays her to build the fence minus the $10 it costs her to build it. When the rancher is *not* liable for damages, the farmer loses $10: the amount she has to spend to build the fence. So the bottom line here is that in the low transaction cost case, the two legal regimes yield outcomes that are equally efficient but not equally equitable.

## II. TRANSACTION COSTS HIGH

Let us now consider what would happen if the rancher and farmer cannot easily negotiate; that is, if transaction costs are high. To keep things simple, let us assume it is literally impossible for them to negotiate. In this case, the end result is completely different when the rancher is liable versus when he is not. In this case, when the rancher is liable, he is staring at a $15 bill for crop damages, but transaction costs make it impossible for him to pay the farmer to build the fence. And, of course, she has no incentive to build it because she is getting compensated for all her damages. *Outcome: fence does not get built.*

So, when transaction costs are high, the fence does not get built, despite the fact that it is the efficient solution. Furthermore, notice that compared to the low transaction cost case, both the rancher and the farmer are worse off. The rancher is on the hook for the entire $15 in damages, not the $12 he would be paying the farmer to build the fence. And the farmer is just breaking even, not pocketing $2 paid by the rancher for building the fence.

The bottom line is that when transaction costs are high, what the law says about the relative rights of the rancher and the farmer matters in three important ways. First, it can determine which outcome we get. Second, it can determine the efficiency of the out-come. And, finally, it can affect the equitableness of the outcome to the negotiating parties, in terms of their net income positions.

This is the reasoning of the Coase Theorem, but to be clear, let me state it explicitly. The theorem says that when transaction costs are high, who enjoys the legal entitlement can affect the final outcome and matters for efficiency and equity. Where transaction costs are zero, who enjoys the entitlement does not matter from the viewpoint of effi-ciency. Negotiations and private self-interest ensure that the efficient outcome always occurs. With high transaction costs, it matters very much who enjoys the rights. Coase believed that transaction costs are generally present in real-world situations. Consequently, he ascribed great importance to the question of who enjoys the rights.

If you think about it, Coase's logic, developed in these simple examples, may apply to a wide range of circumstances. Whenever voluntary exchanges are possible, transaction costs provide a framework for: (1) understanding why those exchanges may not occur; (2) evaluating the efficiency of the resulting outcomes; and (3) evaluating the equity of those outcomes. And the implications extend way beyond markets for goods and services like our produce marketplace and our two simple examples. It has implications for many areas of economic study, including: industrial organization, financial markets, the study of bureaucracies, administrative design, antitrust policy, international trade, monetary theory, political economy and public choice, and, especially important for our purposes, envi-ronmental policy. For a better sense for when and how it applies, let us now conclude this chapter by considering one last foundational concept in the institutional approach to environmental policy: the concept of *property rights*.

## Property rights

In all the areas of economic study that I just mentioned, exchanges of goods and services figure prominently. Now that we are getting used to thinking in legal terms, let us think about the legal meaning of an exchange. It is perhaps most intuitive to think about market exchange, where buyers and sellers exchange goods at agreed-upon prices. In legal terms, a market exchange is a transfer of the property right in the good from the seller to the buyer. Prior to the transaction, the good "belongs" to the seller, who enjoys certain legal rights in how the good is used. For example, a homeowner can live in a house, paint it, remodel it, allow guests into it, raise kids in it, and so forth. When he sells the house to someone else, those legal rights are transferred to the buyer. The former owner can no longer do these things to that house; only the buyer – the new owner of the house – can.

More generally, property rights consist of a series of permissible actions that one can do with a piece of property. A property right begins when you acquire something – by say, purchasing a house or car, or inheriting an estate – and continues until it is transferred to someone else. In between these two end points of a property right, the owner has rights to do various things with the property, typically circumscribed by the law. For example, you may not be able to make an addition on your house that blocks your neighbor's panoramic view. Or you can ride your motorcycle to the big football game, but you cannot use it to run over a fan of the opposing team.

In addition, a property right needs to be enforced while someone is in possession: own-ing a car is not much use to you if anyone can drive off with it. Some simple examples of ways to enforce property rights are: a fence around your property; locks on car doors; and alarms and security systems for your house. But property right enforcement can mean a lot of different things, including police patrols around your neighborhood; legal remedies for burglaries and theft afforded by the legal system; patent protection on inventions; and copyrights on books, movies, and songs. In the coming discussion, it will help to think broadly about the issue of enforcement.

The reason you want to acquire something is, of course, that you expect to get something out of it. Economists call this a stream of benefits, because it often applies to pieces of property that you hold onto for a while; that is, multiple time periods. For example, buying a house may give you the legal right to rent it out and collect rents for as long as you own the house. This stream of benefits determines how much you would be willing to pay for it. So, for example, when you consider buying a house, you expect benefits in each year, over a time span of so many years. These benefits might be the amounts that you collect in rents from a series of tenants. Let us call each of these annual benefits $B_t$, where B is the dollar amount of benefits, and t is the year you realize it. Recalling our earlier discussion of discounting, the total present discounted value of benefits you expect to realize from the house is:

$$PDV = \sum B_t / (1 + r)^t + X_{t+1} \qquad (4.1)$$

In equation (4.1), of course, each annual benefit is being appropriately discounted (depending upon how far in the future it is realized), and $X_{t+1}$ is the expected sale price at the end of your ownership. Presumably, a prospective buyer would be willing to pay no more for the house than the value of this sum.

Now: you may have noticed that some of the examples I used in the discussion of the Coase Theorem do not seem to fall into the category of market exchange. The produce marketplace certainly does. But what about the cattle trampling the crops, and the loud noise from the planes? The idea is that the notion of exchange is actually a lot broader than just what happens in markets. Even though there is no market, we can still imagine negotiations occurring over whether to build a fence, or the noise from airplanes.

In fact, we can imagine negotiations over many things. Teachers may negotiate with school districts over salaries and benefits. Senators may negotiate among themselves over how they will vote on new Supreme Court nominees. Entire countries negotiate with each other over participation in a climate change treaty. You may have negotiated with students living down the hall who are playing their music too loud. In all of these cases, there are things of value being proposed for exchange: pension contributions, votes for particular justices in exchange for support for pet projects, aid packages for low-income countries, and even peace and quiet so you can study for your final exam.

With all this in mind, let us return to the notion of transaction costs and ask the question:

What can get in the way of a successful voluntary exchange? This is a *big* question. So here, let us focus on a few factors, which are considered common sources of transaction costs in a number of environmental issues.

### A. Large numbers of parties

One is when transactions involve a large number of parties. In the Coase example, there were only two parties: the farmer and the rancher. But what if there were dozens of ranchers who were liable for damages and all having an interest in negotiating with the farmer to build his fence? It might be difficult to determine which ranchers' cattle were responsible for causing the damages. Each rancher might then claim that he should not have to contribute to paying for the fence. This might cause a breakdown in negotiations, with the ranchers unable to arrive at an agreement among themselves before even sitting down at the negotiating table.

A similar real-world environmental example is a factory spewing smoke into the air and causing respiratory problems for residents of communities living downwind. Suppose the factory is not liable for damages to the people living in the downwind communities. An agreement to compensate the factory in return for curtailing its smoke pollution might fail because the downwind residents could not agree on how much their relative contributions should be.

### B. Information issues

A second factor to consider is the question of what information potential buyers have about whatever it is they are considering buying. With many goods, there are things hidden from the buyer that it is important for her to know before she forks over her hard-earned cash. Is this car a lemon? Does this house have a mold issue? Furthermore, many questions are about things that the buyer may well know. He may know that a car is a lemon. He may know about the mold in the basement, or the squirrel infestation in the attic. In these cases, we say that there is *asymmetric*, or *private*, *information*. As a seller, you can see that he is probably sorely tempted not to be completely forthcoming about his mold issues, or squirrel damage. He wants to make the sale.

In some cases, it may be possible to get information. If you go online, you can often get ratings from previous customers for books, restaurants, and hotel rooms from companies such as Yelp and Amazon.com. Before buying a car, in many cases you can go on CARFAX to see the car's service record. However, in many cases it may not be possible to get all the relevant information. For example, even a stellar CARFAX report may not reveal all of the hidden flaws of a car.

All of this can cause some potential exchanges not to go through. The buyer may really like the house, but she may wonder whether it has some problems she does not know about. The seller may be waxing poetic about the house, but the buyer knows that he has incentive not to tell her about mold issues. She may wonder what else the seller is not telling her. And, if she can't get independent confirmation that nothing is wrong, she may very well pass up the opportunity, again, even though she really likes the house.

### C. Enforcement costs

A third important way that transaction costs can occur is in the form of *enforcement costs*. The basic idea here is that it may be difficult to enforce rights to the thing being proposed for exchange. That is, the buyer may realize that if she assumes ownership, it may be difficult to keep others from simply taking some of the profits for themselves. If so, this may discourage her from buying it in the first place.

Consider, for example, a fisher who has a license to fish in a regulated ocean fishery. Suppose the license is transferable, meaning that the fisher is permitted under fishery regulations to sell the right to someone else (we will return to this idea in chapter fifteen). Even if the current fisher is bringing in large harvests, the prospective buyer of the license may be reluctant to buy the license if she recognizes that other fishers may come along and deplete the fishery. In this case, we say that there are significant costs of enforcing her right to fish in the fishery. If this occurs then, of course, the license may not be worth much, in terms of conferring benefits on her. This may make her reluctant to buy. Thus, significant enforcement costs may put a damper on the potential transaction in the license.

These three factors – large numbers of parties, private information, and enforcement costs – are key components of transaction costs when addressing environmental issues. Thus, they are key reasons why we cannot necessarily count on private exchanges to result in efficient outcomes. There are two important implications. First, when we decide that transaction costs are high, it then becomes important to focus on which parties we assign property rights to, as Coase suggested. According to this analysis, this will matter both for efficiency and equity reasons.

However, this discussion also implies that the magnitude of transaction costs may vary from situation to situation. The number of parties involved in a particular exchange may range from very large all the way down to two. For some goods, private information about quality will not be a serious issue. And, finally, it may be relatively easy to enforce property rights in certain situations. All of this suggests that there may not be a one-size-fits-all solution to environmental problems. When we decide that transaction costs are low, we may want to be more receptive to private negotiations to address environmental problems. All of this will be a recurring theme as we go through the special topics chapters.

## Key takeaways

- Institutions are the rules of the game under which economic activity takes place, and are implicit in every economic analysis. Institutions provide various incentives for economic activity and reduce uncertainty regarding the outcomes of economic activity.
- Not only do institutions influence economic activity, they are themselves derived from economic conditions and may be subject to institutional path dependence.
- Transaction costs are the costs associated with engaging in exchange with others. These costs include the costs of negotiation, enforcement, and maintaining property rights.
- The Coase Theorem states that when there are significant transaction costs, we cannot necessarily rely on private negotiations to achieve efficient allocation of resources.
- Transaction costs tend to be large when there are large numbers of negotiating parties, when there is private information about the goods in question, and when there are significant costs of enforcing property rights in the goods.
- Since transaction costs can vary, it is unlikely that there is a one-size-fits-all solution to environmental problems.

## Key concepts

- Institutions
- Path dependence
- Transaction costs
- Coase Theorem
- Property rights

## Exercises/discussion questions

4.1. Douglass North distinguishes two types of institutions: *formal* (statutes, constitutions, court rulings) and *informal* (customs, norms, and traditions). Both types of institutions can, and typically do, exist at the same time in a given society. What are their respective roles in providing the rules of the game? Do they provide alternative sources of rules for citizens to follow? Or do they complement each other? And how, if at all, does their relationship change over time as an economy develops?

4.2. Consider a large urn full of ping-pong balls. Half of these balls are red and the other half are blue. Now consider the following experiment. You reach into the urn and pick out one ball. Once you see what color it is, you replace the ball and then add another ball of the same color. You repeat the process over and over. What do you think will happen to the composition of colors in the urn? In what ways is this a path-dependent process?

4.3. Consider a system (such as exists at my college) where many students take courses during their first two years and then declare their majors toward the end of their second year. In what ways is this a path-dependent process?

4.4. In which of the following cases would you expect transaction costs to be large, and why?
   (a)  A power plant polluting the air of nearby communities.
   (b)  Your next-door neighbor in your residence hall is playing his music too loudly and you would like him to turn it down.
   (c)  A beekeeper whose bees fly around and cross-pollinate fruit trees in nearby orchards.
   (d)  Greenhouse gas emissions contributing to climate change.
   (e)  A dam in danger of collapsing and inundating a downstream factory.
   (f)  A farm diverting water from an adjacent stream and depriving a Native American tribe located downstream of water providing habitat for fish.

4.5. Joey has decided to start a tanning business. For his business, he builds a tanning deck in his backyard, which will earn him $4000, as long as he enjoys unobstructed sun. Unfortunately for him, his friend Sally buys up the lot next door to start up a new business. Sally has a choice of two building configurations: a tall, skinny building, which will cost $3000 to build but will bring in total revenues of $6100; or a short, stocky building, which will cost $2800 but which will bring in revenues of $4900. The tall, skinny building will inflict losses of $2000 on Joey's business, unless Sally adds a glass porthole, in which case the building would only inflict losses of $800 on Joey. However, the porthole would cost an extra $1100 to build. The short, stocky building will inflict losses of only $1200 on Joey.
   Assuming that Joey and Sally can enter into costless private negotiations,
   (a)  calculate what would and would not be built, and describe the nature of the negotiations which would occur: (i) If Sally is <u>liable</u> for the damages to Joey's tanning business; and (ii) If Sally is <u>not liable</u> for the damages.
   (b)  <u>If possible</u>, calculate each one's income (i.e., give a specific dollar amount) when Sally is liable for all damages inflicted on Joey, and when she is not. If you <u>cannot</u> give a specific dollar amount but can give a <u>range</u> of dollar amounts, do so. Explain your answer.

4.6. Consider the apple tree in Mark's backyard, which produces 10 bushels of apples per year. Mark, owning his land, has a legal right to those apples. Every year, Mark is able to sell as many apples as he wants for $41 per bushel at the local farmer's market. At the same time, in order to produce those apples, he incurs production costs according to the marginal cost schedule MC = Q, where MC is in dollars per bushel, and Q is the number of bushels.

    The problem is that his backyard is not fenced in and every year hungry deer come by and eat all of the apples off his tree. So he is considering building a fence that, because it uses the latest technology, would be 100% effective in keeping the deer out. Assume that Mark has a discount rate of 20%, a time horizon of only two years (this year and next year), and the fence would be completely effective for two years. To keep things simple, you may assume that he can sell his apples right now, and exactly one year from now. Finally, in answering this question, you may assume that he must incur production costs associated with 10 bushels of output, after which the deer swoop in.

    (a)    What would be the most that Mark would be willing to pay for this fence? Explain.

Suppose the price of apples goes down to $29 per bushel.

    (b)    Reanswer (a).

Now suppose *in addition* that there is another fencing technology that costs $200 less than the latest technology. The problem is that the deer can still get into his backyard to eat his apples but, now, they can only eat half of his apples.

    (c)    Assuming it is worth it to build a fence, which fencing technology would he use, and why?

## Notes

1 Technically, what everyone thinks of as the Nobel Prize in economics is actually the *Sveriges Riksbank Prize in Economic Sciences in Memory of Alfred Nobel* (long story). Because this is such a mouthful, I will simply refer to it as the Nobel Prize in economics, as we encounter other Nobel Prize winners in this book.
2 See also Alchian (1950).
3 For the sake of this example, assume that the rancher does not have access to the fence-building technology.

## Further readings

### I. Institutions

*A couple of readings from the master:*

North, Douglass C. "Economic performance through time," *American Economic Review* 84(June 1994): 359–68.

*Develops the argument for how institutions can contribute to the economic performance of economies and how institutions evolve over time.*

North, Douglass C. "Economic theory in a dynamic economic world," *Business Economics* 30(January 1995): 7–12.

*Provides a critique of standard (neoclassical) economic theory and provides some suggestions for ways economic theory needs to evolve in order to better explain the performance of economies.*

## II. Transaction costs and the Coase Theorem

*The economic literature on the Coase Theorem is vast, but here are a few interesting studies:*

Hoffman, Elizabeth and Matthew L. Spitzer. "The Coase Theorem: Some experimental tests," *Journal of Law and Economics* 25(April 1982): 73–98.

*Provides some experimental evidence regarding the conditions under which the predictions of the Coase Theorem hold.*

Hylan, Timothy R.; Maureen J. Lage; and Michael Treglia. "The Coase Theorem, free agency, and major league baseball: A panel study of pitcher mobility from 1961 to 1992," *Southern Economic Journal* 62(April 1996): 1029–42.

*Uses the Coase Theorem to examine the impact of free agency in major league baseball.*

Kessel, Reuben A. "Transfused blood, serum hepatitis, and the Coase Theorem," *Journal of Law and Economics* 17(October 1974): 265–89.

*Applies the Coase Theorem to understand inefficiencies in the provision of blood for blood transfusions.*

## III. Property rights

*Two empirical studies on the effect of secure property rights:*

Anderson, Terry L. and Dean Lueck. "Land tenure and agricultural productivity on Indian reservations," *Journal of Law and Economics* 35(October 1992): 427–54.

*Argues that insecure land tenure on Indian reservations may have negative effects on agricultural productivity.*

Libecap, Gary D. and Dean Lueck. "The demarcation of land and the role of coordinating property institutions," *Journal of Political Economy* 119(June 2011): 426–67.

*Finds that the rectangular land demarcation system used in the United States provides more secure, transferable property rights, significantly boosting land values.*

# References

Alchian, Armen. "Uncertainty, evolution, and economic theory," *Journal of Political Economy* 58(1950): 211–21.

Coase, Ronald. The problem of social cost. *Journal of Law and Economics* (1960): 1–44.

David, Paul A. "Clio and the economics of QWERTY," *American Economic Review* (May 1985): 332–7:

North, Douglass C. *Structure and change in economic history.* New York: Norton, 1981.

North, Douglass C. *Institutions, institutional change, and economic performance.* Cambridge: Cambridge University Press, 1990.

North, Douglass C. and Barry R. Weingast. "Constitutions and commitment: The evolution of institutions governing public choice in seventeenth-century England," *Journal of Economic History* 49(December 1989): 803–32.

# 5 Common-pool resources

In this chapter, we focus on an extremely important subset of natural resources known as *common-pool resources* (CPRs). Common-pool resources, which I will define shortly, are commonly encountered in environmental economics. This is because many natural resources may be characterized as common-pool resources. Such resources include fisheries, grazing pastures, forests, and local water resources. Analyzing resources as CPRs provides a systematic way to think about this special class of resources and helps explain why we get certain outcomes such as overharvesting or depletion under certain conditions. This allows us to characterize the economic efficiency of those outcomes and to propose sensible policies. All of this is to say that we get a lot of analytical mileage out of this concept, and we will be returning to it repeatedly in the later special topic chapters.

## What is a CPR?

To understand the notion of a CPR, consider a valuable natural resource, like a fishery or a prime grazing area. Because it is valuable, many people would like to tap it for its economic value, by sailing their fishing boats into it, or driving their cattle onto it. For the purposes of analysis, CPRs have two key characteristics. First, the amount taken by me reduces the amount that is available for you. If I catch a fish, you cannot catch that same fish. If I drive my cattle into a grazing area and the cattle consume forage grasses, that means less forage for your cattle. Since you and I are rivals for the resource, we say that our activities are *rivalrous*. Alternatively, you sometimes hear people say that our uses are *subtractable*. That is, my use subtracts from the resource available to you. You should think of these terms as meaning the same thing.

Second, it is difficult to keep every Pam, Dacie, or Harriet from using and benefiting from the resource. For example, it might be extremely difficult to keep anyone with a boat from fishing in a fishery on the open seas. It may also be difficult to fence in a grazing area or a forest, or to keep people from tapping a source of valuable groundwater. In all of these cases, we say it is difficult to exclude users. This is the notion of *non-excludability*.

To avoid confusion, it may be useful to keep in mind a visual image of a resource with boundaries, like a pasture area. Exclusion can refer to either of two sets of users: ones already there – what I will call *current users* – and ones not already there – what I will call *potential users*. Non-excludability can mean either that you cannot keep current users from using the resource as they please, or that you cannot keep potential users from coming in and using the resource. When people refer to CPRs, they admit both of these possibilities.

As it turns out, the notions of non-excludability and rivalry provide a useful way to characterize all goods, not just CPRs, for the purposes of analysis. This is because every

Rivalry of consumption

|  | HIGH | LOW |
|---|---|---|
| HIGH | CPRs | Public Goods |
| LOW | Private Goods | Club Goods |

Difficulty of exclusion

*Figure 5.1* Four types of goods

good can be thought of as some combination of non–excludability and rivalry. To see this, consider Figure 5.1. Here, we categorize different types of goods based on the degree of rivalry of consumption, and the difficulty of excluding others. Standard private goods – the ones that occupy much microeconomic analysis – are in the lower left-hand box. When I eat an apple, that pretty much forecloses your being able to eat that same apple: we both cannot eat the same apple, and you are probably reluctant to try to wrestle the apple out of my hand.

Another type of good is the so-called *public good*, which is characterized by *difficulty* of exclusion but *non-rivalry* in consumption – the exact opposite of a private good. National defense is a classic example of a public good. When the government provides for an army and navy, buys tanks and drones, and sets up a missile defense system – all to ward off foreign invasions – my use in no way reduces the amount you enjoy. And, once it is provided to me, it is really difficult to keep you from enjoying it. If you think about it, this means that every self-interested citizen of a country has no incentive to pay for national defense. That is, each citizen will be saying to her/himself: *Well, why should I pay if I am going to receive it no matter what? If I don't pay, I can get something for nothing!* To introduce a little terminology, we say that everyone has incentive to *free-ride*; that is, to enjoy the benefits without having to pay for them. This is why many public goods like national defense are provided by the government and subsidized by taxes, which basically forces everyone to pay.

CPRs, in some sense, have the worst of both worlds. Rivalrous consumption means that individual users are inflicting costs on other users while, at the same time, it is extremely difficult to keep others from doing the same. This means that there is a real danger that the natural resource will be degraded, and, in extreme cases, destroyed. This argument has major implications for environmental economics, as there are many resources that have CPR characteristics.

The last type of goods are so-called *club goods*, which are characterized by easy exclusion and low rivalry of consumption.[1] An example of a club good might be a large movie theater with plenty of empty seats, but charging admissions at the door. We will encounter all four types of goods in the special topics chapters.

## An important metaphor: The tragedy of the commons

To understand the CPR argument better, let us now consider one of the most famous and influential arguments in environmental history. In 1968, the biologist Garrett Hardin wrote an article for *Science* magazine entitled *The Tragedy of the Commons*. In this article, Hardin argued that many natural resources with the characteristics of CPRs are in danger of being destroyed by individual users acting in their own self-interest. He used the example of a pasture open to everybody. The problem, he argued, is that every herdsman has incentive to continue to add more and more cattle because he derives more benefits from every extra head of cattle than the costs he incurs from degrading the pasture in doing so. The result is the eventual destruction of the pasture. His words, having now been repeated by others countless times, are nevertheless still worth reproducing here:

> The rational herdsman concludes that the only sensible course ... is to add another animal to his herd. And another; and another... But this is the conclusion reached by each and every rational herdsman sharing a commons. There is the tragedy. Each man is locked into a system that compels him to increase his herd without limit – in a world that is limited. Ruin is the destination toward which all men rush, each pursuing his own best interest in a society that believes in the freedom of the commons. Freedom in a commons brings ruin to all. [Hardin (1968), p. 1244]

We can formalize Hardin's argument using a simple game theory construct known as the *prisoner's dilemma*. In the classic prisoner's dilemma game, two accused criminals, accomplices in a crime, are each being interrogated in separate rooms and not allowed to communicate with each other. The police are trying to get each criminal to confess to the crime. If they both confess, they both get sent to prison. If neither confesses, they both receive a lighter sentence for a lesser crime. Finally, each is told that if he does not confess and his accomplice does, he will get the book thrown at him while his accomplice goes free. It turns out that the structure of payoffs induces each player to confess, and both end up going to jail. This is despite the fact that if they both stayed silent, they would both be better off.

To see how this argument applies to Hardin's pasture, consider Figure 5.2, which formalizes this argument in the context of two herders on a common pasture. Here, the two herdsmen are A and B. Each is considering whether to abide by an agreement to share the pasture 50–50. The column on the right-hand side shows the payoffs to A and B. So,

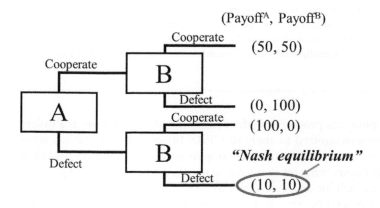

*Figure 5.2* The prisoner's dilemma game

for example, if both cooperate, then they both reap rewards of 50 units (dollars). If neither cooperates (both defect), they degrade the pasture and both receive only 10 units. However, if only one defects, he is able to grab all of benefit of the pasture forage (receiving 100) while his former friend(?) gets none (0).

Under these conditions, each herdsman's best option is to defect (not hold to the agreement) – the pasture gets degraded, and both are worse off than if they had both abided by the agreement. Thus, both herdsmen defecting (defect, defect) is the so-called Nash equilibrium of the game, which is inefficient because it does not maximize the total returns to the players.

Hardin's sobering message has been taken to heart by many biologists, economists, political scientists, and policymakers. Indeed, it is an extremely powerful model, providing insights into a wide range of issues including famine, acid rain, U.S. congressional overspending, urban crime, and international cooperation [Ostrom (1990), p. 3]. However, researchers have also documented a great many real-world examples of apparent CPRs, certainly characterized by rivalry and non-excludability, that somehow manage to avoid destruction or even degradation of the resource. In instance after instance, users tapping forests, grazing areas, fisheries, and local water resources are able to devise sustainable management arrangements. The question is why.

## Interpreting the tragedy of the commons

The answer is that the tragedy of the commons model makes a few implicit assumptions that may not hold in many real-world CPR situations. To understand these assumptions, consider the following metaphor: that tragedies of the commons tend to occur when users behave like *roving bandits* (see Box 5.1).

---

### Box 5.1: Roving bandits in the news

**News item:** "'Roving bandits' depleting fisheries, experts say," *National Geographic*, March 16, 2006.
"Improved technologies – faster boats, better refrigeration – have allowed today's fishers to quickly roam across every corner of the seas. No longer limited to their local waters, some marine-life traders are flouting international law and illegally overharvesting species faster than regulators can respond. Now 15 researchers from around the globe have joined forces to draw attention to the damage to fisheries caused by these law-breaking 'roving bandits.' The traders are central drivers in the overexploitation of the world's oceans, destroying local stocks and evading authorities, the researchers say. ... Due to rapid globalization, people buying and selling marine life-forms can travel almost anywhere in the world within a day. Many species and habitats that were previously too inaccessible to be economically viable targets for fishers are now open to exploitation." [Petherick (2006)]

---

The roving bandits metaphor was proposed by the economist Mancur Olson in a different context, but it has been recognized to apply to CPRs. Imagine mask-wearing, rapacious bandits who recognize profit opportunities, swoop in, scoop up a resource, and then disappear before anybody knows what hit them. Ask yourself: Why exactly would we not expect resources subject to such hit-and-run tactics to be managed sustainably?

Well, first, this is obviously a CPR-type situation, which has rivalrous consumption and non-excludability (with a vengeance!). But think about what is missing from this picture.

The bandits obviously take an extreme short-term view, with no stake in the long-run health of the resource. The bandits have no historical relationship with local users. It is extremely difficult to keep them out. There is no way to monitor and sanction them for their predatory behavior, because they are in and out before anyone even knows they are there. And, even if they could be found, negotiation to try to manage the resource sustainably is both infeasible and pointless, because the bandits are better off just grabbing what they can, while they can. It is really no surprise that these fisheries can be destroyed very easily under these conditions.

By the same token, these missing elements provide us with insight into what it would take to avoid a tragedy of the commons. In many real-world CPRs, the CPR has a distinct boundary, or one can be made fairly easily, which might make it possible to exclude outsiders. Often, local users do have a stake in the long-term health of the resource. In many cases, they know other users, either personally or by reputation. They are often in a position to engage in face-to-face communication with other users. This helps establish trust among the users, which increases the likelihood that they can reach agreement on how to manage the resource, and on a set of sanctions to impose on anyone who violates or free-rides on the agreement. A charismatic local leader may be able to reduce the transaction costs of negotiations among members of the community. Finally, users are likely to abide by existing norms that may penalize self-serving actions at the expense of the community of users.

## Ostrom's design principles

Many economists have been thinking for some time about the CPR problem and what to do about it. But I think it is fair to say that the single person who has done the most to advance our understanding of CPR governance is Elinor Ostrom (see Figure 5.3), and for her work on this topic, she was awarded the Nobel Prize in economics in 2009. Before passing away in 2012, Ostrom had studied real-life CPRs for many years and noticed something that many found surprising: that there seemed to be a wide range of resource outcomes. Some CPRs did indeed get destroyed or severely degraded. But other CPRs did just fine. Intrigued, she looked more closely to see if there were discernible patterns: specific factors that would explain the difference. And after much work and careful thought, she came up with a set of what she called *design principles* [Ostrom (1990), pp. 88–102]. These design principles are factors that consistently seem to spell the difference between CPR success and failure, which are shown in Table 5.1.

So, for example, Ostrom concluded that CPR management tends to be successful when the resource in question has clearly defined boundaries, which makes it easier to keep out roving bandits. In addition, she concluded that governance rules needed to be congruent with local conditions. By this, she basically meant that there is no one-size-fits-all system. Rather, the local rules need to be tied to the specific local conditions. In addition, successful local management is more likely when users can participate in the making of the rules, when the local system can be monitored by individuals who are accountable to the users, and so forth. You can think of these design principles as a way to systematize our understanding of the difference between situations with roving bandits and non-roving bandits. And, therefore, they can shed light on the reasons for the difference between good and bad CPR outcomes. To this day, these design principles are used by practitioners interested in figuring out what is going "wrong" with a particular CPR.

However, it turns out that the design principles by themselves are still insufficient, for several reasons. First, though they are quite useful, they seem to operate at an overly broad

*Figure 5.3* Elinor Ostrom. *Source*: Holger Motzkau, Wikimedia Commons

*Table 5.1* The design principles

| Principle | Explanation |
|---|---|
| Clearly defined boundaries | Clearly defined boundaries around the resource system and its users. |
| Congruence | Governance rules are congruent with local social and economic conditions. |
| Collective-choice arrangements | Individuals governed by the operational rules have standing to participate in modifying them. |
| Monitoring | Presence of individuals who effectively monitor the use and condition of the resource, and these monitors are accountable to the users. |
| Graduated sanctions | Users who violate the rules are subject to a system of graduated sanctions. |
| Conflict-resolution mechanisms | Presence of low-cost conflict resolution mechanisms that users have ready access to. |
| Recognition of the right to organize | Rights of local users to devise their own governance systems, which cannot be challenged by external authorities. |
| Nested enterprises (when CPRs are a part of larger systems) | Governance activities are organized in multiple nested layers. |

level of generality. What exactly does it mean for rules to be "congruent"? And wouldn't it matter exactly how users are allowed to participate in modifying the operational rules? If it is a voting system, for example, what kind of voting system are we talking about, and what are the voting rules?

Furthermore, the list seems incomplete. One would think that a number of other factors may matter, such as the presence of local norms of behavior, the existence of local galvanizing leaders, or the sheer economic value of the resource. And, finally, different factors may interact in complex and unpredictable ways. For example, how much would local norms offset the need for an effective monitoring system? In order to try to fill these gaps, Ostrom and other scholars recognized the need to systematize more fully our understanding of the workings of CPRs. This recognition has given rise to a field of scholarly inquiry centered on the notion of so-called *social-ecological systems*.

## Social-ecological systems

### I. What are social-ecological systems?

You can think of social-ecological systems (SES), sometimes called human–environment systems, as models of natural resource use that explicitly include both social and environmental dimensions. The environmental dimension concerns the natural and physical attributes of the resource itself, while the social dimension concerns the human factors that affect how the resource is used. These human factors include various characteristics of the human actors and the system of governance in which they operate. In SES, all of these factors interact in complex ways to produce resource use outcomes, which can include many possibilities, including growth, sustainable use, gradual depletion, resource destruction, and species extinction. Furthermore, the systems themselves can grow and adapt over time as conditions change and human actors respond to those changes.

Interest in SES has grown in recent years because simple models are often limited in their ability to capture adequately the complex dynamics of resource use. Simple models tend to be parsimonious, consisting of a few variables that interact and evolve in a simple, often linear, way. These models tend to isolate and analyze processes in a larger setting that is unaffected by the outcomes of these processes. In these models, systemic changes occur on the margin in continuous, mostly predictable ways.

By contrast, models of SES tend to be large, containing many variables and interactions. Actions in local populations and communities spread to higher levels through collective behavior, which then feed back in potentially complex ways to influence local behavior. The overall system has emergent properties, meaning that it may behave in ways not predicted by the sum of its component subsystems. This includes the possibility of abrupt systemic changes that may be irreversible. For all these reasons, they are sometimes referred to as *complex adaptive systems* [Levin et al. (2013)].

Generally speaking, the advantage of modeling natural resources as SES rather than simple systems is that they can often better capture complex processes and better predict actual resource outcomes. Some possible examples include climate change, grasslands depletion, deforestation, and the destruction of coral reefs. The downside is that they are much more difficult to model rigorously, involving dynamic optimization of systems of differential equations that also incorporate non-linear feedbacks from outcomes to processes. There may also be a spatial component to modeling the dynamics of the resource system. In addition, such models sometimes try to model strategic interactions among individual actors in the local setting, including allowing for actors to differ in their behaviors (so-called *agent-based modeling*). Finally, in such models there is the possibility of multiple equilibria, path dependence, and outcomes that are irreversible [Levin et al. (2013), pp. 118–24]. Mathematically modeling SES is not for the faint of heart.

## II. The social-ecological systems framework

An alternative approach to modeling natural resources as SES has been proposed by Ostrom and her collaborators, what they call the *SES framework*. This framework emerged as Ostrom thought more about the challenges of modeling CPRs. In developing the SES framework, Ostrom and her collaborators took a very different approach from the mathematical approach just described. Instead, they focus on creating and applying a taxonomy of different variables in order to establish the conditions under which natural resources can be managed sustainably, often focusing on resources with CPR characteristics.[2] The approach thus applies to a variety of resources, such as: watershed management, fisheries, aquaculture systems, forests, grazing pastures, marine ecosystem management, and coastal development [Partelow (2018)].

There are four key components of the SES framework: the *resource system*, the *resource units*, the *actors*, and the *governance system* (see Figure 5.4). Each of these components influences the interactions that determine use of the resource. These interactions result in a potentially wide variety of outcomes regarding the resource: anything from growth to sustainable use to slow degradation to total collapse. In addition, the SES framework allows for feedback from the interactions and outcomes back to the various components. For example, how a resource is used can result in alteration of the rules in place to govern the resource: a groundwater aquifer in critical overdraft may well induce actors to modify the rules for groundwater extraction. The approach thus fits nicely with Douglass North's notion of institutions as responding to various economic and other factors. I should mention that the SES framework finesses the difficult question of characterizing the nature of these feedback mechanisms with any precision.

This basic framework obviously operates at a high level of abstraction so far. To make it useful for analyzing individual CPRs, each of the four components contains a set of variables that capture various features of each component. Ostrom (2007) provides a list of

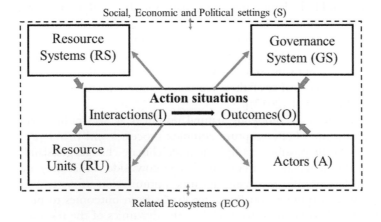

*Figure 5.4* Social-ecological systems framework

these variables, referred to as second-tier variables, which is reproduced in Table 5.2. The idea is that every CPR situation is potentially some unique combination of these second-tier variables, which determines the resource outcome. Depending on the CPR, different variables may play more or less important roles and in some cases, some of these variables may be absent entirely.

To illustrate how to apply the framework, let us consider Hardin's open pasture. Using this approach, it would immediately be assigned a sector under RS1: pasture. The size of the pasture (RS3) and clarity of its boundaries would be noted (RS2), and so on down the list. Of particular relevance for Hardin's pasture are the following facts, which would serve as key assumptions in an analysis of likely pasture outcomes:

- Mobile, identifiable resource units with significant economic value (RU1, RU4);
- Large number of users (A1) given the size of the pasture (RS3), negatively affecting its productivity (RS5);
- The users operate independently, and they make decisions to maximize their short-run returns (A7), in the absence of any behavioral norms (A6);
- No governance system is present (no GS variables).

*Table 5.2* Second-tier variables in SES framework

| Resource (RS) | Governance system (GS) |
| --- | --- |
| RS1 – Sector (e.g., water, pasture, forests, fish) | GS1 – Government organizations |
| RS2 – Clarity of system boundaries | GS2 – Non-government organizations |
| RS3 – Size of resource system | GS3 – Network structure |
| RS4 – Human-constructed facilities | GS4 – Property-right systems |
| RS5 – Productivity of system | GS5 – Operational rules |
| RS6 – Equilibrium properties | GS6 – Collective choice rules |
| RS7 – Predictability of system dynamics | GS7 – Constitutional rules |
| RS8 – Storage characteristics | GS8 – Monitoring and sanctioning processes |
| RS9 – Location | |
| *Resource units (RU)* | *Actors (A)* |
| RU1 – Resource unit mobility | A1 – Number of actors |
| RU2 – Growth or replacement rate | A2 – Socioeconomic attributes of actors |
| RU3 – Interaction among resource units | A3 – History of use |
| RU4 – Economic value | A4 – Location |
| RU5 – Size | A5 – Leadership/entrepreneurship |
| RU6 – Distinctive markings | A6 – Norms/social capital |
| RU7 – Spatial and temporal distribution | A7 – Knowledge of SES/mental models |
| | A8 – Dependence on resource |
| | A9 – Technology used |
| *Interactions (I)* → *Outcomes (O)* | |
| I1 – Harvesting levels of diverse users | O1 – Social performance measures |
| I2 – Information sharing among users | (e.g., equity, efficiency, accountability) |
| I3 – Deliberation processes | O2 – Ecological performance measures |
| I4 – Conflicts among users | (e.g., overharvested, resilience) |
| I5 – Investment activities | O3 – Externalities to other SES |
| I6 – Lobbying activities | |
| I7 – Self-organizing activities | |
| I8 – Networking activities | |
| I9 – Monitoring activities | |

These assumptions yield the prediction that the pasture grasses will be harvested at high levels (I1), leading to severe degradation of the pasture (O2).

SES practitioners emphasize the practicality of this approach, arguing that it can be used in a diagnostic way to determine what is going wrong with a resource system, similarly to the way a doctor can diagnose what is wrong with a patient. To see this, suppose you conducted one hundred studies of different CPRs and in every single case this particular configuration of these variables was associated with destruction of a resource. You could confidently conclude that these values for RU1, RU4, A1, RS3, RS5, A7, A6 and no GS are likely to result in a tragedy of the commons. This would be useful information to have in assessing a CPR situation.

### III. An SES example: Urban lake commons in Bangalore, India

For a better sense of how an SES study is actually carried out, let us consider a specific example: a study by Elinor Ostrom and her co-author, Harini Nagendra [Nagendra and Ostrom (2014)]. The focus of their study was the Indian city of Bangalore, the third largest city in India. In recent years, Bangalore has experienced serious environmental challenges due to rapid economic growth. Perhaps foremost among these challenges is severe water pollution in a series of lakes within the city used for a variety of purposes, including agriculture, fishing, drinking water, and other domestic uses.

Nagendra and Ostrom selected seven lakes for study. These lakes varied in a number of ways relevant to the SES framework, both social and ecological. They used this variation to determine which variables were associated with governance success. The specific variables they focused on were:

- Size of the resource system (RS3);
- Number of actors (A1);
- Whether certain socioeconomic groups were excluded (subset of A2);
- Leadership/entrepreneurship (A5);
- Presence of norms/social capital (A6);
- Dependence on/importance of the resource (A8);
- Presence of operational rules (GS5);
- Networking with government (subset of I8);
- Presence of informal norms for monitoring (I9).

These variables were chosen because previous studies indicated that they were candidates for factors that determine the success of local self-governance.

Important for their research strategy was their ability to eliminate a number of these second-tier variables from their analysis beforehand, because either they did not vary across the different lakes or they did not apply given the nature of the resource. For example, all of the lakes were obviously water and, thus, belonged to the same resource system sector. There was thus no variation in RS1, so they could eliminate this variable from consideration. Similarly, they did not consider resource unit mobility (RU1) or knowledge of SES/mental models (A7), and a number of other variables in the SES framework. In this way, they were able to cull down the variables to a manageable number.

They also needed to define what they meant by governance success. Here, they used two criteria: the extent of collective action, and the environmental condition of the lakes.

So they were interested both in the factors that resulted in lakes being cleaned up, and the use of collective action to achieve these outcomes. Another outcome often used in SES studies – external impacts on other SES – was not considered relevant to the study, which focused on outcomes in individual lakes.

Table 5.3 reproduces part of a table from their study that summarized their findings for five of the seven lakes they examined. Each row represents one of the SES variables, along with their assessment of the value of each variable for each of the lakes. For example, the values in the first row for RS3 (size of the resource system) indicate that Lakes A, C, and D were small, Lake B was moderate-sized, and Lake E was large. The values for Lakes A and B are boldfaced, to highlight the finding that they were the two lakes with positive outcomes for both of the criteria: a high level of collective action (CA), and good environmental quality of the lake (ENV).

These findings paint a complex picture, but Nagendra and Ostrom are able to draw several conclusions. First, it appears that large lakes present more challenges to effective cleanup than small to mid-sized lakes. They argue this is probably because cleanup is more costly and large lakes tended to be stressed more with contaminants such as sewage and industrial effluents. Another large lake, not shown in this table, also experienced issues similar to Lake E.

Second, the findings are suggestive of certain social barriers to effective cleanup of lakes. Both successful lakes had a moderate number of participating actors (A1), suggesting that there may be challenges to having either too few or too many actors. Nagendra and Ostrom interpret this as suggesting there may be a critical mass of actors necessary for successful management, but too many may lead to problems in coordinating local cleanup efforts. The successful lakes also had effective leadership, high levels of social capital, and they did not exclude certain socioeconomic groups from participating in the decision-making process. None of the less successful lakes had all of these things.

Finally, they note that the successful lakes all had operational community rules governing permissible activities, informal norms that supported monitoring activity, and networking with government agencies. All of these they characterized as variables that could be critical for successful water quality management.

*Table 5.3* SES categorization of second-tier variables, Bangalore lakes

|       | Lake A   | Lake B   | Lake C   | Lake D   | Lake E   |
|-------|----------|----------|----------|----------|----------|
| RS3   | **Small**    | **Moderate** | Small    | Small    | Large    |
| A1    | **Moderate** | **Moderate** | Small    | Moderate | Large    |
| A2a   | **No**       | **No**       | Yes      | No       | No       |
| A5    | **Present**  | **Present**  | Present  | Absent   | Absent   |
| A6    | **High**     | **High**     | Low      | High     | Moderate |
| A8    | **Moderate** | **Moderate** | Low      | High     | Moderate |
| GS5   | **Present**  | **Present**  | Present  | Absent   | Absent   |
| I8a   | **High**     | **High**     | Moderate | Low      | Low      |
| I9    | **Present**  | **Present**  | Absent   | Absent   | Absent   |
| CA    | **High**     | **High**     | High     | Low      | Moderate |
| ENV   | **High**     | **High**     | Low      | Low      | Low      |

(Adapted from Nagendra and Ostrom (2014))

### IV. Challenges in using the SES framework

Despite its many advantages, there are a number of challenges presented to practitioners wishing to use the SES framework. One important challenge you may have already noticed is that the results may not be clear-cut in their implications for the combination of variables that would lead to successful self-governance. Referring back to our earlier description of the Hardin commons, the chances of achieving broad unanimity across a large number of studies is really small.

Furthermore, small changes in assumptions may lead to significant changes in outcomes. In the case of Hardin's pasture, for example, would we get the same outcome if the number of actors was smaller or the pasture was larger? How much smaller? How much larger? Would getting this outcome depend on the productivity of the pasture in producing forage grasses? What if the market for cattle crashes, dramatically reducing the value of beef? Would this reduce the pressure on the pasture sufficiently to avert a tragedy of the commons? What if the resource is not a pasture but, rather, an ocean fishery? All of this would make it difficult to know how much the results of one SES study would generalize to other situations.

A second challenge lies in defining the appropriate variables to include in the framework. There is a general sense among many practitioners that the selection of variables to date has perhaps been a bit *ad hoc*. For one thing, the variables included so far vary quite a bit in their level of generality. Furthermore, it is sometimes unclear exactly how particular variables are being defined. For example, different studies may define variables such as norms/social capital (A6), in different ways. This may make it difficult to compare results across the different studies in order to draw general conclusions. Natural scientists have voiced a concern that the variables included to date overly emphasize the interests of social scientists. They want more attention paid to variables suggested as potentially important by studies in the natural sciences.

Third, the SES framework presents a methodological challenge for many economists who are accustomed to hypothesis testing using large ("large-N") data sets in order to advance knowledge. The complexity of the framework, with many potentially important variables, largely rules out standard econometric analysis. Indeed, the vast majority of existing studies using the SES framework are case studies that mostly rely on secondary data sources [Partelow (2018), Table 3]. The method relies on a gradual accumulation of findings over time in order to grow increasingly comfortable with drawing certain conclusions. Different economists will find this approach more and less profitable in terms of its ability to create meaningful knowledge, depending upon their methodological bent.

Future studies are likely to work on refining the choice of variables included in the framework. This will likely include developing theoretical criteria for variable inclusion, establishing a consistent (ontological) basis for creating and grouping different variables, and evaluating empirical evidence for whether particular variables merit inclusion. It is also quite conceivable that variants of the framework for different resources will be developed. Any changes to variable definitions will likely have implications for how to define other variables, especially lower-tier variable subcategories. Finally, the governance system variables are likely to undergo considerable additional refinement in order to forge cleaner connections between variables and resource system outcomes [McGinnis and Ostrom (2014)].

### A final possibility: The anticommons

Before concluding our discussion of common-pool resources, I would like to point out one additional logical possibility. In certain situations, it may be the case that *every* user of a common-pool resource has a right to exclude others from using the resource. Because the pure commons situation is where *no user* can exclude anyone, this case represents the polar opposite of, and stands in stark contrast to, a commons. For this reason, this situation is referred to as an *anticommons*. Like the commons, this idea turns out to have environmental applications.

The term anticommons was coined by the legal scholar Michael Heller, who developed the idea as a way to explain the following strange phenomenon observed in Russia shortly after the breakup of the Soviet Union in 1990 [Heller (1998)]. In Moscow, people noticed that there were a lot of stores sitting empty. At the same time, there were hundreds of kiosks on the sidewalks in front of them, filled with goods and by all accounts doing a bang-up business (see Figure 5.5). So obviously, the explanation for the empty stores could not be lack of demand. Furthermore, this situation went on for years. So why were the stores not being used?

The answer, Heller decided, lay in the peculiar circumstances existing in Russia in the wake of the Soviet breakup, where local and regional government agencies exercised a great deal of power over commercial real estate. However, the new Russian government maintained much control over the leasing terms these agencies could offer to potential store owners. Thus, store owners did not enjoy unfettered right to operate the stores as they wished. Meanwhile, worker collectives who had occupied the stores under the Soviet regime still retained rights of occupancy under existing law.

The overall result was a myriad of competing claims to the use of and control over commercial store property. Store owners, occupants, and local agencies and governments all had an interest in how stores could be used, and each one had standing to object to the ways others wanted to use them. In effect, each of these parties had some semblance of a property right in the stores, which allowed them to exclude others from using them.

*Figure 5.5* Moscow kiosks. *Source*: Gennady Grachev, Wikimedia Commons

So merchants who wanted to actually sell things apparently went ahead and set up cheap metal kiosks, thus bypassing that tangled bureaucratic web, and good riddance.

You can think of the Moscow kiosk example as a metaphor for any situation where multiple parties have standing to object to use of a resource, which is the essence of an anticommons situation. This standing may be manifested in what is essentially veto power, thus stymieing use of the resource. Or that standing may allow them to impose costs on other users. In either case, the predicted outcome is *underutilization* of the resource [Buchanan and Yoon (2000)]. This is, of course, in stark contrast to a commons, where the likely outcome is *overutilization*.

If you think about this argument, transaction costs seem to be present in spades in anticommons situations. If this were not the case, it should be possible for individual users to buy others out, in order to remove objections to resource use. This implies that, among other things, the larger the number of interested parties, the greater the degree of underutilization. In the limit, as the number of parties gets extremely large, the resource may not be used at all (see Box 5.2). And, as you might expect, this would be terribly inefficient.

---

### Box 5.2: Anticommons in medieval times

The Rhine River is one of the great rivers of Europe, starting from its source in the Swiss Alps and flowing north some 770 miles until it empties into the North Sea. In medieval times, it was a major trade route, with trading under the protection of the Holy Roman Emperor. However, over time the Emperor lost power and German barons built castles along the river and started charging traders tolls for passage. It became too expensive for traders to use the river, and so trading dried up. The river had too many "owners," resulting in its underutilization [Heller (2013), pp. 9–10].

---

It turns out that the anticommons idea may apply to a lot of situations. Remember the last time you took a flight? Did you ever wonder why you got stuck in a long line at the airport? Well, ever since the airline industry was deregulated in the mid-1980s, the number of people flying has skyrocketed. However, over this time virtually no new airports have been built in the United States. The reason may be that landowners in the major cities are able to block projects for new airports or airport expansions [Heller (2013), pp. 6–7]. Other resources to which this term may apply are: the broadcast spectrum; federal disaster relief; biomedical research; cyberspace; and water markets.[3] We shall encounter environmental examples of anticommons later in the book.

### Key takeaways

- Common-pool resources are characterized by rivalry in consumption and non-excludability.
- Tragedies of the commons (TOCs) are situations where resources get severely degraded and, in extreme cases, destroyed. TOCs are not inevitable, but occur under specific conditions, as captured in the roving bandits metaphor.
- The SES framework is an attempt to characterize the combination of factors that facilitate successful local management.
- Anticommons are characterized by multiple users having the right to exclude. In these situations, resources tend to be underutilized.

## Key concepts

* Common-pool resources
* Tragedy of the commons
* Roving bandits
* Social-ecological systems
* Social-ecological systems framework
* Anticommons

## Exercises/discussion questions

5.1. Which of the following situations would you expect to result in a TOC or near-TOC? Why?
   * A groundwater aquifer.
   * The Thames River.
   * A non-enclosed small city greenspace.
   * Yellowstone Park.
   * A large city community garden.

5.2. Consider a grazing commons game between two ranchers Smith and Jones. Each rancher has two possible options: to agree to cooperate in grazing the commons or to defect from the agreement, in which case it is every woman for herself. If they cooperate, each receives a payoff of $10. If both defect, they each receive a payoff of $2. If one defects and the other cooperates, the one who cooperates loses $4 while the one who defects gains $15. (a) Calculate the Nash equilibrium of the game. (b) What is the **smallest** penalty that a benevolent government could levy on defectors that you would predict would result in cooperation among these parties? Briefly explain.

5.3. Suppose there were *more than* two players in the prisoner's dilemma game, but with a similar payoff structure for the strategies (defect, cooperate). Do you think that all players would still defect in the (Nash) equilibrium? Why or why not?

5.4. Suppose you apply the SES framework to study a local CPR and you conclude that a strong, charismatic local leader was instrumental in ensuring the ongoing success of the local governance arrangement. You are wondering whether this conclusion applies to other CPRs. What values would *other* second-tier variables have to take to make you more confident that this conclusion would hold for these other CPRs?

5.5. In which of the following situations is an anticommons likely and why?
   * A community swimming pool.
   * The internet.
   * The local siting of a hazardous waste site, subject to public hearings.
   * Congressional oversight of EPA policy.

## Notes

1 These are sometimes called *toll goods* by certain scholars. See McGinnis (2011), p. 174.
2 See, for example, Ostrom (2007); McGinnis and Ostrom (2014).
3 *Broadcast spectrum*: Heller (2013); *Federal disaster relief*: Sobel and Leeson (2006); *Biomedical research*: Heller and Eisenberg (1998); *Cyberspace*: Hunter (2003); *Water markets*: Bretsen and Hill (2009).

# Further readings

## *I. The tragedy of the commons*

Hill, Peter J. "Are all commons tragedies? The case of bison in the nineteenth century," *Independent Review* 18(Spring 2014): 485–502.
*An interesting take on whether the extermination of the Great Plains bison in 19th century America really can be interpreted as a tragedy of the commons.*

## *II. Roving bandits*

Berkes, F. et al. "Globalization, roving bandits, and marine resources," *Science*, New Series, 311(March 17, 2006): 1557–8.
*Brief overview of roving banditry in the world's fisheries.*
Perez, I; M.A. Janssen; A. Tenza; A. Gimenez; A. Pedreno; and M. Gimenez. "Resource intruders and robustness of social-ecological systems: An irrigation system of Southeast Spain, a case study," *International Journal of the Commons* 5(August 2011): 410–32.
*A study of the effect of roving bandits on an irrigation SES in Spain.*

## *III. Social-ecological systems*

Cole, Stroma and Mia Browne. "Tourism and water inequity in Bali: A social-ecological systems analysis," *Human Ecology* 43(June 2015): 439–50.
*Uses Ostrom's social-ecological systems framework to understand degradation of Bali's water resources.*
McGinnis, Michael D. and Elinor Ostrom. "Social-ecological system framework initial changes and continuing challenges," *Ecology and Society* 19(June 2014).
*A valuable examination of the SES framework and its evolution over time.*
Partelow, Stefan. "A review of the social-ecological systems framework: Applications, methods, modifications, and challenges," *Ecology and Society* 23(2018).
*A useful recent summary of the state of SES framework scholarship.*

## *IV. Anticommons*

D'Agata, Antonio. "Geometry of Cournot-Nash equilibrium with application to commons and anticommons," *Journal of Economic Education* 41(April/June 2010): 169–76.
*This article provides a nice unified analysis of the commons and anticommons and their implications. For the mathematically inclined.*

# References

Bretsen, Stephen N. and Peter J. Hill. "Water markets as a tragedy of the anticommons," *William and Mary Environmental Law and Policy Review*, 33(2009): 723–83.
Buchanan, James M. and Yong J. Yoon. "Symmetric tragedies: Commons and anticommons," *Journal of Law and Economics* 43(April 2000): 1–14.
Hardin, Garrett. "The tragedy of the commons," *Science* 162(December 13, 1968): 1243–8.
Heller, Michael A. "The tragedy of the anticommons: Property in the transition from Marx to markets," *Harvard Law Review* 11 (1998): 621–88.
Heller, Michael A. "The tragedy of the anticommons: A concise introduction and lexicon," *The Modern Law Review* 76(January 2013): 6–25.
Heller, Michael A. and Rebecca S. Eisenberg. "Can patents deter innovation? The anticommons in biomedical research," *Science* 280(1998): 698–701.

Hunter, Dan. "Cyberspace as place, and the tragedy of the digital anticommons," *California Law Review* (2003).

Levin, Simon et al. "Social-ecological systems as complex adaptive systems: Modeling and policy implications," *Environment and Development Economics* 18(April 2013): 111–32.

McGinnis, Michael D. (2011) "An introduction to IAD and the language of the Ostrom Workshop: A simple guide to a complex framework," *Policy Studies Journal* 39(2011): 169–83.

McGinnis, Michael D. and Elinor Ostrom. "Social-ecological system framework initial changes and continuing challenges," *Ecology and Society* 19(June 2014).

Nagendra, Harini and Elinor Ostrom. "Applying the social-ecological system framework to the diagnosis of urban lake commons in Bangalore, India," *Ecology and Society* 19(June 2014).

Ostrom, Elinor. *Governing the commons*. Cambridge: Cambridge University Press, 1990.

Ostrom, Elinor. "A diagnostic approach for going beyond panaceas," *PNAS* 104(September 25, 2007): 15181–7.

Partelow, Stefan. "A review of the social-ecological systems framework: Applications, methods, modifications, and challenges," *Ecology and Society* 23(2018).

Petherick, Anna. "'Roving Bandits' depleting fisheries, experts say," *National Geographic*, March 16, 2006.

Sobel, Russell S. and Peter T. Leeson. "Government's response to Hurricane Katrina: A public choice analysis," *Public Choice* 127(2006): 55–73.

# 6 Public choice and the politics of environmental policy

You may recall the distinction between positive statements and normative statements that we discussed briefly in chapter two. To refresh your memory, positive statements are about what things happen. Like hitting your finger with a hammer causes pain. Normative statements are about assessing those outcomes, like being extremely unhappy when you are in pain. Similarly, we know what happens in markets when, say, consumer incomes increase. Demand tends to increase, resulting in increased prices and quantities sold (positive). And under certain conditions we really like markets because they generate efficient outcomes (normative). These two kinds of analysis often go hand in hand when economists consider real-world situations.

It is also useful to think about this distinction when we are considering a government setting environmental policy. When economists prescribe a policy like, say, setting aside some habitat for endangered species, we are making a normative statement: *This is how the government should behave in order to maximize social welfare.* However, we should also be asking a more fundamental positive question: *What would we actually expect the government to do?* Is it reasonable to expect the government to actually create a program that sets aside habitat? More generally, completely aside from the question of what we would *like* the government to do, what do we think it will actually end up doing?

To begin to answer these questions, think about the way the government actually operates in the real world. Upon closer inspection, it is not at all clear that real-world governments would set aside habitat in order to obtain efficient outcomes. Governments are messy things. In many countries, elected legislators – like your congressman or senator – are primarily responsible for choosing environmental policies. These legislators represent different constituencies – everyone from Green New Dealers to climate deniers – who may possess a wide variety of preferences on environmental outcomes. Legislatures produce policy by trying to get a majority of legislators with different parties and ideologies to agree. And this brief description does not even begin to do justice to the complexities of legislative policymaking. The point is that policies have to undergo a long, involved political process before they see the light of day. This means we need to ask ourselves the question: Why on earth would we expect an efficient policy to emerge?

All of this means that environmental policymaking is very much a political exercise. Thus, we have to pay attention to the political process and the political factors that can lead to one policy outcome rather than another. This is not to argue that economics will not matter. Rather, we need to think about how economics and politics interact in order to produce environmental policy. Exploring this idea is the basic objective of this chapter.

# What is public choice?

The economist Gordon Tullock has defined public choice as "the use of economic tools to deal with traditional problems of political science" [Tullock (2008)]. Notice that Tullock is saying that the economic approach can be applied to provide insights into political decisions. What is the economic approach? As we have seen, it entails thinking in terms of scarcity, tradeoffs, and incentives. But, in addition, it generally involves the behavioral assumption that individual actors are, broadly speaking, essentially rational and self-interested, meaning that they purposefully pursue actions that benefit themselves, often at the expense of others. As we put it in chapter three, people act in order to maximize their own utility, to the extent they can.

We also saw this assumption in chapter two, for example, when we considered a profit-maximizing monopolist. However, though it was not explicitly developed there, we were also assuming that consumers and competitive firms take purposeful steps to pursue their own interests when they interact in markets. Following Tullock, in this chapter we will develop the idea that voters, politicians, and agency administrators also behave in a largely rational manner to pursue their own interests. The question is: What are the implications for environmental policy if they do?

Let us start with a simple example that illustrates the public choice approach to politics, an argument made a number of years ago by the economist Anthony Downs. In 1957, Downs published the book *An Economic Theory of Democracy*. This highly influential book was one of the first books ever to take the public choice approach, assuming rational, self-interested actors. One of the questions Downs raised was one that is fundamental to the working of democracies everywhere: *Why do people vote?*

If you are used to thinking like an economist and you ponder this question, the answer may not be entirely obvious. After all, applying the economic approach, we should think that people would vote only if the benefits of voting outweigh the costs. The difficulty is that the costs of voting seem larger than the benefits. In order to vote, you may have to drive over to the nearest polling station, wait in line, maybe take time off from work or miss classes, and so forth. In some cases, you may even have to endure long lines and it could be raining, snowing, or blazing hot that day.

On the other hand, think about the benefit you derive from voting, in terms of what is presumably your reason for voting: to help elect your favored candidate, or to help pass a local environmental initiative, or something like that. In a situation where the outcome depends upon a majority in favor, ask yourself the following question: What are the chances that your vote would tip the balance from no to yes or the other way round? I mean: in an election, you may be one of thousands, or even millions, of voters. So, even if you *really* want your candidate to win, or you *really* want your environmental initiative to pass, the likelihood that your single vote will change the outcome is probably really small. This means that, in expected terms, you derive virtually no benefit from going out and casting your vote.

And yet many people vote. There have been two reactions to this unquestionable fact. The first is that there must be something wrong with the notion that voters behave in a rational, self-interested way; that is, there must be something wrong with public choice theory itself. As one famous political scientist put it, the fact that many people vote is the "paradox that ate rational choice theory" [Fiorina (1990)]. Other, less polite, commentators have called the public choice perspective "nonsense on stilts," like an edifice ready to crumble [Grofman (2008)].

The other reaction has been to accept the public choice approach, but to argue that the narrow notion that people only care about the election outcome may be incomplete. Some people may consider it their civic duty to vote. For others, voting may serve to send a message about who they are politically or as a person, or to show group solidarity with others they identify with. None of this is to argue that there is no self-interested element but, simply, that for certain types of behavior like voting, there may also be other important factors. I would like you to keep in mind this idea as you progress through this book: that public choice theory explains a lot, but it may not explain everything.

## Special interest politics

One of the key insights of the public choice approach is the way we understand legislative outcomes as the result of competition among *special interest* groups. In the basic special interest model, individual voters combine to form interest groups based on some common objective. Why do they do this? To increase their political influence, since their individual voting behavior is unlikely to matter much (as we have just seen). Interest groups then compete for favorable policies by providing things of value to legislators. These things of value include votes, campaign contributions, bribes, and the prospect of lucrative employment after their legislative career is over. All of these things reflect their willingness to pay for policies. So, in a real sense, interest groups "demand" policies, which are then "supplied" by legislators (see Figure 6.1). The key questions are then about the relative strength of competing demands by different groups, and the "costs" to legislators of supplying different policies.[1]

Key to this approach is the assumption that interest groups can exercise influence over their elected representatives through what political scientists have called the *electoral connection* [Mayhew (1974)]. The term electoral connection refers to the desire of legislators for re-election, which explains much of their legislative behavior, including wanting to keep their constituents happy by pushing for policies that favor them. These could include environmental policies of various kinds, like stiffer air quality regulation, funds for toxic waste disposal, or remediation of local groundwater supplies. So we start to see the positive origins of environmental policy, through the actions of interest groups and their influence on legislators.

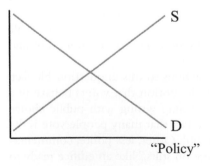

*Figure 6.1* The political market for policy

To expand on the special interest approach to legislation, consider the demand of an interest group for a policy that favors it. This demand will depend upon the benefit it expects to gain from the policy. So, for example, suppose the policy confers $1 million in benefit on the members of the group. This implies that the group should be willing to pay up to $1 million for the policy. More generally, the greater the benefit it expects to receive, the greater the group's stake in the outcome, which translates into greater willingness to pay. In the terminology of public choice, this willingness to pay for favorable policies gives rise to *rent-seeking* behavior: political actions aimed at securing rents (benefits) from the political process.

In general, there will be competition among interest groups for favorable policies, as a policy that benefits one group may impose costs on another group. For example, policies to limit greenhouse gas emissions are supported by environmental groups and opposed by the fossil fuel industry. So enacting policies often becomes a question of competing demands, a kind of tug-of-war among interest groups with the highest bidder winning. The question then becomes: In addition to group stakes, what factors will determine which group will win?

### I. Interest group size

One factor is the relative size of the interest groups, and here we get an interesting result. Given that many legislative systems are based upon majority voting, one might think that legislators would tend to favor groups that are larger; that is, containing more voters. After all, one would think that such groups would have larger sway in elections, by virtue of their sheer numbers. As it turns out, this is often not the case, and this is one of the major contributions of public choice theory. There are two basic reasons why.

To understand the first argument, let us turn to another classic of public choice theory, a book published in 1965 by the economist Mancur Olson entitled *The Logic of Collective Action*. In this book, Olson argued that larger interest groups suffer from an important disadvantage when in competition with smaller groups. Namely, members of larger groups have greater incentive to sit back and let others in the group put in the effort to pressure their representatives to serve the group's interests. This is because they will receive the benefits of the group's efforts – say, enactment of a piece of environmental legislation – regardless of whether they put in any effort on their own. So, since effort is costly, each member of the larger group has more incentive to sit on the sidelines. If this argument sounds familiar, it should. This is the idea of *free-riding* that we encountered in chapter five.

Since this is an extremely important argument, let me clarify one key point so there is no misunderstanding. Every individual in an interest group would be most well-off if everyone in the group worked hard; that is, if nobody engaged in free-riding. If this happened, the group would have more influence and would be more likely to secure the most favorable policies for itself. However, every individual also has individual incentive to free-ride. So, overall, every member wants to free-ride while hoping nobody else does. But since they all have the same incentive, free-riding occurs and it tends to occur more with larger groups. This is why they tend to be less politically effective, other things equal.

### II. Individual member stakes

The second reason that larger groups are not necessarily more politically influential is that members of larger groups often have less at stake individually, making them less likely to push for favorable, or to oppose unfavorable, policies. This would be the case, for example, if the cost of a new policy was paid for out of general taxation or some other way that spread the cost over the rest of society. Smaller groups would tend to push for policies that benefited their group and, therefore, their individual members, greatly. But if the cost of these policies were spread out over many other people, they may not get much political pushback.

Let me give you an example using numbers, to make the argument concrete. Suppose a new policy benefits a group of 100 people to the tune of $1 million. But suppose the $1 million cost of this policy is spread out over the entire society of 100,000 people through a general tax. Each member of the group receives $10,000 in benefit from the policy, making each of them very likely to push hard for it. At the same time, if the cost is spread evenly, each one of the 100,000 people will only have to pay $10 in additional taxes. This is unlikely to arouse much political opposition, making the group likely to get its way.

A real-world example of this phenomenon of *concentrated benefits and dispersed costs* is protectionist trade policy such as tariffs on imported consumer goods, which is designed to protect a domestic industry from international competition. Such policies tend to raise consumer prices, but, since the effect on prices tends to be relatively small, consumers have little incentive to oppose the tariffs, while the protected industry has huge incentive to support them. The real-world result is often that tariffs get imposed. Real-world environmental examples include: tax breaks for the fossil fuel industry, and federal subsidies to support a local mass transit system, hydropower plant, or irrigation system. These types of projects are sometimes unflatteringly referred to as *pork barrel* projects [Ferejohn (1974)].

I should mention that this process can also work in reverse, in cases where there may be *concentrated costs and dispersed benefits*. In this case, a narrow interest group with much to lose may successfully block an environmental policy that benefits many people. For example, in the 1970s and 1980s, representatives of states that produced high-sulfur coal successfully fought tightened pollution standards for sulfur dioxide ($SO_2$) because those policies would have imposed costs on them.[2] According to the political scientist James Wilson, it is no surprise that interest groups are often able to block new policies:

> It is remarkable that policies [with concentrated costs and dispersed benefits] are ever adopted, and in fact many are not. ...The framers arranged things ... so that a determined minority has an excellent chance of blocking a new policy. ...The opponent [of such a policy] has every incentive to work hard; the large group of prospective beneficiaries may be unconvinced of the benefit or regard it as too small to be worth fighting for. [Wilson (1992), p. 436]

### III. Interest group heterogeneity

So far, the discussion of interest group effectiveness has depicted interest groups as monolithic: that is, that all members of the group share the same interests, ideologies, preferences for policy, and magnitude of stakes in the policy outcome. However, this will not be an accurate description of many interest groups, whose members may not be the same in various important respects. The question then becomes: How will these differences affect the ability of the interest group to effectively exert demand for its favored policies?

The answer turns out to be a bit complex, as it depends upon precisely how – that is, along which of these dimensions – the members differ from each other. I will be unable to give you a comprehensive answer to this question, but let me give you a couple of factors to keep in mind.

Let us begin with an argument made by Olson himself: that heterogeneity of an interest group might actually make the group more effective politically. What he was thinking of was a case where a few group members have substantially larger stakes in a policy outcome than others. If so, this might enable the group to overcome the free-rider problem. The argument is essentially that these larger-stakes members would be highly motivated to push for the policy and could provide a critical mass of support, while smaller-stakes members free-ride on their efforts. Olson called this "exploitation of the great by the small" [Olson (1965), p. 29]. A number of subsequent studies have investigated varying stakes and found the effect to be complex, as it depends on the way stakes vary, and the cost of lobbying effort.[3]

A second way in which individual interest group members could be heterogeneous is if they differ in their political goals and objectives. If members of a group differ in important ways relating to their goals and objectives, this can prevent them from presenting a united front on behalf of a policy. This will tend to make the interest group less effective.

For example, consider two real-life environmental groups, the Sierra Club and Ducks Unlimited. The Sierra Club champions a wide variety of environmental causes. If you visit their website, for example, they invite donations so they can: protect wilderness areas and endangered species, help keep air and water clean, work toward clean energy, combat climate change, protect the oceans and fragile coastal environments, and promote safe and healthy communities. Ducks Unlimited, on the other hand, has a much narrower focus: wildlife management and habitat protection. All this makes members of the Sierra Club much less likely to be on the same page when trying to agree on which policies to support.

One way to think about the relative situations of the Sierra Club and Ducks Unlimited is that the Sierra Club is likely to experience higher transaction costs of negotiation among its members. Sierra Club members may, for example, be at odds over which issues to prioritize and champion and internal discussions over strategy may be contentious. These transaction costs will tend to reduce the willingness to pay of the Sierra Club for favored policies. So, for example, if a favorable policy provides $1 million in benefits but the Sierra Club experiences $100,000 in transaction costs over internal negotiation, they would only be willing to pay up to $900,000 for the policy. This might put them at a bidding disadvantage relative to another group with lower transaction costs, like Ducks Unlimited.

None of this is to argue that Ducks Unlimited would always get its way in a head-to-head competition with the Sierra Club. The Sierra Club is bigger and has more resources, and other factors might affect their relative demands for favorable policy. The point is that, other things being equal, more heterogeneous interest groups will tend to experience greater difficulties in coming to agreement on policies, which will tend to reduce their willingness to pay.

*IV. Political entrepreneurs*

One final factor concerns the possibility of what we call *political entrepreneurs*. Political entrepreneurs are individuals or organizations who can galvanize political support around a particular policy issue. A good example of such a person is Ralph Nader, who gained national recognition in the United States beginning in the 1960s for his strong advocacy for consumer protection and other environmental issues. Political entrepreneurs like

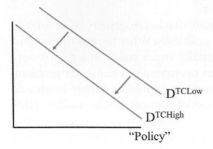

*Figure 6.2* Effective interest group demand

Nader reduce the transaction costs of negotiation among the members of an interest group. In Nader's case, it seems less likely that consumers as a broad group would have effectively organized to push for consumer protection without Nader playing a role to focus attention on the issue and to push strongly for reform.

In analyzing political outcomes, you can think of group stakes, group size, individual member stakes, interest group heterogeneity, and political entrepreneurs as shifting the demand for particular policies. For example, in Figure 6.2, we see the effect of high transaction costs in reducing the demand of an interest group for a favorable policy. We call $D^{TCHigh}$ the *effective demand* for the policy.

## An expanded public choice perspective on the Coase Theorem[4]

Now that we understand the concept of transaction costs in the context of public choice, let us return briefly to the Coase Theorem discussed in chapter four. You may recall the key result there: in the presence of significant transaction costs, there is no necessary presumption that free exchange will result in the efficient outcome. Furthermore, the assignment of property rights – say, under a rule of liability for damages – could matter greatly in determining the efficiency of the outcome. However, if transaction costs are low, it should not matter who is assigned the property rights, assuming we are interested in efficiency. Of course, it could matter a great deal from the viewpoint of equity.

During that discussion, did you ever wonder about the issue of how the property rights were assigned? In fact, there was an important unspoken assumption. Namely, that the property rights were assigned in a simple, uncontentious manner: say, by a court just sitting down and assigning liability. Importantly, of all of the negotiations that took place in that described setting, none was about the assigning of the property rights in the first place.

Now that we are in the public choice world, we have to seriously reconsider that assumption. The practical effect of many environmental policies is to award property rights to one group or another. These include the right to clean air, clean water, and the like, all of which are analogous to the right of Coase's farmer to untrampled crops. But we have just seen that the special interest model is based on the notion that policy is determined by the interplay of competing rent-seeking groups. All sorts of resources – bribes, lobbying efforts, negotiations within interest groups to settle on a policy position – go into those efforts to influence policy. In other words, for many environmental policies the assigning of property rights may be drawn-out, costly, and contentious.

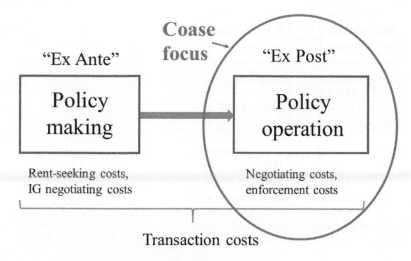

*Figure 6.3* Coase and environmental policy

Figure 6.3 places the Coase argument within the broader context of environmental policy. Coase was largely concerned about negotiations that occurred after policy was set and property rights had been assigned. This occurs in the *ex post* world (that is, given the fact that the policy exists) on the right-hand side. In this ex post world, what is central to the analysis of efficiency are the costs of negotiating over, exchanging, and enforcing property rights *that have already been assigned* under a policy. Figure 6.3 adds the fact that there is a previous stage, prior to (*ex ante*) the operation of the policy, during which all of the interest group interplay occurs.

The key point here is that even if transaction costs are low in the policy operation stage, they could be considerable in the policymaking stage. Indeed, *policymaking* transaction costs could be considerably greater than *policy operation* transaction costs. More generally, environmental policymaking may want to consider tradeoffs between the ex ante and ex post stages, in order to minimize overall transaction costs. We will see applications of this framework in later chapters.

## The supply of policy under the public choice framework

So far, we have been focusing on the demand side of public choice theory, by focusing on the demanders of policy: voters and interest groups. Let us now turn to the supply side, meaning the elected officials who make policy.

In order to provide an organizing framework for the discussion to come, consider the simplified description of the political system of the United States shown in Figure 6.4.[5] This illustrates a *bicameral* system, where there are two legislative bodies. In the United States, both the Senate and the House of Representatives must pass a bill in order for it to become law. Legislation can originate in either the House or the Senate, and it must pass that chamber before moving on to the other one.

Within each chamber, legislators can introduce bills, which are then referred to a committee with appropriate jurisdiction. This committee decides, by majority vote of

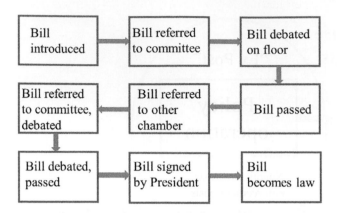

*Figure 6.4* Simplified bicameral legislative procedure

committee members, whether the bill can move to the floor of the chamber, to be debated and voted on. If a majority of legislators vote to approve a bill in a floor vote, it then moves on to the other chamber, where basically the same procedure is followed. In a final step before it becomes law, the bill must also be signed by the president. The president has the power to veto a bill, which signifies that she/he does not approve it being enacted into law.

There are basically four ways in which bills do not become law. First, the committee may decide not to move a bill to the floor. Second, a bill may be voted down by a majority of legislators in either chamber. Third, a bill may fail if the second chamber revises the bill submitted by the first one and the two chambers are unable to hammer out a common version in a so-called *conference committee* consisting of members of both chambers. Finally, a bill vetoed by the president can become law only if legislators vote to override the veto, which requires a two-thirds majority of both the House and Senate.

A final piece of the picture is the fact that once a law is passed, its provisions are commonly administered by a government agency. In the case of environmental laws in the United States, for example, the Environmental Protection Agency, the Fish and Wildlife Service, and state and local governments are charged with administering policies regarding air and water quality and endangered species protection. Generally speaking, laws contain procedures for legislatures to exercise oversight of administrative agencies, such as requiring them to prepare annual reports, or requiring agency officials to appear when committees hold public hearings.

Much of the public choice scholarship has focused on three steps in this process: legislative voting, the importance of committees, and legislative oversight of agencies. Let us consider each of these in turn, beginning with legislative voting.

### I. Legislative voting

Majority voting systems play an important role in most legislative systems. Under majority voting systems, typically any majority voting in favor of a bill results in its passage. It does not matter whether the bill passes 50.1% to 49.9%, 60% to 40%, or unanimously: it is enacted in any case.[6] A question that has been asked by many public choice scholars is: Under a majority voting rule, what legislative outcome would you expect? This question is obviously relevant for us if we are considering environmental policy set by legislatures operating under majority voting rules.

It turns out that this question is not easy to answer, as the institutional details matter a lot. Policies are often complex, involving many factors. Different policies may have very different distributional consequences, resulting in very different political dynamics. The procedures used by legislatures can make all the difference in whether bills get passed: by determining when, how, and even whether bills get considered. In this section, it will not be possible to have a complete discussion of a vast and nuanced scholarly public choice literature on majority rule voting outcomes. Instead, we will focus on a few key results of that literature.

### A. The median voter theorem

One extremely important result in the public choice literature ascribes a lot of importance to the political preferences of the so-called *median voter* [Black (1958); Krehbiel (2004), pp. 114–15]. The median voter is the ultimate moderate, the voter whose political views are smack dab in the middle of the distribution of political views, in the following sense. Imagine you could measure political views and order them along a line that runs from *extreme liberal* on one end to *extreme conservative* on the other end (see Figure 6.5). Now think of every voter as occupying a point along that line. In Figure 6.5, there are seven such voters, A, B, C, D, E, F, and G. Here, D is the median voter.

Why would voter D be so important? Well, imagine that each voter, though preferring the position represented by her/his own letter, has preferences over other outcomes too. So D might not mind C's position: she just does not consider it to be as good as D. Furthermore, assume that each voter dislikes positions that are farther away from their ideal position. So D prefers C to B, which is in turn preferred to A. Similarly, E is preferred to F, which is preferred to G. One way to represent all of this is to imagine each position represents a different amount of utility to each voter, where utility is maximized at one's ideal position and decreases in both directions. Figure 6.6 represents this assumption with a series of utility hills in the simplified case of three voters: A, D, and G. For perhaps obvious reasons, these voters are said to have *single-peaked* preferences.

Now, let us refer back to Figure 6.5 and ask ourselves: Which outcome, from extreme liberal to extreme conservative, would be likely to muster majority support among these seven voters? Well, first of all, neither of the extreme positions A or G, right? Even though these positions are loved by the extreme voters, they are disliked by everyone else. So, for example, G would fail if he tried to get his bill passed, because voters A through F all have less conservative bills that they prefer to G. In a face-to-face runoff between G and any bill more liberal than (to the left of) G, G would be defeated. And the same argument goes

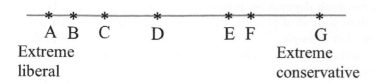

A  B  C     D     E  F     G
Extreme                  Extreme
liberal                  conservative

*Figure 6.5* The median voter model

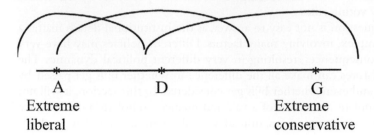

*Figure 6.6* Single-peaked preferences

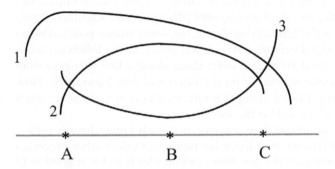

*Figure 6.7* Voting with non-single peaked preferences

for A, in the opposite direction of course. So neither the extremely liberal nor the extreme conservation bill is likely to be adopted.

But B and F are likely to fare no better, right? In each case, other bills would be preferred by a majority in a face-to-face vote. Indeed, the *only* bill that would not be defeated under a majority voting rule is D. So the predicted outcome under a majority voting rule is that the moderate median voter position is the one that gets enacted. This remarkable result is known as the *median voter theorem*.

It is important to keep in mind the assumptions being made in deriving this median voter result. First, there is only one policy dimension, in our case liberal/conservative. However, voting preferences are generally not this simple. Think of all the policy issues you are concerned with, which may include some of the following: climate change, abortion rights, income inequality, gun reform, and health care reform. It may be difficult to reduce your preferences on all of these issues down to one single dimension, let alone try to compare positions across different voters.

Second, the median voter result depends on preferences being single-peaked. That is, the farther you get away from your ideal position, the less you like the policy. However, suppose this is not the case. For example, consider Figure 6.7, which graphs the preferences of three voters 1, 2, and 3 over three policies A, B, and C. *Before going on, can you determine which policy would win under a majority voting rule?*

In this case, it turns out that there is no clear prediction of a winner. To see this, let us take the information in Figure 6.7 and translate it into a table of preferences over the three policies. This is done in Table 6.1. For example, voter 1 likes policy A best, then B, then C. Voter 2 likes B best, then C, then A. And similarly for voter 3. In this example, notice that the preferences of Voter 3 are not single-peaked, because she derives less utility from B than from A, which is further away from her ideal policy.

To see that there would be no clear majority winner, consider head-to-head matchups between different policies. According to Table 6.1, if A is pitted against B, A would win 2 to 1 because voters 1 and 3 prefer A while only voter 2 prefers B. If B goes up against C, B would win 2 to 1 because voters 1 and 2 prefer B. Finally, if A goes up against C, C would win 2 to 1. In this case, where there is the possibility of non-single peaked preferences, there is no predicted winner. Instead, we have the phenomenon of *cycling*, where we get different winners depending upon which matchup we select. This is a disconcerting result because, after all, the choice of which policies to match up against each other in this simple example is arbitrary.

Finally, the median voter model implicitly assumes that there are no procedural restrictions in terms of how bills can be introduced or amended, or what happens when bills are defeated. To make the point simply, let us pare down the world to two policy options, F and G. In the above example, when we were comparing G to F, we assumed that F was an actual option. But suppose that there was a procedural restriction that kept F from even being considered by the legislature for a vote. G might lose to F, but G might be preferred to nothing at all!

The other issue in this example is: In this case – F non-existent – what if G gets voted down anyway? The median voter model says nothing about what might happen in this case. The bottom line is that the median voter result holds when policies can be reduced to a single dimension, when preferences are single-peaked, and when there are no procedural restrictions on voting. Of course, there is no guarantee that these conditions will hold in the real world. We will revisit this issue in the next section on the role of committees.

It turns out that a number of economists have used the median voter theorem to try to better understand environmental policy. Let me give you an example. Have you ever thought about the relationship between environmental policy and income inequality? For example, would you expect pollution standards to be stricter or weaker when income is more unequally distributed among a nation's citizens? Or would you not expect it to make any difference? Different economic models predict different relationships, so it becomes an empirical question.

As we have seen, under the median voter theorem, the preferences of the median voter decide what policy we will get. When there is greater inequality, there will tend to be a greater gap between the median income in a society and the average income. So, if pollution abatement declines as this gap widens, then you can conclude that growing inequality

*Table 6.1* The possibility of cycling

| Voter | | |
| --- | --- | --- |
| (1) | (2) | (3) |
| A | B | C |
| B | C | A |
| C | A | B |

may be less conducive to strict environmental policy. Some studies have found this to be the case [Eriksson and Persson (2003); Kempf and Rossignol (2005)].

### B. Logrolling

In the basic legislative voting model, we encounter the same issue that we saw earlier with the general electorate: that a legislator's individual vote may have very little impact. Indeed, in the median voter model, only the median voter's vote makes a difference to the outcome. This is an important reason why legislators often get together and form coalitions to gain passage of bills. One way they accomplish this is by trading votes, commonly known as the practice of *logrolling*.[7]

The idea behind logrolling is intuitive and can be illustrated with a simple example. Suppose there are three congresswomen, each of whom represent different constituencies. Congresswoman A, who represents an agricultural district, may be interested in enacting ethanol subsidies. Congresswoman B, who represents a city, may be interested in subsidizing mass transit. Congresswoman C, who represents a suburb, may be interested in subsidizing creation of more green spaces. On their own, each one may have little chance of gaining passage of her favored bill. After all, each has only one vote. Under logrolling, congresswomen agree to vote for each other's bill. Then, if enough trades occur, bills can muster majority support and gain passage.

To make sure we understand the example, let us use some numbers, shown in Table 6.2. Each entry in the table represents the net benefit derived by each congresswoman from each policy. For example, Congresswoman A, who represents a rural area, derives −200 in net benefit from mass transit (MT) subsidies. Each congresswoman derives positive net benefits from her preferred policy but negative net benefits from the others because her constituents will have to pay taxes in order to pay for those policies. Assuming that negative net benefits for constituents would cause a congresswoman to vote no, you can see that none of these policies would be able to command majority support. If none of the policies are enacted, each congresswoman earns zero net benefits for her constituents.

However, consider the possibility of logrolling. In this case, Congresswoman A might approach Congresswoman C and offer support for her bill in return for C's support for hers. A is willing to do this because she really wants the ethanol subsidies, while C should agree because she really wants the subsidies for green spaces. In this case, both green spaces and ethanol subsidies would garner majority support because both A and C would be better off: A would benefit by 400 (400 − 100), while C would benefit by 200 (300 − 100).

Notice also that this logrolling is efficient. Even though this deal hurts B by 300, the combined net benefits to A and C (600) more than make up for this. Total net benefits to society have increased. However, it is important to note that logrolling may not always be

*Table 6.2* A logrolling example

|   | MT | GS | ETH |
|---|---|---|---|
| A | −200 | −100 | 500 |
| B | 400 | −100 | −200 |
| C | −100 | 300 | −100 |

*Table 6.3* Inefficient logrolling

|   | MT | GS | ETH |
|---|-----|------|------|
| A | −200 | −100 | 300 |
| B | 400 | −100 | −400 |
| C | −100 | 200 | −100 |

efficient. Consider Table 6.3, which modifies the numbers in Table 6.2 slightly. Here, A and C benefit from the logrolling by a total of 300, but B is hurt by 500. In this case, logrolling actually reduces total net benefits to society.

Logrolling is obviously a form of exchange and, like all exchanges, is potentially subject to transaction costs. Transaction costs arise for several reasons. First, it may be the case that payoffs are not well-known in advance. Tables 6.2 and 6.3 mask this issue by just assuming numbers that we could take as true. However, in the real world, it may not be clear exactly how much benefit suburbs derive from green spaces or rural areas derive from ethanol subsidies. This might especially be problematic if the stream of benefits stretches off into the future and future events are uncertain. This might be the case with many ongoing programs, unlike a one-shot deal to build green spaces, a hydropower dam, or a bridge. If so, it might simply be difficult for legislators to know how much a trade will benefit them, which may discourage some vote trading.

Another potential source of transaction costs is associated with the problem of enforcing logrolling deals. This problem can especially arise when the votes for the different policies are not contemporaneous, or the benefits occur at some future point in time. So if one vote takes place first and A gets her ethanol subsidies, she then loses her incentive to vote for C's green space subsidies. This may be true even in a "repeated-play" situation where she may be concerned that reneging will tarnish her reputation, if she can argue that changing conditions legitimately make it difficult for her to stick to the deal.

Furthermore, when policies are enacted, it is often difficult to enforce those policies over time. Future legislatures, which may be subject to different political pressures and may not even contain the same set of legislators, may seek to amend or ignore agreements from previous legislative sessions[8]

However, there are a couple of important ways that legislatures may be able to reduce the transaction costs of logrolling. One is through rules that allow various policies to be bundled together into one package before they are voted on, producing so-called *omnibus bills*.[9] This ensures that votes for different policies are simultaneous, solving the problem of reneging on deals by individual legislators once they get their way. However, while this solution addresses the problem for one-time policies such as the contents of many pork-barrel projects, it does not address the issue of ongoing enforcement.

Another way to reduce transaction costs is through the committee system. Let us now take a closer look at that important feature of many legislative systems.

## II. Role of committees

In many legislative systems, committees play an important role in the legislative process. In the U.S. Congress, there are a number of standing committees, each of which oversees a particular policy area. For example, standing committees in the Senate include committees overseeing agricultural policy, foreign relations, homeland security, environmental policy, veterans' affairs, and appropriations for discretionary spending. Most of these committees also have subcommittees that handle sub-areas of policy concern. There are also a few special, or select, committees, mostly formed to address general issues, like aging, intelligence, or ethics. However, we shall not focus on these because the argument really applies mostly to standing committees that have jurisdiction over areas of special economic concern to their constituents.

When bills are introduced into either the House or the Senate, they are immediately referred to a committee with relevant jurisdiction. Bills are discussed in committee, which sometimes then passes them on for consideration by ("advances to the floor of") the larger legislative body. However, committees can also choose not to advance bills, at which point they are effectively dead for that legislative session. Committees thus have effective veto power over whether a bill can even be considered by the full legislative body. For this reason, committees are believed to have a great deal of influence in the legislative process.

The power of committees has two important implications for the discussion in this chapter.

## A. Effect on median voter outcomes

First, even under many of the assumptions of the median voter model, adding committees to the picture can result in outcomes that differ significantly from the median voter position within a legislative setting. They accomplish this by imposing additional structure on the way the bills are considered.

To see this, consider a simple model that resembles the median voter model except that it adds three features. First, and most importantly, it assumes there is an *agenda setter*: a person or entity that can determine what bills are considered, indeed whether they are considered at all, by the legislature. That is, this party sets the agenda for what can be considered by the legislature, hence the name. This is sometimes referred to as the *gatekeeper* role of committees. As applied here, the committee plays the role of the agenda setter.[10]

Second, it assumes the agenda setter can propose a so-called *take-it-or-leave-it* bill to the legislature. This means that any bill advanced to the legislature cannot be amended once it is there. That is, the legislature operates under a *closed rule* with regard to committee referrals. Finally, if the agenda setter's proposed bill is not passed by the entire legislature, then it assumes we go back to the drawing board and simply maintain the status quo.

Under these assumptions, the committee as agenda setter may have incentive to choose a bill to advance to the floor for consideration by the entire legislative body that is different from the median voter position. To see this, consider Figure 6.8, which shows two utility functions, one for the agenda setter and the other for the median voter. Each function peaks at the respective ideal positions, $A^\star$ for the agenda setter and $M^\star$ for the median voter. Under the median voter theorem, the predicted outcome is $M^\star$, the ideal position of the median voter [see Krehbiel (2004), p. 116].

However, notice how the additional assumptions change things. Instead of $M^\star$, the agenda setter will have incentive to propose $X^\star$. Why? Because this is the best he can do and still ensure that the median voter will prefer it to the status quo. That is, the median voter will vote for any position to the right of $X^\star$ because he prefers any of them to the status quo. So all these positions will command a majority vote. So the agenda setter will choose $X^\star$ – or more accurately, as close to $X^\star$ as he can get and still gain a majority in

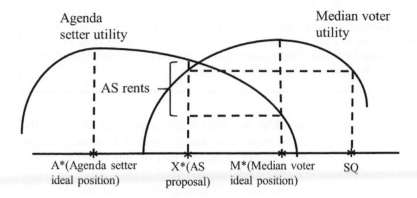

*Figure 6.8* The power of the agenda setter

favor – because that is the one that maximizes his utility while gaining majority support. And, at that position, he will be better off than he would be at the median voter position, in the amount labeled "AS rents." This is the height of his utility function at the chosen position X★ minus the height of his utility function at the median voter position.

### B. Effect on logrolling

The second effect of committees is that they can lower the transaction cost associated with logrolling on the floor of the legislature. They can accomplish this by discouraging legislators who try to renege on logrolling deals. For example, consider two legislators who have had their bills passed in a logrolling deal. Once they get their goodies, each has incentive to form a coalition to repeal the other's subsidy. In order to do so, however, they must introduce another bill, which must be referred to a committee. This committee controls whether the bill advances to the floor for a vote and thus has power to kill the attempt to overturn the previously agreed-upon subsidy.

In order for this system to succeed in keeping legislators from reneging on deals, the committee must serve the interests of the subsidized group. However, the very life-and-death power that committees have over legislation provides legislators incentive to serve on those very committees with jurisdiction over issues important to their constituents. For example, legislators representing rural districts will want to serve on committees with jurisdiction over farming. By doing so, they put themselves in a position to help advance legislation favorable to farming interests while helping to forestall unfavorable legislation (see Box 6.1). This includes attempts to overturn previously agreed-upon deals.

---

### Box 6.1: The importance of congressional committees

Democrat Collin Peterson represents the Seventh District, a heavily Republican district in rural Minnesota, and he has done so for the last thirty years. In 2016, Donald Trump out-polled Hillary Clinton in the district by a margin of thirty-one points. Yet Peterson keeps winning. Why? Well, one big reason is that Peterson has the staunch support of the agricultural sector, including the Minnesota Farm Bureau and representatives of many farming industries such as the powerful sugar

> beet and dairy industries. Peterson's long tenure in Congress makes him one of the more senior members of Congress, putting him in a position to chair the House Agriculture Committee if the Democrats control the House in 2020. This powerful position would enable him to play a highly influential role in advocating for farmer interests, a fact not lost on farm interests, who contribute millions of dollars to keep Peterson in office [Schneider (2020)].

You might reasonably ask the question: Why wouldn't legislatures place limits on the ability of committees to act as agenda setters for new legislation? After all, in principle, legislatures based on majority voting rules should be able to alter or eliminate procedures by a majority vote. It seems strange that they would set and keep in place procedures that effectively permit themselves to be consistently held hostage by special interests with incentive to milk the system as much as they can.

Part of the answer may be that committees have more specialized expertise in the areas under their jurisdiction than do those in the larger legislature. After all, committees have incentives to go out and collect information that will enable them to serve their constituencies better. Committee members may be more willing to collect information and convey it to the legislature, the more they can benefit their constituents. So, for the larger legislature, delegating agenda-setting to committees may reflect a tradeoff: compromising on their ideal majority position in order to obtain a more informed policy [Gilligan and Krehbiel (1987; 1990)].

### III. Legislative oversight of agencies

Many environmental policies are administered by government agencies. In the United States, for example, there are a variety of such federal agencies, including the Environmental Protection Agency, the Fish and Wildlife Service, the Forest Service, the National Park Service, and the Department of Energy. In addition, there are many administrative agencies at the regional and state level. Most states have a department of environmental protection or environmental quality, and a number have separate departments of natural resources. There are also a number of regional agencies responsible for managing resources of various kinds like coastal fisheries, and interstate rivers.

The performance of environmental policy depends upon the performance of these agencies, in how well they carry out their legislative mandates. Thinking like economists, their performance should in turn depend on the incentives experienced by members of those agencies. This very point has generated a debate among public choice scholars regarding how to interpret agency behavior. Scholarly positions on bureaucracy vary greatly, and you can think of them as being arranged on a spectrum.

At one extreme is the view that agencies are simply non-responsive bureaucracies, full of paper-pushers who twiddle their thumbs all day, wasting taxpayers' money. At the other extreme is the view that agencies are full of idealistic civil servants, whose fondest desire is to work 80-hour weeks serving the public. Both of these extreme views are stereotypes and, like all stereotypes, each has a modicum of truth, which helps explain their persistence in people's imagination. However, neither tells anything like the complete story [Weingast and Moran (1983)].

We can gain insights into agency behavior by thinking about the issue of legislative oversight. If you think about it, legislators have every incentive to engage in agency oversight, for two reasons. First, agency officials may have very different objectives than energetically executing policy. They may be more interested in exerting power within the agency, amassing resources, not rocking the boat, or hanging onto their jobs. Second, legislators have constituencies back home that they want to keep happy. In order to do this, they must be able to show results so that they can claim credit to their constituencies. If agencies do not do their job, legislators risk getting booted out of office during the next election cycle.

For these reasons, when Congress passes a new environmental law and designates an agency to carry it out, it typically provides for congressional oversight. This includes things like: requiring agency heads to submit annual reports and to testify periodically before an oversight committee, and threatening to cut agency funding if agencies do not do their job (the so-called *power of the purse*) [Fenno (1966); Ting (2001)]. In principle, these powers impose discipline by keeping agency officials accountable to the legislature.

In practice, the amount of discipline imposed on agencies by legislative oversight probably varies a lot. Much occurs in an agency that legislators cannot observe and, therefore, can do little to correct. This includes everything from disruptive internal personnel decisions to staff members playing computer games on the job. Furthermore, there may be practical limits to what legislators can do to penalize agency officials for not vigorously pursuing legislative objectives. It is difficult to fire career civil servants, for one. And factors like public pressure and political infighting among legislators may stymie attempts to slash funding.

In order to see the argument here, consider a model – the so-called *principal-agent model* – that economists have proposed to capture and understand situations like this. In this model, there are two parties, a principal and an agent. The principal has a task that she wants done and hires the agent to do it. So the agent is hired by the principal to work on her behalf. The *principal-agent problem* occurs when the principal and the agent have different objectives and the principal cannot observe what the agent does. You can probably see why this might be a problem, especially for the principal, who may have a tough time getting her task done.

To illustrate: in the classic example, the principal is a firm and the agent is a worker hired by the firm for a fixed salary. Of course, the firm wants the worker to work hard on its behalf. The problem is that the worker does not like to work hard, and he knows that if he does not work hard, he will still get paid. So he has incentive to goof off and, if he does, the firm suffers. The worker is said to be engaging in *shirking* behavior. Under these conditions, the firm's ability to make money depends on its ability to monitor the worker's behavior, and if he goofs off, to impose sanctions, like docking wages or even termination of employment.

The key here is that the firm cannot fully observe ("has imperfect information on") the worker's effort level, making shirking possible. This provides the worker with the opportunity to satisfy his own objectives rather than those of the firm. If possible, the firm might try to write a different type of contract to elicit more effort from the worker. For example, instead of a fixed salary, the firm might want to offer a "piece-rate" contract where the salary is tied to output. However, in many industries this might be difficult. A piece-rate contract might work for manufacturers of hub caps, but it might not for universities, law firms, consulting firms, or firms providing consumer services of various kinds, where output is much harder to measure.

If you think about it, the relationship between a legislature and an agency is very similar to the relationship between a firm and a worker on a fixed salary. The legislature has a task that it wants done: the implementation of a policy. It designates an agency to perform this task. But, as we mentioned earlier, agency officials may have different objectives and it may be difficult for the legislature to observe their behavior. The end result may be shirking by agency officials and ineffective policy implementation [Weingast(1984)].

The advantage of using the principal–agent model is that it tells us an important reason for ineffective agency behavior. It often comes down to ineffective oversight, where the legislature is unable to adequately observe the actions of the agency officials. Or, if it can, it may have few weapons at its disposal to discipline the agency for not doing its job. In later chapters, we will encounter various examples of administrative agencies charged with carrying out environmental policy delegated by a legislature. When evaluating the effectiveness of the policy, one important question will be: How effective was the legislative oversight?

## Key takeaways

- Public choice is the use of economic tools to deal with traditional problems of political science.
- Under the public choice approach, policy outcomes occur largely as the result of competing demands by special interest groups, and supply by legislators.
- Several factors determine the political effectiveness of an interest group, including the size of the group, the collective stakes of the group in a policy outcome, the stakes of individual members of the group, the heterogeneity of the group, and the presence of a political entrepreneur.
- Large interest groups may not necessarily be more political effective than small interest groups, because they may be subject to an internal free-rider problem.
- Interest groups that are less heterogeneous and that have higher stakes in legislative outcomes tend to be more effective politically.
- Political entrepreneurs can galvanize political pressure either for or against a set of policies.
- There can be transaction costs in policymaking; that is, in the assigning of property rights, even prior to the actual operation of policies.
- Under certain restrictive conditions, under majority voting rule systems the predicted policy is that preferred by the median voter.
- In legislatures, vote-trading, or *logrolling*, supports coalition building to effect legislative outcomes. Vote trades may be subject to difficulties of legislators credibly committing to uphold their votes, but some legislative mechanisms may be designed to promote successful logrolling.
- Legislative committees may serve an important agenda-setting function in determining legislative outcomes.
- Legislatures set in place oversight procedures to attempt to ensure that agency administrators carry out their legislative mandates. The effectiveness of agency oversight depends upon the ease with which legislatures can observe administrator behavior and impose effective sanctions.

## Key concepts

- Paradox of voting
- Electoral connection
- Free-riding
- Concentrated benefits and dispersed costs
- Interest group heterogeneity
- Political entrepreneurs
- Policymaking vs. policy operation transaction costs

- Median voter theorem
- Single-peaked preferences
- Logrolling
- Omnibus bills
- Power of the agenda setter
- Principal-agent problem

## Exercises/discussion questions

6.1. In your opinion, is individual voting behavior inconsistent with rational choice?

6.2. Under what conditions might large interest groups be more politically effective than small interest groups? Can you give an example of an issue for which this might be true?

6.3. As we have seen, large interest groups suffer from free-riding problems, like the provision of national defense, which we have characterized as a public good. Does this imply that lobbying legislators on behalf of the interests of a group is a public good? What, if anything, is the conceptual difference between a country providing national defense and an interest group providing lobbying activity?

6.4. In the discussion of policymaking and policy operation transaction costs, it was suggested that the courts assigning liability to establish property rights in an environmental dispute would entail low transaction costs. Would this always be the case? More generally, suppose there were two ways to assign property rights in an environmental dispute: the courts making a ruling or a legislature enacting a statute. Where would the policymaking transaction costs be greater, and why?

6.5. Can you think of an issue where voters may not have single-peaked preferences? What issue might this be, and why?

6.6. The Superfund program enacted by Congress in the wake of the Love Canal disaster has been characterized by various observers as a pork barrel program. One feature of the program is that it calls for toxic waste sites to be created in all fifty states, thus spreading the benefits around. In your opinion, does this make it not a pork barrel program?

6.7. What would be the effect of each of the following factors on the political effectiveness of an environmental interest group?
- A successful membership drive.
- A coalition with another environmental interest group.
- An endorsement by the president.

### Notes

1 A few of the classic studies that develop this approach are Stigler (1971); Peltzman (1976), Romer and Rosenthal (1978); Becker (1983); and Weingast and Moran (1983).

2 See Joskow and Schmalensee (1998), pp. 44–5. See also the later discussion in chapter ten.

3 See, for example, Heckathorn (1993); Baik (2008); Epstein and Mealem (2009); Nitzan and Ueda (2014).

4 For more on political transaction costs and the Coase Theorem, see Jung et al. (1995); Krutilla and Krause (2010); Krutilla and Alexeev (2014).

5 Many countries around the world operate under a so-called *parliamentary* system. Parliamentary systems are similar to the system in the United States, but there are some important differences. Under parliamentary systems, the executive and legislative branches are not separate, the executive is a member of the legislature, and political parties tend to have more control over the elective process. In many

countries, the legislature consists of only one body, a *unicameral* system. In general, keep in mind that there are many subtle differences in structure and procedures across political systems.

6  Of course, the more lopsided is the vote in favor of passage, the more likely it is that any veto by the executive will be overridden.

7  For a classic treatment of logrolling, see Buchanan and Tullock (1962).

8  For a discussion of the points in the last three paragraphs, see Weingast and Marshall (1988).

9  For more on omnibus spending bills, see Oleszek (2007).

10  For classic example of committees as agenda-setters, see Romer and Rosenthal (1978); Denzau and Mackay (1983); and Shepsle and Weingast (1987).

# Further readings

## *I. Demand for policy*

Stigler, George J. "The theory of economic regulation," *Bell Journal of Economics and Management Science* 2(Spring 1971): 3–21.

*A classic reading in public choice, this influential article pushes the notion that economic regulation derives largely from special interest interaction, not public interest objectives.*

Olson, Mancur. *The rise and decline of nations.* New Haven: Yale University Press, 1982.

*A follow-up to Olson's Logic of Collective Action. Argues that there is a consistent political dynamic across different countries that special interest groups tend to accumulate over time, which reduces growth and results in economic inefficiency. This is the famous "institutional sclerosis" theory.*

## *II. The electoral connection*

Ringquist, Evan J. and Carl Dasse. "Lies, damned lies, and campaign promises? Environmental legislation in the 105th Congress," *Social Science Quarterly* 85(June 2004): 400–19.

*Examines, using roll-call voting data, the extent to which members of Congress actually keep their campaign promises.*

## *III. The median voter theorem*

Frederiksson, Per G. and Le Wang. "Sex and environmental policy in the U.S. House of Representatives," *Economic Letters* 113(December 2011): 228–30.

*Use the implications of the median voter theorem to conclude that female members of the House of Representatives favor stricter environmental policies than their male counterparts.*

## *IV. Committees*

Rabe, Barry G. "Legislative incapacity: The congressional role in environmental policy-making and the case of Superfund," *Journal of Health Politics, Policy and Law* 15(Fall 1990), 571–89.

*Argues that the committee structure in the House has contributed to fragmentation of federal policy regarding regulation of hazardous waste sites. Makes a number of recommendations for reform of the congressional committee structure.*

## *V. Legislative oversight*

Lazarus, Richard J. "The neglected question of congressional oversight of EPA: Quis custodiet ipsos custodies (Who shall watch the watchers themselves)?" *Law and Contemporary Problems* 54(1991): 205–39.

*Argues that there has been excessive and hostile oversight of the EPA by congressional committees, which has made it difficult for EPA to do its job.*

McCubbins, Mathew D. and Thomas Schwartz. "Congressional oversight overlooked: Police patrols versus fire alarms," *American Journal of Political Science* 28(February 1984): 165–79.
*Rebuts the common perception that the U.S. Congress neglects its oversight responsibility. Argues that it oversees agency behavior by setting down rules and procedures to respond to episodes of agency shirking, so-called fire-alarm oversight. This is as opposed to a more easily observed form of oversight where Congress routinely monitors agency behavior through regular channels, like field observations and hearings, or what it calls police patrol oversight.*

# References

Baik, Kyung Hwan. "Contests with group-specific public-good prizes," *Social Choice and Welfare* 30(January 2008): 103–17.

Becker, Gary. "A theory of competition among pressure groups for political influence," *Quarterly Journal of Economics* 98(August 1983): 371–400.

Black, Duncan. *The theory of committees and elections*. London: Cambridge University Press, 1958.

Buchanan, James M. and Gordon Tullock. *The calculus of consent*. Ann Arbor: University of Michigan Press, 1962.

Denzau, Arthur T. and Robert J. Mackay. "Gatekeeping and monopoly power of committees: An analysis of sincere and sophisticated behavior," *American Journal of Political Science* 27(November 1983): 740–61.

Downs, Anthony. *An economic theory of democracy*. New York: Harper and Row, 1957.

Epstein, Gil S. and Yosef Mealem. "Group specific public goods, orchestration of interest groups with free riding," *Public Choice* 139(June 2009): 357–69.

Eriksson, Clas and Joakim Persson. "Economic growth, inequality, democratization, and the environment," *Environmental and Resource Economics* 25(2003): 1–16.

Fenno, Richard F. *The power of the purse: Appropriations politics in Congress*. Boston: Little, Brown, 1966.

Ferejohn, John A. *Pork barrel politics: Rivers and harbors legislation, 1947–1968*. Stanford: Stanford University Press, 1974.

Fiorina, Morris. "Information and rationality in elections," in *Information and Democratic Processes*, John Ferejohn and James Kuklinski (eds.) Urbana: University of Illinois Press, 1990: 329–42.

Gilligan, Thomas and Keith Krehbiel. "Collective decision-making and standing committees: An information rationale for restrictive amendment procedures," *Journal of Law, Economics, and Organization* 3(1987): 287–335.

Gilligan, Thomas and Keith Krehbiel. "Organization of informative committees by a rational legislature," *American Journal of Political Science* 34(1990): 531–64.

Grofman, Bernard. "Anthony Downs," in *Readings in public choice and constitutional political economy*. Boston: Springer-Verlag, 2008.

Heckathorn, Douglas D. "Collective action and group heterogeneity: Voluntary provision versus selective incentives," *American Sociological Review* 58(June 1993): 329–50.

Joskow, Paul L. and Richard Schmalensee. "The political economy of market-based environmental policy: the U.S. acid rain program," *Journal of Law and Economics* 41(April 1998): 37–84.

Jung, Chulho; Kerry Krutilla; W. Kip Viscusi; and Roy Boyd. "The Coase Theorem in a rent-seeking society," *International Review of Law and Economics* 13(1995): 259–68.

Kempf, Hubert and Stephane Rossignol. "Is inequality harmful for the environment in a growing economy?" *Nota di Lavoro* No. 5(2005), Fondazione Eni Enrico Mattei (FEEM), Milano.

Krehbiel, Keith. "Legislative organization," *Journal of Economic Perspectives* 18(Winter 2004): 113–28.

Krutilla, Kerry and Alexander Alexeev. "The political transaction costs and uncertainties of establishing environmental rights," *Ecological Economics* 107(2014): 299–309.

Krutilla, Kerry and Rachel Krause. "Transaction costs and environmental policy: As assessment framework and literature review," *International Review of Environmental and Resource Economics* 4(2010): 261–354.

Mayhew, David R. *Congress: The electoral connection*. New Haven: Yale University Press, 1974.

Nitzan, Shmuel and Kaoru Ueda. "Intra-group heterogeneity in collective contests," *Social Choice and Welfare* 43(June 2014): 219–38.

Oleszek, Walter J. *Congressional procedures and the policy process.* Washington: CQ Press, 2007.

Olson, Mancur. *The logic of collective action.* Cambridge: Harvard University Press, 1965.

Peltzman, Sam. "Toward a more general theory of regulation," *Journal of Law and Economics* 19(August 1976)): 211–40.

Romer, Thomas and Howard Rosenthal. "Political resource allocation, controlled agendas, and the status quo," *Public Choice* 33(Winter 1978): 27–43.

Schneider, Gabe. "Why the agriculture industry is all in on sending Collin Peterson back to Congress," *MinnPost*, September 15, 2020.

Shepsle, Kenneth A. and Barry R. Weingast. "The institutional foundations of committee power," *American Political Science Review* 81(March 1987): 85–104.

Ting, Michael M. "The 'power of the purse' and its implications for bureaucratic policy-making," *Public Choice* 106(2001): 243–74.

Tullock, Gordon. "Public choice," *The New Palgrave Dictionary of Economics Online.* 2008.

Weingast, Barry R. "The congressional-bureaucratic system: A principal agent perspective (with applications to the SEC)," *Public Choice* 44(1984): 147–91.

Weingast, Barry R. and William J. Marshall. "The industrial organization of Congress; or, Why legislatures, like firms, are not organized as markets," *Journal of Political Economy* 96 (February 1988): 132–63.

Weingast, Barry R. and Mark J. Moran. "Bureaucratic discretion or congressional control? Regulatory policymaking by the Federal Trade Commission," *Journal of Political Economy* 91(October 1983): 765–800.

Wilson James A. *American Government* (5th edition). Lexington: Heath, 1992.

# 7 Population

Let us begin our discussion of special topics by considering perhaps the most important one of all. You may be aware that in the early 2010s, the global population surpassed seven billion people. This sounds like a lot of people, and it certainly is. That many mouths to feed. That many people needing increasingly scarce land to live on and farm. That many people encroaching on natural areas, resulting in deforestation and loss of endangered species. That many people using energy to drive cars and heat their homes. That many people in need of safe and clean drinking water. That many people engaging in activities that generate greenhouse gases, thus contributing to climate change. And global population will continue to increase, as it is projected to rise to nine or ten billion people by mid-century. Not surprisingly, it is common to hear references in the popular media to the "population crisis."

## Overview of population

To understand population, it helps to take the long view. Figure 7.1 does precisely that, showing data compiled by *Our World in Data* on total world population at 20-year intervals, beginning in 1750 and projected into the future all the way to 2100. In 1800, for example, there were fewer than one billion people on earth. By 2015, this number had increased to 7.4 billion. And by 2100, this number is projected to increase to 11.2 billion.

This is, of course, only one projection into the future, which is based on a certain set of assumptions regarding fertility and mortality. Under other assumptions, these numbers can change somewhat, and they become increasingly uncertain as we move further into the future. But all credible projections show the same basic pattern shown here.

The other thing that is interesting about Figure 7.1 is the time trend in the annual growth *rate* of the population, the solid line. Growth rates were fairly stable and low for most of this historical period until about 1920. They then shot up dramatically, reaching a peak of 2.1% in 1960. Since then, they have fallen precipitously, to about 1.2% in 2015. Moving forward into the future, the growth rate is projected to continue to decline, to close to zero by 2100.

To help interpret these numbers consider the *Rule of 72*. This rule states that, roughly speaking, 72 divided by the annual growth rate gives the number of years it takes for something to double in size. So, at an annual growth rate of 2%, population would roughly double in 36 years. When the growth rate falls to 1%, it would take twice as long, or 72 years. So the difference between 1% and 2% is pretty important in determining how quickly something like population will grow.

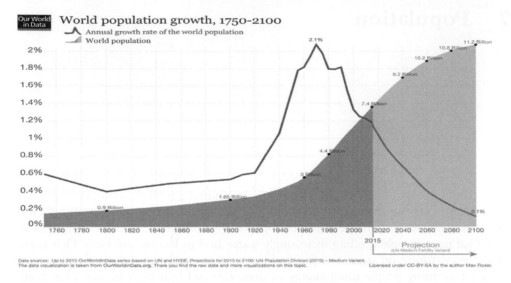

*Figure 7.1* World population growth, 1750 to 2100. *Source*: Our World in Data: Licensed under CC-BY-SA by Max Roser

Figure 7.1 shows the situation for the whole world. However, you should keep in mind that population growth varies dramatically across different regions of the world. In low-income countries, populations grow more rapidly than in high-income countries. Currently, the rate of population growth is highest in Africa. In 2015, the population of Africa stood at 969 million. By 2050, the United Nations projects it to rise to 2.168 billion. If this occurs, it would represent an annual growth rate of 2.3%. Nigeria alone could have a population of 400 million by 2050, which would make it the third most populous country in the world. In 2017, in 22 countries the average number of children per woman was at least five. Of these 22 countries, 20 are in Africa and two are in Asia [United Nations (2017)].

At the other extreme, certain countries of the world are actually experiencing shrinking populations. China's annual population growth rate, for example, is forecast to become negative before 2030 [McKirdy (2019)]. Similarly, South Korea's population will probably also begin to shrink before that date. The population of Japan is already shrinking: this trend began in 2005 [Feyrer et al. (2008), p. 3]. As of 2018, the population of Japan was 127 million. By 2049, at least one official projection has it falling below 100 million.[1] Other countries whose populations are projected to start shrinking soon, or which already have, are Spain, Italy, Germany, Hong Kong, and Singapore.

It will be helpful to remember that fast-growing countries and slow-growing/shrinking countries face very different population-related economic issues. Fast-growing countries like ones in Africa tend to have low levels of income. For them, the main population-related issues are feeding and housing the population, judicious use of scarce natural resources, and environmental degradation. Slow-growing or shrinking countries tend to enjoy higher levels of income. Because they grow slowly, they tend to have older populations, with a higher ratio of elderly to young. For these countries, the main population-related issues are about having a sufficiently large and skilled workforce, and providing for the needs of an aging population.

# The Becker model of fertility

To understand the discussion to come, let us begin by considering a model of fertility behavior developed by the economist Gary Becker [Becker (1960)]. This model has become the dominant framework that economists use to understand fertility behavior.

Becker's model is based on the notion that decisions made by parents regarding children can be analyzed in roughly the same way as the decisions they make regarding other things they value. In the standard consumer choice model, consumers choose a bundle of goods in order to maximize utility, subject to a budget constraint. This generates demand for all of the goods they purchase, which was implicit in the demand analysis in chapter two.

In Becker's model, parents "demand" children in the same way. They derive utility from children, but children are "costly" to consume, in terms of baby food and diapers to start with, and, later on, pizza parties and music lessons. The money they spend on their children has other potential uses, to purchase other things they value, but they have only so much money to spend. This generates a tradeoff between having children on the one hand, and dinners out or summer vacations on the other. As the Rolling Stones remind us, you cannot have everything you want, and neither can parents.

In the simplest version of the Becker model, you can think of parents having the following utility function:

$$U = f(N, G) \tag{7.1}$$

Where N is the number of children they have and G is all other goods they consume. Both N and G have prices, and parents have so much income to spend. This model generates certain predictions, like increases in the "price" of children will tend to reduce the number of children they have, and increases in income will tend to increase that number. Just like other goods.

Becker was troubled, however, by the fact that the world does not quite behave as this model predicts. Specifically, he noticed that parents with more income tended to have *fewer* children. So Becker reasoned that there is more that parents value in children than just how many they have. They want them to be healthy. They want to provide them a good education. And doctor visits and schoolbooks are other things that compete for scarce household resources. So Becker modified the model slightly, to come with the following version:

$$U = f(N, Q, G) \tag{7.2}$$

Where Q is the so-called "quality" (i.e., healthier, better educated) of the children. And, in the household budget constraint, both the number of children and the quality of children have prices.

The nice thing about this model is that it permits the modeling of tradeoffs between the number of children and the quality of the children. If you have more children, for example, you may have less to spend on healthcare or schoolbooks for each child. If the costs of pregnancy and delivery go up, one might have fewer children and decide to spend more on education.[2] And the model allows for the possibility that the desired number of children could go down as income increases. Higher-income parents may decide to use their extra income not to have more children, but to invest more in each child.

## A model of population change: The demographic transition model

Now that we have the main facts and a way to think about them, let us turn to the question of how we understand population growth and population dynamics more generally. For this, we turn to a very useful framework for understanding population change: the well-known *demographic transition model*. This is a model used by demographers to help them think about the dynamics of population change.

Demographers have long noticed a consistent pattern in population growth as countries go through various stages of development, from low-income to high-income countries. At extremely low levels of income, birth rates and death rates both tend to be high and, as a result, population sizes tend to be stable. But as countries start to develop, death rates start to fall while, at first, birth rates stay high. As a result, population levels start to rise. As economic growth continues, however, birth rates start to decline and population growth starts to moderate. Then, birth rate reductions catch up with death rate reductions and populations stabilize at higher levels. In a final stage, birth rates fall still further and go below death rates, and so population may actually start to decline. In sum, as economies develop, they go through a transition consisting of several distinct periods. Hence, the name *demographic transition model*.

These stages of the demographic transition process are shown in Figure 7.2. In moving from high birth and death rates in the first stage to low birth and death rates in the fourth stage, the key factor driving population growth is the fact that declines in the birth rate lag behind declines in the death rate. And depending upon how quickly death rates decline and how much falling birth rates lag behind this decline, population levels can increase extremely quickly.

Figure 7.3 shows the actual demographic transition pattern for five countries for the period 1820 to 1910, as compiled by Max Roser. These countries – Germany, Sweden, Chile, Mauritius, and China – could not be more varied in terms of culture and economic development. Yet each one shows a variant of the demographic transition model. In each case, at some point the death rate declined, followed eventually by a decline in the birth

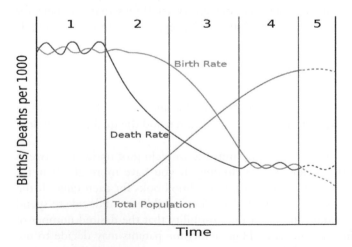

*Figure 7.2* The demographic transition model. *Source:* Wikimedia Commons (public domain)

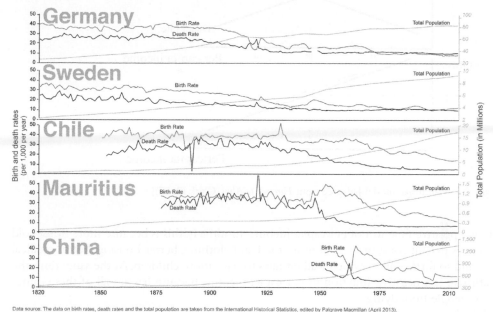

*Figure 7.3* Demographic transition in five countries, 1820 to 1910. *Source: Our World in Data*, by Max Roser

rate. The overall effect was for the population to increase, followed by a levelling-off as the birth rate approached the death rate.

The demographic transition also seems to apply if you compare countries of the world cross-sectionally. Low-income countries tend to have high birth rates and high death rates. High-income countries tend to have low birth rates and low death rates. And rapidly growing countries have commonly caused their populations to grow rapidly by reducing their death rates.

## Stage 1: The Malthusian trap

Let us now dig a little deeper to understand what is going on in each of these stages of the demographic transition, beginning with Stage 1. Here I will introduce the ideas of the English economist Thomas Malthus, who has had a tremendous influence on how economists think about population and its economic consequences. In 1798, Malthus wrote one of the most famous economic arguments ever written: *An Essay on the Principle of Population*. In this essay, Malthus argued that there were built-in stabilizing influences on population growth that operated through their effect on income. Some of these influences cause populations to increase, while others cause them to decrease.

To see the argument, consider Figure 7.4.[3] The top panel graphs the birth rate and death rate in a society as a function of per capita income. Here, it is assumed that the birth

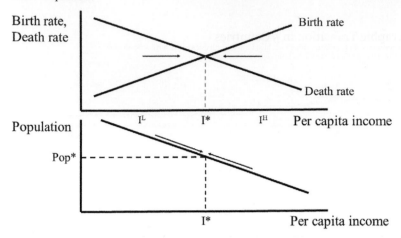

*Figure 7.4* Malthusian model [adapted from *Farewell to Alms*, Clark (2007)]

rate is a positive function of per capita income. Intuitively, this is because more household resources permit consumption of more food and clothing, better housing, better medical care, and so forth – all of which lead families to have more children. At the same time, the death rate is a negative function of per capita income, because having all of these things keeps people from dying.

The bottom panel graphs population as a function of per capita income. Here, there is a negative relationship between population and per capita income. The intuition is that for economies with fixed resources – that is, economies everywhere – having a greater population means less wealth for each member of the economy.

Now consider what you would expect to happen in equilibrium. When the birth rate exceeds the death rate, there will be a tendency for the population to grow. When the opposite is true, there will be a tendency for the population to shrink. So, at relatively high per capita income levels like $I^H$ where birth rates exceed death rates, the population will grow, pushing per capita incomes down. And this will be true as long as birth rates exceed death rates.

On the other hand, when per capita incomes are low, like $I^L$, death rates exceed birth rates, tending to decrease the population size, which will exert upward pressure on per capita incomes. The size of the population is thus driven to an equilibrium level where it is stable, Pop★, and an associated per capita income level I★. I★, the income level at which the population size is stabilized, is called the *subsistence income*.

To be clear, the economy depicted here could experience departures from this equilibrium. For example, a series of good harvest years could lead families to have more babies, pushing the birth rate above the death rate. But, if this occurs, it will increase the population size, in turn lowering incomes and tending to push the population size back down. Or a drought resulting in famine could result in an increase in the death rate. But, in this case, the size of the population will fall, in turn raising incomes, which will tend to push the population size back up.

Thus, so-called *shocks* to a system, such as bumper crops and famines, may result in temporary departures from this equilibrium. However, the economy eventually settles back down to the equilibrium population and income levels. This seemingly inexorable process has come to be known as the *Malthusian trap*.

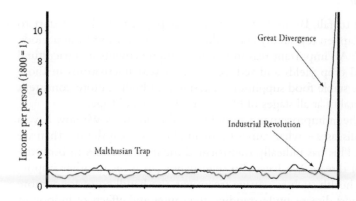

*Figure 7.5* The Malthusian trap. *Source*: *A Farewell to Alms*, by Greg Clark, p. 2.

The Malthusian trap is shown in Figure 7.5. It depicts the long-term history of per capita income, over the course of centuries. It shows, for the vast majority of human history, humankind trapped at fixed, low income levels with high birth rates and death rates. Thus, Stage 1 of the demographic transition model has been the reality for most of human history.

Figure 7.5 also shows, however, that the Malthusian trap has turned out not to be inevitable. At some point in history – in particular, starting in the late 18th and early 19th centuries with the coming of the Industrial Revolution – we began to escape from the Malthusian trap. And, since that time, per capita income, and standards of living more generally, have increased dramatically.

## Stages 2–4: Patterns of economic development

So, how did we manage to escape from the Malthusian trap? In a nutshell, we accomplished this great feat by finding ways to grow the economy more quickly than we grew the population. In the Malthusian world, population growth drags down standards of living. Because no sooner has humankind managed to get its head above water with a little extra productivity then it starts to produce more offspring, who eat up all the gains we have made. In principle, these population increases offset, and in practice may be large enough to reverse, the increases in standards of living.

In general, there are two ways to free ourselves from this Malthusian rat race. One is by keeping down the increases in population. The other is by increasing production of goods and services enough to stay ahead of the increases in population.

---

**Box 7.1: Did the Black Death help us escape from the Malthusian trap?**

In the mid-14th century, the Black Death struck Europe, killing ¼ to ½ of the population of the continent. This was a huge shock to European countries and, obviously, a tragic event of mammoth proportions. However, it triggered a significant increase in real wages for the surviving laborers. And it was a big enough shock that it may have begun a process whereby parts of Europe – specifically, England and the Netherlands – started to emerge from the Malthusian trap. For more, see Haddock and Kiesling (2002); Voigtlander and Voth (2013), in *Further readings*.

## I. Lowering population growth

In Stage 2, death rates start to fall. By itself, this means that population should start to rise, and we have seen this happen repeatedly across different countries and time periods. Why have death rates fallen? An important reason has been improvements in agricultural practices that have increased crop yields and reduced year-to-year fluctuations in those crop yields. Larger and more stable food supplies have resulted in higher, more consistent caloric intake, reducing mortality at all stages of life, from infancy to old age.

A second key reason has been improvements in public health practices. We now have a far better understanding of diseases – what causes them and how to treat them – than we did in even the recent past. This has radically transformed medical practices in both the treatment and prevention of disease. We have also seen major advances in other important areas relating to public health such as sewage treatment, the handling of food, more widespread vaccinations against disease, understanding the causes and effects of indoor air pollution on health, the spread of indoor plumbing, and so forth.

As it turns out, reductions in the death rate have probably occurred much more rapidly in the last several decades than at any other time in human history. During the 18th and 19th centuries, for example, reductions in the death rate were gradual, resulting from steady but gradual advances in science and technology relating to public and human health. However, in recent decades reductions in death rates have occurred through transfers of existing knowhow to areas of the world with high death rates. This has enabled death rates to be brought down much more quickly. The result has been the "explosion" of population in many areas of the world.

The reductions in the rate of population growth has occurred in Stage 3 for a variety of reasons. Among the key reasons is the development of contraceptive technologies and family planning services. Birth control pills, condoms, and diaphragms, and how to use them effectively, have permitted families to have greater control over the number and timing of offspring. However, these factors alone would probably not have had much of an effect on reproductive decisions without other important changes in economic systems, culture, and society.

One important change occurs when economies come to be based less heavily on agriculture. In agricultural economies, children serve an important economic function by performing various roles on farmsteads. This particular value of children declines as parents increasingly take on manufacturing jobs, giving them less incentive to have children. In terms of the Becker model, the value of having more children decreases relative to the value of child quality. Fertility also tends to fall with the rise of various social programs associated with higher levels of development, including retirement security and compulsory education. Finally, changes in the social status of women that promote their pursuit of education and employment opportunities increase the opportunity cost of having children and tend to lead to reductions in fertility.

The pace at which economic growth is accompanied by these changes varies considerably across economies and cultural contexts. As a result, the rate at which fertility rates decline also varies across economies and cultures. To the extent that these changes occur more quickly, declines in birth rates will lag less behind declines in death rates, which will tend to moderate Stage 3 population increases.

## II. Increasing production

The Malthusian model assumes that production of food and other necessities of life will never be able to outstrip population growth. And this was apparently true for centuries of human history. However, the experience of the last couple of centuries has revealed that, in fact, we can increase output faster than we reproduce. We have just seen one of the reasons: our ability to curb fertility behavior. The other reason is that the ability of humankind to increase production of food and other goods and services has grown by leaps and bounds.

There are many reasons for this, some of which we will explore in a later chapter on agriculture in the context of food production. However, one key factor helps us understand how we have managed to stay ahead of population growth in recent years. That factor is *technological change*, which has made workers more productive, increasing production yields. For example, new inventions for threshing and reaping, improved farming techniques, improvements in farm organization, more disease- and drought-resistant plant hybrids, and so forth all permit farmers to produce more output. There is considerable evidence that the effects of technological advance on production are huge, not just in farming but in all sectors of the economy. Furthermore, the effects of technological advance have probably swamped the effects of adding capital, labor, and even land.

But even this is not the end of the technology story. In the longer term, growth in population may itself serve as a catalyst for technological change to occur. This provocative idea was advanced by the Danish economist Ester Boserup in 1965 in an influential book *The Conditions for Agricultural Growth*. Boserup argued, in essence, that agricultural hardship forces people to find new ways to produce more food. That is, Malthusian population growth, rather than simply causing people to die off when conditions become too harsh, actually plants the seeds for technological advance to mitigate the very hardships caused by that growth. Some have called the Boserup thesis: "Necessity is the mother of invention."

Figure 7.6 illustrates the Boserup effect. The two lines represent food production and population growth as they progress over time. In this dynamic, increases in food production provoke spurts in population growth à la Malthus. But those very population spurts provoke technological change that results in new and better ways to produce food. The resulting increase in food production encourages more population growth, which in turn provokes more technological change, and so forth. As a result, food production and population growth co-evolve together over time, with food production steadily increasing sufficiently to feed ever-growing populations. We will encounter the Boserup effect again in chapter thirteen, when we discuss various theories of agricultural production.

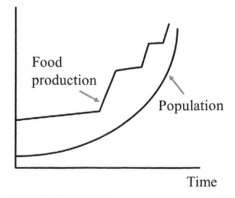

*Figure 7.6* The Boserup effect

So who is right, Malthus or Boserup? The answer is probably *yes*. More accurately, there are probably conditions that are more conducive to the creation and adoption of new technologies to expand production. When these conditions are present, we are more likely to observe the Boserup effect. However, when they are not present, Malthusian forces may dominate.

To understand these conditions, it will help to be concrete about *why* one might expect population growth to produce more technological change. Boserup herself made the following argument, specifically about agriculture. Growth of the population in a farming area pressures farmers to squeeze more output out of the land they have. More intensive use of land increases the rate of soil depletion, necessitating new methods to deal with declining soil productivity. This may lead to the development and adoption of new technologies and better ways of managing the land [Boserup (1965); Goldman (1993), p. 47].

However, it is not simply a matter of being forced to be ingenious by being in dire straits. In addition, more people means more sources of new ideas and inventive activity [Kremer (1993); Johnson (2000), pp. 6–7]. If only one in 10,000 people is capable of coming up with an idea for a brilliant new invention, a population of 1,000 will generate no new ideas, whereas a population of one million people will generate 100. Furthermore, population growth tends to result in greater population densities, meaning more opportunities for people to come in contact with each other, sharing ideas, capitalizing on shared interests, and collaborating on joint ventures.

However, whether these ideas, opportunities, and collaborations result in actual increases in productivity also depends on whether they can be turned into innovations that generate revenue streams. In other words, like everything else in life, we will tend to see more inventive activity when people have more incentive to engage in it.

Here, good institutions that provide support for farmers to profit from engaging in crop production appear to be key, for several reasons. First, secure private property rights in land encourage farmers to work the land harder and to make investments to increase productivity, including adoption of new methods and technologies. This is because they expect to be able to enjoy the fruits of their labor.

On this point, the nature of the government and legal systems matters greatly. Governments and legal systems that protect property rights, support appropriate pricing of resources, and promote local arrangements to manage resources will all tend to promote productive use of lands [Cruz and Gibbs (1990)]. At the other extreme, governments that expropriate lands and do not respect private property will tend to have the opposite effect. Similarly for situations of major political and social instability, extended local violence, and civil conflict [Koren and Bagozzi (2017)].

Second, being able to market and sell one's farm products also tends to encourage farmers to work and invest in their farms. Marketing farm products is facilitated by the presence of well-developed infrastructure such as roads, railways, bridges, and local supply and distribution centers that enable farmers to get their products to markets.[4] We will return to some of these issues in the later chapter on agriculture, where we will take a more in-depth look at the economics of agriculture and some key policy issues.

## Population policies

Countries all over the world have adopted a variety of policies that target fertility behavior. Some are designed to reduce birth rates, others are designed to increase birth rates. These policies are sometimes referred to as *antinatalist* and *pronatalist* policies. Example of antinatalist policies include:

- Providing access to low-cost, safe, and effective contraception;
- Integrating family planning into primary healthcare systems;
- Improving education and employment opportunities for females;
- China's One Child Policy;
- Raising the minimum legal age for marriage.

Examples of pronatalist policies include:

- Paid parental leave;
- Flexible work schedules;
- Baby bonuses;
- Tax incentives of various kinds.

Not surprisingly, countries with high fertility rates have tended to adopt antinatalist policies, and countries with low fertility rates have tended to adopt pronatalist policies.

### I. Antinatalist policies

Let us first focus on antinatalist policies, which tend to be used in high-fertility, lower-income countries. The objective of these policies is to either reduce the demand for children or raise the cost of having more children. In the Becker model, this should tend to reduce the size of families and result in fewer children for the economy as a whole. A famous example of an antinatalist policy is the *One Child Policy* instituted by China in 1980, which dramatically increased the cost to Chinese families of having a second child (see Box 7.2).

---

**Box 7.2: China's One Child Policy: The Becker model at work**

For years, China enforced a strict national policy of essentially allowing families to have only one child. This policy was designed to limit population growth, which was seen as necessary to improving living standards. Additional children could not attend public school or receive public healthcare. This One Child Policy worked: it dramatically reduced the growth of population. But it perhaps worked too well. China's population is predicted to peak in the late 2020s and then begin declining. By mid-century, one-third of the population will be over the age of 65, and the working-age population will have fallen precipitously.

   Acknowledging these obvious downsides, the government of China reversed course in 2016, declaring that families were now encouraged to have at least two children. But this change has failed to have the desired effect of encouraging Chinese families to have more children. Instead, families are devoting their resources to spending more on their one child, with art classes, public speaking classes, piano lessons, English language lessons, and the like. Now the government of China is considering taking additional measures to encourage people to have more children, like lowering the legal marriage age, providing living expenses, and even free delivery for the second child. [Fifield (2019)]

---

For the purposes of this discussion, let us focus on family planning and contraception, which have been the subject of much scholarly study. When you examine economic studies, on the surface there appears to be wide disagreement about the effects of family planning on fertility behavior. Some studies have found large, statistically significant effects, while others have found little or no effect. However, some of this disagreement is attributable to possible issues of methodology used in some of the studies. The more reliable studies use better data and larger samples, and they design their studies more carefully to allow them to rule out alternative explanations for their findings.

With all this in mind, a few patterns emerge from the studies in this literature. Table 7.1 lists a number of representative studies from the family planning literature with an emphasis on low-income countries (LIC denotes low-income countries). There are a few takeaway messages from Table 7.1. First, family planning does appear to have an effect on fertility behavior, though there is somewhat less consensus on the magnitude of the effect.[5]

There are a few other things to note about the results shown in Table 7.1. First, the education level of the mother seems to matter: multiple studies have found that family planning appears to be more successful at lowering fertility rates among more highly educated women. Second, incomes levels may also matter, with higher-income women tending to have fewer children in response to family planning programs. In terms of the Becker model, we may interpret both of these findings as factors that contribute to increases in the opportunity cost of having children.

Finally, one study found that family planning, even though it may have no effect on the total number of children borne by a mother, may still have an economic impact through its effect on the *timing* of the children's births. That is, if family planning helps women postpone when they start their families, it may allow them to finish school, postpone their marriages, and start their careers, all of which improves their economic outcomes over the long term [Goldin and Katz (2002); Bailey (2006); Miller (2009)]. Furthermore, studies have shown that this results in positive outcomes for offspring as well [Miller (2009)].

A final implication of all this is the potential importance of economic development as a means of controlling population growth. If more education and higher incomes are associated with lower fertility rates and greater responsiveness to family planning programs, policies that promote economic growth and development should encourage reductions in fertility. This is why family planning programs are often not considered in isolation but, rather, in conjunction with other ways of raising standards of living.

*Table 7.1* Effect of family planning, studies of (mostly) low-income countries

| Study | Country | Effect | Affected group/finding |
|---|---|---|---|
| Koenig et al. (1992) | Bangladesh | Yes | Does not distinguish |
| Schultz (1994) | Panel of LIC | No | Education matters |
| Schultz et al. (1995) | Rural China | Small | Education matters, income may matter |
| Pritchett (1994) | Multiple LIC | Small | -- |
| Bloom et al. (2003) | Ireland | Yes | -- |
| Sinha (2005) | Bangladesh | Yes | -- |
| Miller (2010) | Colombia | No | Effect on birth timing, life outcomes |
| Pop-Eleches (2010) | Romania | Yes | Education matters |
| Mello (2012) | Rwanda | Yes | Education matters |
| Becker et al. (2013) | Prussia | -- | Education matters |
| Jones (2015) | Ghana | Yes | Income, rural location matter |

## II. Pronatalist policies

At the other extreme are countries and regions of the world experiencing declining populations. Low birth and death rates mean fewer babies and longer lives, which combined
mean a larger ratio of elderly to young. Virtually all high-income, low-fertility countries have in place retirement systems to support people in their old age. The financial
solvency of such systems requires a sufficiently large population of (younger) taxpayers.
Furthermore, as more people retire, these countries may experience shortages of skilled
workers. Low fertility rates may make it more difficult for countries to achieve economic growth. So low birth rates have potentially important ramifications for economic
performance.

In theory, pronatalist policies have the exact opposite effect of antinatalist policies. They
shift the demand and/or supply curves of children outward, not inward. In doing so, they
encourage parents to have more children.

The first set of pronatalist policies we will consider are ones that attempt to influence
women's fertility behavior by offering families direct financial incentives to have children.
Governments have tried policies of this kind for decades. In 1966, the government of
Romania banned abortion and imposed a celibacy tax on women who were childless by
the age of 45. In 2007, a region in Russia declared September 12 the *Day of Conception*
and gave couples time off from work to procreate, even offering valuable prizes to successful couples, including money, SUVs, refrigerators, and washing machines ["Russians
offered day off"]. More recently, Hungary, another low-fertility country, has announced
a program of tax incentives to encourage women to have more children. This included
permitting families with four or more children to not pay any income tax [Taylor (2019)].

Do programs like these work? Perhaps. For example, in Russia, births in the regional
hospital reportedly shot up nine months later ["Russians"], and local authorities subsequently turned it into an annual event. However, it is not clear whether these types
of programs have any lasting impact, enough to make a serious dent in declining birth
rates. In some cases, programs like these have a novelty effect that may wear off over time
[Demeny (1986), p. 351]. And some of the apparent short-term effect may even be illusory.
The grand prize winners in the first Russian Day of Conception contest told reporters
"they were planning to have another child anyway" when they heard about the contest
["Russians"].

Economists who have examined the effect of financial pronatalist policies implemented
through the tax code and other programs have found mixed evidence that they encouraged families to have more children. Some studies have found that increases in various tax
provisions – including the personal exemption, child tax credits, and family allowances –
all have a positive effect on fertility. One study has found that a government program in
Quebec, Canada that paid families up to $8,000 (Canadian dollars) for an additional child
had a significant effect on fertility. On the other hand, studies of the effect of welfare payments on fertility have concluded that the effect is somewhat mixed.[6]

The nice thing about policies offering financial incentives is that they offer families
flexibility in how they make decisions regarding child-bearing and child-rearing. However,
programs offering targeted financial incentives are probably extremely expensive. When
Australia introduced a "baby bonus" policy in 2001, the policy apparently ended up costing, in total, $126,000 per new baby [Demeny (1986), p. 353; Hammond (2018)]. This is
why economists have examined the effectiveness of other types of pronatalist policies not
involving direct financial incentives.

Central to recent developments in many high-income, low-fertility countries are pronatalist policies to reduce the cost of birthing and child care. These policies include paid parental leave and access to childcare. Studies have found that extended maternity leave and greater availability of childcare facilities can have a significant positive impact on fertility.[7] Policies to promote these measures are probably a good deal less costly than targeted financial incentives and may be able to have a significant impact on fertility behavior.

## Key takeaways

- Population growth varies dramatically between low-income and high-income countries.
- In the economic model of fertility, parents select family size by maximizing utility subject to constraints on their resources. This results in their trading off having children against other uses of their household resources.
- Parents make specific tradeoffs between the number of children they have and their desire to spend more on each child. The result is that higher-income families tend to have fewer children.
- In a Malthusian economy, both birth rates and death rates are high, and income levels are kept low.
- In the course of economic development, countries generally undergo a characteristic pattern of decline in birth rates and death rates, with declines in birth rates lagging declines in death rates.
- Death rates have fallen as crop yields have increased due to technological advance, and with advances in public health services and practices.
- Birth rates have fallen as economies transitioned from being heavily agricultural, with the development of contraceptive methods and family planning services, and with the wider adoption of various social programs such as retirement security and compulsory education.
- Population growth may itself serve as a catalyst for technological advance that increases food production (Boserup effect).
- Good institutions, including secure private property rights and stable governments, encourage agricultural production.
- There are a variety of population policies, which can be divided up into antinatalist and pronatalist policies.

## Key concepts

- Becker model of fertility
- Demographic transition model
- Malthusian economy
- Malthusian trap
- Boserup effect
- Antinatalist vs. pronatalist policies

## Exercises/discussion questions

7.1.   Consider the Becker model of fertility. What would be the predicted effect, if any, of each of the following events on (a) family size and (b) expenditures per child?

- One of the spouses in a two-breadwinner family receives an unexpected promotion and a large raise.
- The non-working spouse in a two-breadwinner family receives a large unexpected inheritance.
- The government provides contraception and family planning services.
- The government enacts tuition-free public education.

7.2. Can you see any ways that escaping from the Malthusian trap may have elements of path dependence?

7.3. Discuss ways in which educational subsidies and public health policies could be used to enhance the Boserup effect.

7.4. Can you see ways that China's failure to escape from the antinatalist effects of its One Child Policy may have elements of path dependence?

7.5. Other than the ones listed in the discussion, can you think of other examples of pronatalist policies? Antinatalist policies?

## Notes

1 National Institute of Population and Social Security Research. http://www.ipss.go.jp/pp-zenkoku /e/zenkoku_e2017/pp_zenkoku2017e_gaiyou.html#e_zenkoku_II, retrieved August 2, 2019.

2 The simplifying assumption being made here is that pregnancy and delivery costs are a function only of the number of children, and not the quality of children. See Guinnane (2011), p. 597.

3 This figure is adapted from Clark (2007), p. 22.

4 See, for example, Goldman (1993) for the case of Kenya.

5 The study of Prussia focused on the 19th century and thus did not investigate family planning effects.

6 *Positive effect of tax provisions*: Whittington (1992, 1993); Zhang et al. (1994). *Canada payment program*: Milligan (2005). *Welfare payments*: Hoynes (1997); Moffitt (1998).

7 Blau and Robins (1989); Buttner and Lutz (1990); Winegarden and Bracy (1995); Lalive and Zweimuller (2009).

## Further readings

### I. The Becker fertility model

Caceres-Delpiano, Julio. "The impacts of family size on investment in child quality," *Journal of Human Resources* 41(Fall 2006): 738–54.

*Examines the effects of larger family size, through giving birth to twins, on parental investment in their children. Finds a reduced likelihood that older children attend private school, and increased likelihood that parents divorce.*

Black, Sandra E.; Paul J. Devereux; and Kjell G. Salvanes. "The more the merrier? The effect of family size and birth order on children's education," *Quarterly Journal of Economics* 120(May 2005): 669–700.

*Examines relationship between family size and children's educational attainment level. Finds that the negative relationship between the two disappears when they control for birth order. The main impacts on education and other life outcomes is for the younger children, especially females.*

### II. Contribution of the Black Death to rising per capita incomes

Haddock, David D. and Lynne Kiesling. "The Black Death and property rights," *Journal of Legal Studies* 31(June 2002): S545–87.

*A study of the effect of the Black Death in Europe, which raised the value of labor and human capital, leading to an erosion of feudal institutions, especially serfdom.*

Voigtlander, Nico and Hans-Joachim Voth. "Gifts of Mars: Warfare and Europe's early rise to riches," *Journal of Economic Perspectives* 27(Fall 2013): 165–86.

*Argues that the rise of Europe in the Middle Ages was accentuated by constant warfare and religious strife, which dramatically reduced population pressures and led to unusually high income levels.*

### III. Determinants of fertility behavior in low-income countries

Ainsworth, Martha; Kathleen Beegle; and Andrew Nyamete. "The impact of women's schooling on fertility and contraceptive use: A study of fourteen sub-Saharan African countries," *World Bank Economic Review* 10(January 1996): 85–122.
*A detailed examination of the relationship between fertility behavior and mothers' educational attainment for a number of African countries.*

Jones, Kelly M. "Contraceptive supply and fertility outcomes: Evidence from Ghana," *Economic Development and Cultural Change* 64(October 2015): 31–69.
*Examines the effect of loss of contraceptive services on fertility behavior in Ghana. Finds significant differences in the effect on conception in rural vs. urban areas, and between women of different income classes.*

### References

Bailey, Martha J. "More power to the pill: The impact of contraceptive freedom on women's life cycle labor supply," *Quarterly Journal of Economics* 121(February 2006): 278–320.

Becker, Gary S. "An economic analysis of fertility," *Demographic and Economic Change in Developed Countries*, Gary S. Becker(ed.) Princeton: Princeton University Press, 1960.

Becker, Sascha O.; Francesco Cinnirella; and Ludger Woessmann. "Does women's education affect fertility? Evidence from pre-demographic transition Prussia," *European Review of Economic History* 17(February 2013): 24–44.

Blau, D.M. and P.K. Robins. "Fertility, employment, and child-care costs," *Demography* 26(1989): 287–9.

Bloom, D.E. and D. Canning. "Contraception and the Celtic Tiger," *Economic and Social Review* 34(2003): 229–47.

Boserup, Ester. *The conditions of agricultural change: The economics of agrarian change under population pressure.* Chicago: Aline, 1965.

Buttner, Thomas and Wolfgang Lutz. "Estimating fertility responses to policy measures in the German Democratic Republic," *Population and Development Review* 16(September 1990): 539–55.

Clark, Gregory. *A Farewell to Alms.* Princeton: Princeton University Press, 2007.

Cruz, Wilfrido and Christopher Gibbs. "Resource policy reform in the context of population pressure: The Philippines and Nepal," *American Journal of Agricultural Economics* 72(December 1990): 1264–8.

Demeny, Paul. "Pronatalist policies in low-fertility countries: Patterns, performance, and prospects," *Population and Development Review* 12(1986): 335–58.

Feyrer, James; Bruce Sacerdote; and Ariel Dora Stern. "Will the stork return to Europe and Japan? Understanding fertility within developed nations," *Journal of Economic Perspectives* 22(Summer 2008): 3–22.

Fifield, Anna. "Beijing's one-child policy is gone. But many Chinese are still reluctant to have more," *Washington Post*, May 4, 2019.

Goldin, Claudia. and Lawrence F. Katz. "The power of the pill: Oral contraceptives and women's career and marriage decisions," *Journal of Political Economy* 111(August 2002): 730–70.

Goldman, Abe. "Agricultural innovation in three areas of Kenya: Neo-Boserupian theories and regional characterization," *Economic Geography* 69(January 1993): 44–71.

Guinnane, Timothy W. "The historical fertility transition: A guide for economists," *Journal of Economic Literature* 49(September 2011): 589–614.

Hammond, Samuel. "Born in Hungary," *National Review*, August 8, 2018.

Hoynes, Hilary M. "Work, welfare, and family structure," in *Fiscal Policy: Lessons from Economic Research*, Alan J. Auerbach(ed.) Cambridge: MIT Press, 1997.

Johnson, D. Gale. "Population, food, and knowledge," *American Economic Review* 90(March 2000): 1–14.

Jones, Kelly M. "Contraceptive supply and fertility outcomes: Evidence from Ghana," *Economic Development and Cultural Change* 64(October 2015): 31–69.

Koenig, Michael A. et al. "Contraceptive use in Matlab, Bangladesh in 1990: Levels, trends, and explanations," *Studies in Family Planning* 23(Nov. – Dec. 1992): 352–64.

Koren, Ore and Benjamin Bagozzi. "Food access and the logic of violence during civil war," *New Security Beat*. Wilson Center, May 15, 2017. https://www.newsecuritybeat.org/2017/05/food-access-logic-vi olence-civil-war/, accessed 8/2/2019.

Kremer, Michael. "Population and technological change: One million B.C. to 1990," *Quarterly Journal of Economics* 108(August 1993): 681–716.

Lalive, Rafael and Josef Zweimuller. "How does parental leave affect fertility and return to work? Evidence from two natural experiments," *Quarterly Journal of Economics* 124(August 2009): 1363–1402.

Malthus, Thomas R. *An essay on the principles of population*. Oxford: Oxford University Press, 1993[1798].

McKirdy, Euan. "Study: China faces 'unstoppable' population decline by mid-century," *CNN Online* (January 7, 2019).

Mello, Steven. "Do changes in condom availability affect short-term fertility? Evidence from Rwanda," *American Economist* 57(Fall 2012): 154–70.

Miller, Amalia. "Motherhood delay and the human capital of the next generation," *American Economic Review* 99(May 2009): 154–8.

Miller, Grant. "Contraception as development? New evidence from family planning in Colombia," *Economic Journal* 120(June 2010): 709–36.

Milligan, Kevin. "Subsidizing the stork: New evidence on tax incentives and fertility," *Review of Economics and Statistics* 87(August 2005): 539–55.

Moffitt, Robert A. "The effect of welfare on marriage and fertility," in *Welfare, the family, and reproductive behavior*, Robert A. Moffitt(ed.). Washington: National Academy Press, 1998.

Pop-Eleches, Cristian. "The supply of birth control methods, education, and fertility: Evidence from Romania," *Journal of Human Resources* 45(Fall 2010): 971–97.

Pritchett, Lant H. "Desired fertility and the impact of population policies," *Population and Development Review* 20(March 1994): 1–55.

"Russians offered day off, prizes to procreate," *NBCNews.com*, August 14, 2007.

Schultz, T. Paul. "Human capital, family planning, and their effects on population growth," *American Economic Review* 84(May 1994): 255–60.

Schultz, T. Paul and Yi Zeng. "Fertility of rural China: Effects of local family planning and health programs," *Journal of Population Economics* 8(November 1995): 329–50.

Sinha, N. "Fertility, child work, and schooling consequences of family planning programs: Evidence from an experiment in rural Bangladesh," *Economic Development and Cultural Change* 54(2005): 97–128.

Taylor, Adam. "Hungary's Orban wants to reverse his country's shrinking population through tax breaks. That's so much easier said than done," *Washington Post*, February 12, 2019.

United Nations. Department of Economic and Social Affairs. Population Division. "Government policies to raise or lower the fertility level," *Population Facts* PopFacts No. 2017/10 (December 2017).

Whittington, Leslie A. "Taxes and the family: The impact of the tax exemption for dependents on marital fertility," *Demography* 29(1992): 215–26.

Whittington, Leslie A. "State income tax policy and family size: Fertility and the dependent exemption," *Public Finance Quarterly* 21(1993): 378–98.

Winegarden, C.R. and Paula M. Bracy. "Demographic consequences of maternal-leave programs for industrial countries: Evidence from fixed-effects models," *Southern Economic Journal* 61(April 1995): 1020–35.

Zhang, Junsen; Jason Quan; and Peter van Meerbergen. "The effect of tax-transfer policies on fertility in Canada, 1921–1988," *Journal of Human Resources* 29(Winter 1994): 181–201.

# 8  Fossil fuel energy

At present, the world economy is heavily dependent upon the burning of fossil fuels to meet its various energy demands. Fossil fuels are what we call "exhaustible" energy resources deposited in the ground over geologic time and then extracted, refined, and combusted to provide energy for power plants, transportation, and various residential and industrial uses. They are an important option for all economies requiring energy to meet their economic objectives; in other words, economies everywhere. And they are a key driver of economic growth and source of enormous economic value.

A simple look at the data reveals humanity's heavy reliance on fossil fuel energy. Currently, oil, coal, and natural gas account for over 80% of world supply of energy [International Energy Agency (2017)]. Figure 8.1 shows, for the year 2017, the breakdown of total energy supply among the seven main categories of energy. Oil accounts for nearly one third of energy supply, while coal accounts for about 27% and natural gas accounts for about 22%. Meanwhile, the other three energy categories – biofuels, nuclear, hydropower, and renewable energy – account for roughly 19% combined. Clearly, fossil fuels are an important piece of the overall world energy picture.

At the same time, production and consumption of fossil fuels come with all sorts of environmental concerns, not the least of which is global climate change. The burning of fossil fuels releases greenhouse gases (GHGs) and various other pollutants into the atmosphere, which cause climate change and other threats to the environment and to human health. In addition, periodic oil spills like the Santa Barbara spill in 1969, the Exxon Valdez spill in Prince William Sound in 1989, and the Deepwater Horizon spill in the Gulf of Mexico in 2010 have dumped tens of thousands of tons of oil into marine environments. As one more example, the recent ascendance of hydraulic fracturing (fracking) in various regions in the United States has taxed local water supplies and threatens the quality of local drinking water. One of the important themes of this chapter is the tradeoff between economic growth and value on one hand, and negative environmental impacts on the other hand.

In considering the development and use of fossil fuels, a number of institutional issues arise. As we have seen, the economic approach involves the careful weighing of trade-offs across competing objectives. It turns out that institutions can importantly influence tradeoffs in the way fossil fuels are developed and used. For example, later in this chapter we shall see that the way we define a property right to an oil field can alter the incentives to develop it efficiently. The transaction costs of negotiation can affect the ability of individual companies to internalize the externalities associated with pumping from a *common-pool resource* (CPR). And given the economic importance of fossil fuels, and the enormous environmental impacts associated with their production and use, you will not

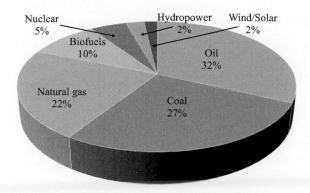

*Figure 8.1* Total world energy supply, 2017. *Source*: International Energy Agency, *Data and statistics*

be surprised to hear that heated political battles have broken out pitting proponents of energy development against environmental advocates. All of this will be developed more fully in this chapter.

## Modeling fossil fuel production and use

As a natural resource, all fossil fuels fall into the category of exhaustible resources, a concept that we encountered earlier in chapter two. Exhaustible resources like fossil fuels do not naturally replenish themselves, at least not within a time frame that the human race cares about. Thus, the key to sensibly exploiting an oil field is to consider its value both now and in the future and to allocate it accordingly; that is, develop and use it when it is more valuable to society.

To analyze this situation, let us return to the simple two-period dynamic model we encountered in chapter two. Figure 8.2 shows the current and future demand curves for a fossil fuel that I will call oil, $D_0$ and $D_1$, as well as the discounted future demand $D_1/(1 + r)$. As always, we assume that both $D_0$ and $D_1$ are declining functions of the amount of oil used in each time period. To keep things simple, we assume that it is costless to extract the oil from the ground (see exercise 8.3 for the case when this is not true). Under competitive market conditions, we would predict the quantity $Q^\star$ to be sold in the current period at a price of $P_0^\star$, and the remainder to be sold in the future period at the price $P_1^\star$. In this equilibrium situation, total discounted net benefits to society are maximized. In other words, the market equilibrium is *dynamically efficient*. Again, this is the virtue of markets, as applied to exhaustible resources such as oil and other fossil fuels.

Now let us think more carefully about what is happening to the price of the resource in equilibrium. Whereas the price in the current period equals $P_0^\star$, the price in the future period is $P_1^\star$, which equals, if you think about it, $P_0^\star$ *times* $(1 + r)$. This means that between the two periods, the price of the resource is growing over time at a rate precisely equal to the discount rate. Extending this same logic to multiple time periods, we can see that, over time, the price of fossil fuels should grow at a rate precisely equal to the discount rate r, shown in Figure 8.3. The resulting price path over time is known as the *Hotelling price path*, after the economist Harold Hotelling, who first derived the result. The Hotelling price path shows the basic theoretical result that the prices of exhaustible

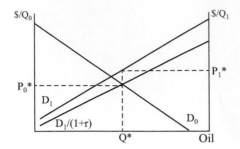

*Figure 8.2* Dynamic market equilibrium

resources like fossil fuels are predicted to increase (roughly) exponentially, as they become steadily depleted over time.

## Real-world oil markets

You should think of the analysis summarized in Figures 8.2 and 8.3 as a baseline set of predictions under simplified assumptions about the world. You might be wondering how these predictions comport with our real-world experience. Figure 8.4 shows the nominal (that is, not adjusted for inflation) world price of oil since the 1940s through 2019.[1] As you can easily see, the actual price path for oil over time bears little resemblance to a Hotelling price path. The difference reflects the complexities of the real world and the many factors that can affect oil prices including new discoveries, non-competitive markets, short-term fluctuations in demand (see Box 8.1), and supply disruptions such as oil embargoes. It also includes wars that have periodically broken out in the Middle East, a major source of world oil supplies.

---

**Box 8.1: A negative price of oil**

The sudden collapse in world demand for oil in the early stages of the COVID-19 crisis in 2020 temporarily led to a negative price of crude oil in U.S. markets. On April 20, the so-called benchmark futures price for a barrel of oil briefly fell to minus $37.63. This means that if you could have taken delivery of a thousand barrels of oil a month later, someone would have paid you $37,630. This strange situation occurred because demand fell faster than supply, and there were no places to store the surplus oil. [Irwin (2020)]

---

However, this uneven price history has not quelled fears that we are "running out" of oil. Such fears were particularly pronounced in the 1970s when they were exacerbated by geopolitical events in the Middle East that sent the world price of oil skyrocketing in 1973, and again in 1979. During this time, many scientists began to express growing concerns that we might have reached so-called "peak oil," the point of maximum oil production, after which production would steadily dwindle over time.[2] This raised the specter that fossil fuel supplies were becoming depleted and encouraged the development of various policies designed to combat depletion, such as residential energy conservation

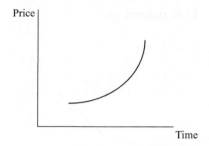

*Figure 8.3* Hotelling price path

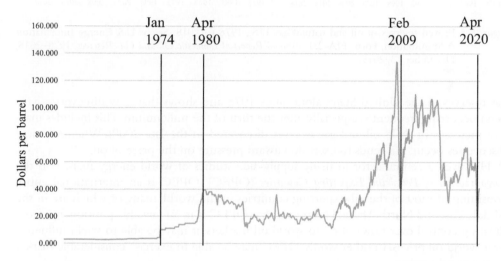

*Figure 8.4* World oil prices, 1960 to 2020. *Source*: Federal Reserve Bank of St. Louis, Economic Research Division

programs, solar energy tax credits, and fuel efficiency standards for automobiles. Currently, we are just as likely to see accounts downplaying this idea, such as a 2018 story in *Forbes* entitled: *What ever happened to peak oil?* [Lynch (2018)] However, periodic lulls in world oil markets have by no means extinguished concerns that we are running out of oil.

One way to view the uneven price history shown in Figure 8.4 is that the long-term depletion effects shown in the two-period model are currently being dominated by short-term market forces. This is in fact the position of most economists. Two main positions have emerged: one that emphasizes supply-side factors, and another that emphasizes demand-side factors.

### I. Supply-side factors

Supply-side studies focus on a variety of factors that have affected the supply of oil to world oil markets. These include, for example, discoveries of new oil reserves, which are additions to world supplies of oil in the long term. Figure 8.5 shows proved oil and natural

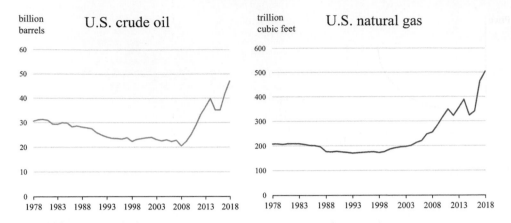

*Figure 8.5* Proved reserves of oil and natural gas, U.S., 1978 to 2018. *Source*: U.S. Energy Information Administration, Form EIA-23L, *Annual Report of Domestic Oil and Gas Reserves*, 1978–2018, December 13, 2019.

gas reserves in the United States alone since 1978 and shows that new discoveries have been occurring frequently, especially after the turn of the millennium. This includes massive increases in shale oil and gas reserves discovered in the late 2000s. Whenever new discoveries occur, this tends to exert downward pressure on the price of oil.

However, a central actor in many supply-side studies of world energy markets is the *Organization of Petroleum Exporting Countries* (OPEC). OPEC is an international entity consisting of most of the oil-exporting countries in the world, many of which are in the Middle East and North Africa. This means that OPEC is in an enviable position: by coordinating control over flows of oil to world oil markets, it may be able to wield influence over world oil prices. In other words, OPEC may be able to exercise considerable market power in oil markets.

This was in fact the dominant view of OPEC in the 1970s, when it seemed as if OPEC was flexing its monopoly muscle in world oil markets [Gately (1984), pp. 1101–4]. As Figure 8.4 shows, there were two large price spikes in the 1970s: in 1974 and 1979. The first spike occurred in the wake of an embargo levied by OPEC in late 1973, which lasted for several months. This embargo was partly in retaliation for U.S. support of Israel during the Yom Kippur War. The second spike was associated with the Iranian Revolution of 1979 and then the Iran-Iraq War, which dramatically reduced production in both countries and, hence, exports into world oil markets.

### II. Demand-side factors

However, the current consensus among economists is that we can better explain the price experience of the past 50 years as reflecting cyclical demand side factors.[3] Broadly speaking, expansions in economic activity during upturns in the business cycle increase oil demand and exert upward pressures on oil prices. When economic activity slows down or, worse yet, economies go into recession or even depression, oil demand drops and prices tend to fall. The swings in oil prices can be extreme, because the supply of oil tends to be quite price-inelastic in the short-term. Figure 8.4 reveals two massive drops in world oil prices, reaching rock bottom in February 2009 and April 2020. These drops occurred in response to the financial meltdown beginning in late 2008 and the COVID-19 crisis in spring of 2020, both of which dramatically decreased worldwide demand for oil.

## Environmental impacts of fossil fuel use

The demand for oil and other fossil fuels comes, of course, from the wide variety of things we use fossil fuels for, like power production, automobiles, heating and cooking in our homes, a whole host of industrial uses, and so forth. When fossil fuels are combusted to provide energy for these activities, this inevitably generates waste by-products, in the form of various pollutants like carbon dioxide ($CO_2$), methane, sulfur dioxide ($SO_2$), nitrous oxides ($NO_x$), and solid particulate matter. When these pollutants enter the environment, they create all sorts of environmental hazards. Production of $CO_2$, methane, and $NO_x$ contribute to growing climate change. $SO_2$ and $NO_x$ can make rain and snow increasingly acidic, causing damages to lakes, forests, and buildings. $NO_x$ and solid particulate matter – basically, airborne particles of ash and dust – contribute to air pollution, which has various effects on human health, including exacerbating respiratory and heart conditions.

We can take various steps to reduce the amount or harmfulness of these pollutants. These include:

- Installing catalytic converters in automobiles;
- Making and driving vehicles that use less gasoline or diesel;
- Burning coal with lower sulfur content;
- Installing so-called smoke scrubbers in power plants.
- Burning natural gas instead of coal;
- Injecting aerosols into the atmosphere;
- Using negative emissions technologies to physically remove $CO_2$ from the atmosphere (plant more trees, or direct air capture);
- Switching to renewable energy sources like wind, solar, and geothermal; and
- Conserving energy in our homes and businesses.

We will encounter all of these options over the next four chapters. And, when we do, we will examine them more closely to understand how they would work, and explain how economists think about their advantages and disadvantages.

## Economic analysis of fossil fuel consumption

Let us now bring these environmental impacts into the economic analysis in order to provide guidance for a general discussion of policy. When pollutants are produced by burning fossil fuels, they are released into the environment, where they can generate smog, acidify lakes, contribute to global climate change, exacerbate respiratory ailments like asthma, and so forth. In economic terms, they generate costs that are incurred by others. The result is a social marginal cost (SMC) curve that exceeds the private marginal cost curve for users of fossil fuels. If no mechanism exists that forces fossil fuel users to bear these costs, we would predict that goods produced using fossil fuels, like electrical power, will tend to be overproduced.

The analysis is shown in Figure 8.6, where the market equilibrium quantity of power is $Q^\star$, where supply equals demand, which is incidentally where private marginal costs

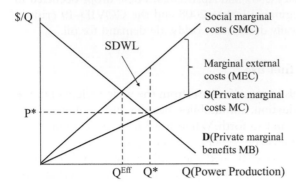

*Figure 8.6* Market inefficiency in equilibrium

(MC) equals private marginal benefits (MB).The efficient amount of power is, however, $Q^{Eff}$, where marginal benefits equals *social* marginal costs (SMC). In efficiency terms, the interests of society would be better served if the power companies produced less than they actually do. But they produce more than the efficient quantity because they are given no incentive to consider the external costs they are imposing on the rest of society.The amount society loses in net benefits is the triangle denoted SDWL, or social deadweight loss.

For some added insight, let us return to the Coase argument that we encountered in chapter four. Rather than think about this situation as the power companies inflicting costs on the nearby residents, Coase would argue that the power companies and nearby residents are inflicting costs on each other.This means, first of all, no necessary presumption on efficiency grounds that the power companies should be the ones penalized: for causing the problem, so to speak. Rather, efficiency requires avoiding the greater cost. How to do this depends on, among other things, the magnitude of the transaction costs of negotiation among the parties involved.

### I. Transaction costs low

To apply the Coase argument, first note that, in this situation, there are potential gains from trade from negotiating.To see this, re-examine Figure 8.6.At the equilibrium quantity $Q^\star$, the marginal benefit to the power companies of producing the last unit – the price $P^\star$ that they can get on the power market for selling it – is exactly equal to their (private) marginal cost of producing it.This tells us that there is no gain to the power companies' bottom line, in terms of extra net benefit, from producing that last unit.

On the other hand, the cost being incurred by the downwind neighborhoods from that last unit is clearly greater than zero.This implies that there are potential gains from trade: in principle the neighborhoods and power companies should be able to strike a deal under which the neighborhoods pay the power companies to produce less. If it were costless for the two sides to negotiate, one would expect negotiations to occur, which would drive down the amount of power production all the way to $Q^{Eff}$. In this case, there would be no rationale for government policy, at least not on efficiency grounds.And you could address equity concerns by deciding which parties to provide with legal rights. In the simplest case, this question might revolve around whether or not to make the power companies liable for damages. If this is unclear, you might want to review the farmer-rancher example in chapter four, which explains the logic at greater length.

## II. Transaction costs high

The more likely case is that there are transaction costs, potentially large ones. That transaction costs are large is suggested by the fact that we generally do not see private negotiations between power companies and affected neighborhoods to reduce power output, even when the neighborhoods are incurring large costs. This suggests that it is indeed costly to negotiate, perhaps because of the typically large numbers of parties in the affected neighborhoods. Or, there may be other sources of pollution, making it difficult to determine who is responsible for what pollution. And, of course, the power companies have every incentive to claim that they are not the ones responsible for the pollution.

With large transaction costs, then, there is the potential for government policies to improve resource allocation, because we cannot rely on private negotiations to address the inefficiencies. Then, once we are considering government policies, we have to keep in mind two additional questions: What *should* the government do? And what is it *likely* to do, given political realities? These are the normative and positive views of government policy.

Regarding what the government *should* do, for economists there are two key issues: efficiency and equity. If negotiations are unlikely to occur, efficiency dictates placing the legal right in the hands of the lower-cost option. For example, this may mean taxing the power companies if they are inflicting massive costs on the nearby neighborhoods. But it may not if the costs of abating the pollution are greater than the costs to the neighborhoods. Here is where equity concerns might well enter the picture, which economists are more wimpy about weighing in on. You may well have strong opinions about who should bear the cost and, if so, I encourage you to express them. Economists may respond with efficiency arguments but are much less likely to dispute your equity concerns.

Regarding what the government is *likely* to do, we are then suddenly in the realm of public choice. Here, the particular circumstances matter a great deal; for example, the size, stakes, and heterogeneity of relevant interest groups, the existence of political entrepreneurs, the particular makeup of key legislative committees, the power of committees to oversee administrative agencies charged with implementing a program, all that stuff. That is, all of the factors that we encountered in chapter six. You can probably see that the question of what the government is likely to do can only be answered on a case-by-case basis because every issue is different. The first extended public choice example in this book will be seen later in this chapter when we encounter the important new issue of fracking. But, before we turn to that discussion, let us first examine an interesting set of institutional issues relating to oil and gas production.

## Local solutions: The common-pool problem

The policy discussion so far has related to the external impacts of fossil fuel consumption. However, there are also some important policy issues related to fossil fuel production, particularly production of oil and natural gas. These issues stem from the fact that much oil and natural gas is extracted from oil and gas fields where multiple private companies are operating. Oil and gas fields can be quite large, extending for literally tens of thousands of acres. To take just one example, the Slaughter oil field in Texas extends for 87,000 acres, or nearly 136 square miles. Generally, if a company wants to tap such a field, all it has to do is buy or lease land from an existing landowner, which gives it the right to pump the underlying oil and gas. You can imagine that an oil field extending for 136 square miles might attract many companies.

In such a situation, however, there is the strong possibility that inefficiencies in production may occur, for two related reasons. First, every company realizes that, if it does not pump the oil and gas, some other company may beat it to the punch. So every company has incentive to pump more rapidly than it would if it had an ironclad guarantee that the oil and gas would still be there tomorrow.

Second, an actively pumping company reduces pressure underground and induces the lateral flow of oil and gas in its direction. Drawing away oil and gas from other companies reduces underground pressure under their sites, which makes it more costly for these other companies to extract oil and gas themselves.

The problem is that no company has incentive to take these impacts on other companies into consideration. Thus, each company tends to pump too much too quickly. Furthermore, since companies know that their oil and gas may migrate to other properties, they may undertake investments to keep this from happening. For example, they may drill *offset injection wells*, through which they can inject gas at strategic locations in order to reduce oil migration off their lands. You will notice that this is basically the exact same problem we discussed earlier in chapter five: the problem of the common pool. Whenever individuals claim a resource in common with others, each has strong incentives to take what it can, while it can, while making unnecessary investments in order to do so.

Let us view, however, what is going on in these oil and gas fields from a slightly different perspective. In a real sense, each of these multiple companies tapping a field is claiming a property right to the oil and gas that it pumps. This right consists of the right to pump the oil and gas and bring it to market, where the company has the right to sell it for as much as it can get for it. If oil and gas were like automobiles, everything would be fine. When you buy a new car, you pay your money to obtain title, and use the car however you want until you sell it to someone else. None of this imposes costs on others.

With oil and natural gas, however, the very act of obtaining possession (by pumping from the ground) inflicts costs on everyone else pumping from the same field. The law gives no one any incentive to consider the effects on others because it generally applies the so-called *rule of capture*. The rule of capture simply says: reducing possession of a resource to your own personal use grants a right. You can see, for example, how this rule is applied when someone goes fishing. Catching a fish gives you the right to possess and do what you like with it, including eating it or throwing it back. Pumping oil and gas from a field confers the same sort of right.

So, what is the solution? Two distinct strategies have received a lot of attention, what I will call the *fiscal approach* and the *property rights approach*.

### I. Fiscal approach

Under both approaches, the overpumping problem can be addressed in a seemingly simple way: make everyone bear at least part of the costs that their extraction inflicts on others. Under the fiscal approach, this is accomplished by imposing a tax on extracting the oil and gas from the ground, commonly referred to as a *severance tax*. Severance taxes are levied at the individual state level. Usually, the tax revenues are then placed in the state's general fund to support general expenditures, though there has been a recent trend to dedicate revenues specifically to mitigate externalities associated with extraction, including impacts on local communities.[4]

As you might expect, severance taxes have been politically controversial, with support often divided along party lines and, in general, they have been bitterly opposed by oil and gas interests. However, political opposition can be defused when much of the extracted oil and gas are exported from the state. In this case, much of the incidence of the tax can be passed through to consumers in other states, since demand tends to be price-inelastic. This is illustrated in Figure 8.7, where under inelastic demand conditions, the severance tax is mostly reflected in an increase in the price paid by consumers and little of the tax is borne by producers. This analysis helps explain why states like Alaska and Wyoming, which produce much oil and gas, have relatively high severance taxes, since almost none of their oil and gas is consumed within the state.

### II. Property rights approach

Under the property rights approach, oil and gas producers are made to bear the external costs of extraction in a different way. Instead of imposing severance taxes on each company, all companies' operations are placed under the control of a single company, which I will call *The Firm*. Under this so-called *unitization* arrangement, The Firm is given sole responsibility for developing the oil and gas field. In essence, this company is given the keys to the car and is permitted to drive it anyway it pleases, as long as it satisfies certain rules (like obeying the speed limit and not crashing into others' cars), which are agreed to ahead of time.

The rationale for unitization may be clear: The Firm, by having the entire field under its control, suddenly has incentive to consider the effect of one company's operations on others. Company A, acting on its own, may have incentive to drill offset injection wells, just to keep oil from migrating to a neighboring tract. However, The Firm has no such incentive. It knows that if any oil migrates across the border between two properties, it can simply pick it up on the other side. No need to waste money drilling those

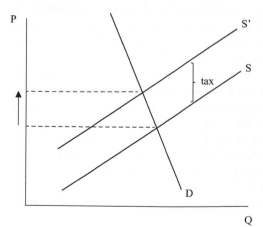

*Figure 8.7* Price effect of a severance tax

wells, whose only purpose is to keep oil from migrating and which do not aid in actual extraction one bit.

It may be clear why unitization is called a property rights approach. In essence, The Firm is given the property right to the entire oil field. Not necessarily in narrow legal terms, in the sense of actually having a paper deed that declares it is the owner. But rather, in the economic sense that it has been given free rein to develop the oil field as it sees fit, and it is entitled to receive some returns for its effort. Just as you, by enjoying the property right to your car, can drive it as you see fit and enjoy yourself while doing so.

In theory, complete unitization of an oil and gas field should permit the efficient development of that field. The effect of unitization is to eliminate all external impacts of extraction, because no portion of the field is external, from the viewpoint of The Firm. All costs are internalized: that is, borne by The Firm itself. Thus, The Firm does not worry that if it does not pump today, someone else might get the oil first. The oil will still be there tomorrow, because what happens tomorrow is under its control. And, more generally, it will have incentive to weigh the benefits of pumping in one location – perhaps the market value of that oil it brings to market – against the costs of that pumping, in terms of reduced extraction elsewhere. All of this should result in efficient development of the oil field.

In practice, however, a number of factors may make it difficult for unitization to occur. To see why, consider what is required. All of the companies in an oil field must sit down and agree on voting rules; arbitration procedures; a formula to share revenues and operating costs among the companies; and, importantly, procedures for monitoring The Firm and for applying sanctions if it does not abide by the procedures agreed upon by everyone [Libecap (1999)]. Furthermore, all of this must be agreed upon ahead of time: that is, in advance of when production begins. This is because oil production typically causes changes in the oil field that affect different companies unpredictably. To be willing to participate, companies must be given assurance that these changes will not harm them.

For various reasons, coming to an agreement in advance on all of these things and more may present major challenges to any group of companies wanting to set up a unitization arrangement. To put it differently, there may be significant transaction costs involved in negotiating such an arrangement.

Why might there be transaction costs impeding the creation of unitization? It is unlikely that the reason is the sheer number of companies, as the number of companies operating in an average oil field tends not to be large, the Slaughter oil field example notwithstanding [Wiggins and Libecap (1985), p. 369]. Rather, transaction costs arise for two related reasons having to do with imperfect information about the oil field and the effect of production activity under the unitization agreement on the field.

First, it is often unclear exactly how much oil and gas there is in an oil field, and under whose land it is located. This is an important issue because the ability to come to a unitization arrangement depends crucially on achieving agreement in advance on the sharing of the proceeds from unitized production. If the sharing formula leaves some companies worse off, they will be unwilling to participate in the arrangement. If it is difficult to tell how much oil and gas each company has, it will be difficult to create a formula that compensates each company appropriately. Companies that turn out to have less oil than expected will tend to be paid too much, and companies that turn out to have more oil than expected will tend to be paid too little. Knowing all this, companies will hesitate before agreeing to join.

Second, every company possesses information regarding how much oil and gas can be pumped from its wells, and this information is not available to other companies. This is important because shares in field proceeds must be based on estimates of the size of the oil supplies of different companies. Thus, every company has incentive to overstate the size of its oil supplies. Since every other company knows this, it may be difficult for them to agree on a sharing formula.

As it turns out, despite its efficiency virtues being known for some time, relatively few unitization arrangements exist. For example, by the mid-1970s, unitized oil field production accounted for well below half of production in Oklahoma, and only a fifth of production in Texas, two of the largest oil-producing states in the United States [Libecap and Wiggins (1985). This surprisingly meager reliance on unitization, despite its well-known efficiency advantages, suggests that transaction costs have presented significant obstacles to unitization.

At the same time, we do know some things about specific factors that tend to reduce transaction costs and make unitization more possible. Older fields are more likely to be able to unitize, because more is generally known about the characteristics of the field, making it easier to arrive at agreement. Also, fields containing holdings that are more similar in productivity should also be easier to unitize, as firms with highly productive fields will tend to not want to include them in the unitization plan [Wiggins and Libecap (1985)]. Finally, unitization appears to be easier on federal lands, because of federal rules that tend to facilitate unitization. These include rules that allow agreements to be reached before any drilling occurs, which reduces the amount of private information on productivity possessed by any company. This makes it less likely that any one company will hold out for a better deal.[5]

## Hydraulic fracturing

A notable recent development in world oil and natural gas markets is the emergence of hydraulic fracturing (fracking) as a major source of oil and gas currently and, possibly, for some time to come. Its role in world energy markets may be huge. The World Energy Council, for example, has estimated that total global shale oil reserves may be more than three times as large as oil from conventional sources [World Energy Council (2016), p. 6].

To date, no country has developed its shale oil and gas reserves more than the United States, and it is not even close. In 2015, for example, total shale gas production in the United States far outstripped that of all other countries *combined* [Statista (2020)]. Figure 8.8 shows that this occurred in the wake of massive discoveries of shale gas deposits beginning in the mid-2000s. From 2006 until 2018, while proved reserves of non-shale gas reserves actually declined, shale reserves increased *23-fold*. In the mid-2000s, fracking was providing only a small fraction of total natural gas production in the U.S [U.S. EIA, "Today in energy"]. In sharp contrast, by 2015 fracking was accounting for some two-thirds of all domestic natural gas production (and about half of all domestic crude oil production!). Every indication is that fracking will be a major component of fossil fuel production in the United States for some time to come.

Expansion of fracking promises to offer major economic benefits, as well as some environmental benefits.[6] The discovery of reserves of shale oil and gas both increases energy output and reduces prices now and in the future, as shown in Figure 8.9. Here, the new added shale oil and gas reserves are represented by increasing the length of the horizontal axis, while not changing either current or future demand. The result is an increase in the current equilibrium quantity from $Q^\star$ to $Q'$ and a lowering of the equilibrium prices in both periods to $P_0'$ and $P_1'$. It should be easy for you to verify that the future equilibrium quantity increases as well.

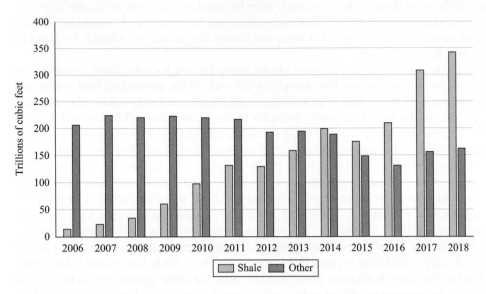

*Figure 8.8* U.S. natural gas proved reserves, 2006 to 2018. *Source*: U.S. Energy Information Administration, Form EIA-23L, *Annual Report of Domestic Oil and Gas Reserves*, 2011–18.

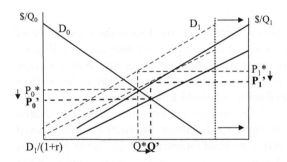

*Figure 8.9* Effect of adding shale oil and gas reserves

The environmental benefits derive from the fact that both oil and natural gas are cleaner fossil fuels than coal, which we currently rely on heavily to produce electricity. Switching from coal to natural gas should reduce greenhouse gas emissions, which will help combat climate change.[7] In addition, switching from coal-fired power plants will address other pollution-related concerns such as acid rain, asthma, and other respiratory ailments.

However, there are also certain negative environmental impacts associated with fracking. To see these, consider the way fracking is done. Fracking shale deposits lie deep underground (see Figure 8.10). In the fracking process, a deep well is drilled into the shale layers and, once there, it is extended horizontally in order to access more of the shale field. Then a mixture of water, sand, and chemicals is injected through the well at high pressures. This causes the shale deposits to fracture, releasing the gas and oil, which then flows back

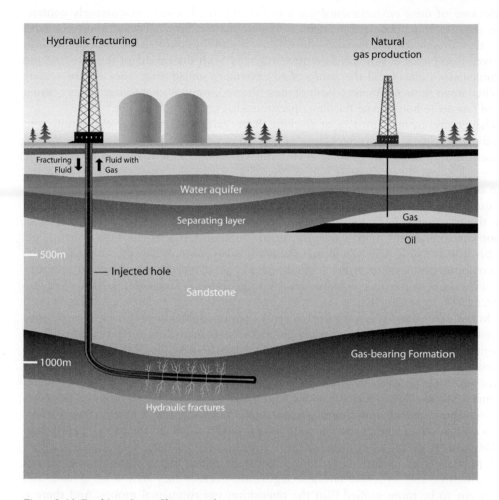

*Figure 8.10* Fracking. *Source*: Shutterstock

up through the well to the surface. In addition, the process produces wastewater – water containing sand and chemicals – that must be disposed of. Sometimes the wastewater, or so-called *produced water*, is recycled to produce more shale gas or oil, and sometimes it is sent to water treatment plants [Hall (2013), pp. 48–9]. Most commonly, however, disposal occurs by drilling wells deep into geologic formations and injecting the produced water into the wells.

This description of fracking indicates a number of ways in which environmental damages could occur. First, fracking requires large amounts of water, which may place onerous demands on local water supplies. Second, chemicals used in fracking may leak into local groundwater. Third, the produced water is commonly laced with chemicals and must be disposed of. In some cases, the processes of fracking and disposal of produced water have actually produced local earthquakes.[Seismological Society of America (2019)] Finally, there may be increased air, traffic, noise, and light pollution in the local community when the well is being drilled and when it is actively producing oil or gas.

Because of these environmental impacts, fracking has become an extremely contentious issue. Highly publicized incidents of contamination have occurred, and some have been tied directly to local fracking efforts [Brown (2007)]. In 2010, the fracking industry received two pieces of unwanted negative publicity, with the release of the anti-fracking documentary *Gasland* and the airing of an extremely unflattering story on the widely watched news show *60 Minutes*. Both of these showed, among other things, water coming out of faucets in houses near fracking operations that could be lit on fire.[8]

The environmental impacts of fracking have galvanized political opposition into a national coalition to ban fracking, *Americans Against Fracking*, which includes among its members a number of national environmental groups such as Friends of the Earth, Greenpeace, and the Rainforest Action Network, as well as other groups on both the state and federal level [Americansagainstfracking.org]. In the 2020 Democratic presidential campaign, both Bernie Sanders and Elizabeth Warren came out in favor of such a ban, though eventual Democratic nominee Joe Biden did not. However, a number of major environmental groups have not joined the coalition, including the Sierra Club, the Environmental Defense Fund, and the National Resources Defense Council. So the environmental movement in the United States is a bit split in its opposition to fracking, with more mainstream environmental groups showing some ambivalence [See also Spence (2013), pp. 156–8].

On the other hand, it is not surprising that continued fracking has been strongly supported by representatives of the oil and gas industry, manufacturers of goods who would benefit from lower gas prices, and many landowners who have leased their lands to fracking companies. In 2011, a group of pro-fracking groups – including chambers of commerce and trade associations of the oil and gas industry – sent a letter to then-President Obama that strongly supported fracking, emphasizing its potential to promote job creation, economic growth, lower energy prices, and energy security. It also portrayed fracking as safe and cited studies documenting its "proven track record for safety" ["Hydraulic fracturing jobs," (2011)].

A public choice analysis of fracking regulation might predict that fracking interest groups will be able to secure favorable legislation because they have a lot at stake and they seem to be more unified than the opposition environmental groups. And there is some evidence that supports this prediction. Pro-fracking groups have managed to secure a largely hands-off approach by the federal government, which has deferred all regulatory responsibility to the states. Perhaps the most well-known example of this is the fact that fracking companies have been exempted from certain provisions of the federal Safe Drinking Water Act, which regulates waste disposal in injection wells in order to protect groundwater supplies. This exemption was built into the 2005 Energy Policy Act and is often called the *Halliburton Loophole*. This is because the exemption was pushed by Dick Cheney, vice president under George W. Bush and a former CEO of Halliburton, a large oil and gas company [Warner and Shapiro (2013), pp. 479–80].

But it is not simply a question of being able to secure this favorable federal policy. After all, this does not necessarily get them off the hook from being regulated at lower levels: by state or even local governments. However, studies have shown that interest groups are often able to wield more influence at the state level in securing their favored policies [Warner and Shapiro (2013), p. 475–6]. As a result, interest groups may prefer to be regulated not by the federal government but by the state, where they believe they are more likely to be politically successful. This is an example of the phenomenon that political scientists call *venue-shopping*. Under this interpretation, in securing the federal exemption the fracking interest groups successfully shopped for the more favorable venue. Venue-shopping represents an added wrinkle in public choice analysis when you have multiple levels of jurisdiction, as occurs under the U.S. federalist system, or where there are other arenas in which political advantage can be secured, like the courts or administrative agencies.[9]

It is of course not true that regulation at the state level would always be preferred by interest groups. Various factors will determine the relative costs and benefits of selecting one venue over the other. These include: the perceived strength and intensity of opposing groups, which may vary across venues; constituent preferences on choice of venue; and whether a group has a relationship with a particular legislator, especially a powerful and connected one [Holyoke (2003), p. 326]. In the case of fracking, the sheer economic value of fracking activity may be quite large in a state like Texas or Wyoming that happens to be blessed with abundant shale oil and gas reserves. This may make fracking interest groups able to wield greater influence over the legislators of those states than at the federal level.

It is also telling that the companies that signed that letter to Obama expressed a clear preference for state regulation, arguing that;

> Because they understand the regional and local geological conditions, [state regulators] are in the best position to protect groundwater and drinking water sources.
> ["Hydraulic fracturing jobs" (2011)]

There is likely some truth to this argument, as state regulators may well have better information on local conditions in general. On the other hand, state regulators may not have the regulatory resources that the federal government has in order to engage in effective regulation. In any case, I will leave it to you to decide whether you find it more plausible that fracking companies were more interested in protecting water resources or avoiding being regulated.

So the reality to this day is that regulation of fracking is the responsibility of state governments, not the federal government. Given the argument so far, one might expect there to be little or no state regulation, if the fracking industry is so politically powerful at the state level. However, it turns out this is not the case: Every single state with shale production has imposed regulations on its fracking industry [Krupnick et al. (2015), p. 53]. And since every state is free to make its own rules, the result is that rules vary widely across different states. And furthermore, it is not at all clear that the regulations are designed to help the fracking industry by keeping entrants out, as a public choice model might predict.

In an interesting set of studies, the economist Alan Krupnick and his coauthors have carefully examined and compared the regulatory regimes of different states with fracking activity [Krupnick et al.(2015); Richardson et al. (2013)]. These studies do not find evidence of a clear negative correlation between the amount of fracking activity and the stringency of fracking regulation, which you might expect if a more powerful fracking interest group would more easily be able to stave off regulation. This is true even when you control for other relevant factors. They conclude that the special interest model explains only partly, at best, the observed pattern of regulation across the states.

So what *does* explain fracking regulation at the state level? Krupnick speculates that some of the differences may be attributable to differences in initial regulatory conditions, which have persisted over time in a path-dependent process [Krupnick et al. (2015), p. 66]. However, an additional possibility is that there is some component of regulation in the interests of consumers. In Krupnick's data, stringency of regulation protecting groundwater, for example, seems to be positively related to the extent of a state's reliance on groundwater for its water supplies [Krupnick et al. (2015), p. 60].

Many economists have been loathe to completely rule out public interest considerations in explaining regulation. How would this arise? Well, one possibility is that public interest regulation – say, to protect drinking water supplies – occurs when there is a highly publicized incident, like tap water set on fire. Some have argued that major shocks – what we often call crises – can stimulate policy actions because of the perceived need to "do something." [Jones and Baumgartner (2005); Grossman (2012)] In the case of fracking, such events may mobilize self-interested legislators to enact legislation because the public is suddenly galvanized. But, overall, it is fair to say the jury is still out on how best to interpret fracking regulation at the state level.

## Key takeaways

- In the baseline case of market competition with zero extraction costs, the price of fossil fuel energy is predicted to rise exponentially over time.
- Actual price paths probably reflect a combination of short-term supply-side and demand-side factors. The supply-side factors include new discoveries of fossil fuels and market power exercises by OPEC. The demand-side factors include upturns and downturns in the business cycle. Most energy economists believe that the recent price experience over the past several decades is more attributable to demand-side factors.
- Development and consumption of fossil fuels produces a variety of environmental contaminants, not the least of which is greenhouse gases.
- The large transaction costs associated with negotiating over the production of various pollutants likely results in overproduction and social deadweight loss.
- Extraction processes in local oil and gas fields likely results in common-pool impacts on the local level, resulting in overpumping of oil and gas. Two possible approaches for dealing with this are the fiscal approach and the property rights approach.
- Under the fiscal approach, severance taxes are imposed on oil and gas extractions.
- Under the property rights approach, oil and gas fields are subject to unitization. In principle, unitization should result in efficient development of an oil and gas field. In practice, there are likely significant transaction costs associated with unitization.
- In recent years, the United States has discovered and developed large amounts of shale oil and gas, through the process known as hydraulic fracturing, or fracking.
- Fracking has some potential environmental benefits, to the extent its development enables us to switch away from burning coal, a relatively dirty energy source. On the other hand, there are also numerous environmental costs, which have been highly visible and resulted in much negative publicity for the fracking industry.
- Environmental regulation of the fracking industry has largely been pursued at the state level, where oil and gas interests are predicted to be politically influential. Despite this, most states have imposed regulations on fracking, which may have arisen from public interest concerns about the environmental damages of fracking.

## Key concepts

- Hotelling price path
- Severance taxes
- Rule of capture

- Unitization of oil and gas fields
- Hydraulic fracturing
- Venue shopping

## Exercises/discussion questions

8.1. *[Basic exercise in applying the intertemporal allocation model]* Consider the two-period model, as applied to oil. Suppose that in each of the two periods, oil is produced under conditions of competition. In each period, the market demand for oil can be expressed algebraically as: $P = 120 - 0.12Q$; where $P$ = price of oil, in dollars per barrel; and $Q$ = quantity of oil, in millions of barrels. Suppose that oil stocks total 1,450 million barrels, oil extraction is costless, and the discount rate is 20%. How much oil is sold in each period, and at what prices?

8.2. Assuming zero extraction costs and competitive markets, the price of fossil fuels is predicted to rise over time at the interest rate r (the Hotelling price path). What is your best intuition for why this would be the case?

8.3. Assume, as in exercise 8.1, the market demand for oil in each of two periods can be expressed as $P = 120 - 0.12Q$. However, assume now that oil extraction is not costless, and, in particular, that marginal extraction costs in each period equals \$20 per barrel of oil. Calculate the new dynamic market equilibrium prices and quantities sold. Does the Hotelling price path hold? Give your best intuition for this result.

8.4. The dynamic market analysis implicitly assumes that current and future market demand are known with certainty. Suppose that future market demand is not known with certainty. How do you think this would affect the price path over time?

8.5. *[Exercise to see the effect of new discoveries on current and future production and prices]* Assume the conditions of exercise 8.1. Now suppose that new discoveries of shale oil increase total stocks of oil by 110 million barrels. Calculate the impact of these discoveries on the price of oil in each of the two periods, and the quantities of oil sold in each period. Confirm that the Hotelling price path still holds.

8.6. Reanswer 8.5 assuming that new discoveries of shale oil increases total stocks of oil by 550 million barrels. How do you explain this result?

8.7. Suppose that a state adopted a relatively lax regulatory environment toward fracking. How might this tend to perpetuate itself over time in a path-dependent process?

8.8. The discussion suggests that both special interest and public interest considerations have shaped state regulation of fracking. Under what political and economic conditions would one of these explanations dominate? And what evidence would you seek to support your argument?

## Notes

1 This data series is West Texas Intermediate (WTI) crude oil prices, measured in dollars per barrel. There are many different prices of oil, reflecting different sources and grades of oil. WTI is the considered the "benchmark" price for oil in the United States.

2 The idea of peak oil is generally attributed to the geologist M. King Hubbert, who wrote about the idea in the 1950s. So sometimes you hear it referred to as Hubbert's Peak.

3 See, for example, Baumeister and Kilian (2016), pp. 141–2 and the sources cited there.

4 For good discussions of severance taxes, see Light (1981); Rabe and Hampton (2005), pp. 390, 402–3.

5  For a clear discussion of the contracting challenges regarding unitizing oil fields, see Libecap (1989), pp. 93–114.
6  One carefully done recent study, for example, estimated that expanded shale gas production from fracking resulted in roughly $48 billion per year in added net benefits for the United States between 2007 and 2013 [Hausman and Kellogg (2015)].
7  This seems to be the consensus view (see, for example, Nordhaus (2013), pp. 158–60). However, some recent studies have concluded that when you take associated methane production into account, natural gas does not do nearly as well. See Krupnick et al. (2015), p. 64.
8  *Gasland*; *60 Minutes*, "Shaleionaires," (2010). See also: "Sparks fly," (2011)
9  Venue-shopping can take a number of different forms. For example, interest groups can attempt to secure advantages not only at different levels of legislative jurisdiction, but also in the courts, in regulatory agencies, and in legislative committee. See Holyoke (2003), p. 325–6.

## Further readings

### I. Prospects for fossil fuel consumption

Covert, Thomas; Michael Greenstone; and Christopher R. Knittel. "Will we ever stop using fossil fuels?" *Journal of Economic Perspectives* 30(Winter 2016): 117–37.
*Argues that we will not without, significant government intervention.*

### II. Public choice in energy policy

Frederiksson, Per G. and Herman R. J. Vollebergh. "Corruption, federalism, and policy forma-tion in the OECD: The case of energy policy," *Public Choice* 140(July 2009): 205–21.
*Investigates the effect of corruption on energy policy. Finds that greater levels of corruption tends to reduce the stringency of energy policy, but that having a federalist system mitigates the impact.*

### III. Hydraulic fracturing

Chyong, Chi Kong and David M. Reiner. "Economics and politics of shale gas in Europe," *Economics of Energy and Environmental Policy* 4(March 2015): 69–84.
*Examines the economics and politics of hydraulic fracturing in Europe.*

## References

Americansagainstfracking.org.
Baumeister, Christiane and Lutz Kilian. "Forty years of oil price fluctuations: Why the price of oil may still surprise us," *Journal of Economic Perspectives* 30(Winter 2016): 139–60.
Brown, Valerie J. "Industry issues: Putting the heat on gas," *Environmental Health Perspectives* 115(February 2007).
*Gasland* (2010), written and directed by John Fox.
Gately, Dermot. "A ten-year retrospective: OPEC and the world oil market," *Journal of Economic Literature* 22(September 1984): 1100–14.
Grossman, Peter Z. "The logic of deflective action: U.S. energy shocks and the U.S. policy process," *Journal of Public Policy* 32(April 2012): 33–51.
Hall, Keith B. "Recent developments in hydraulic fracturing regulation and litigation," *Journal of Land Use & Environmental Law* 29(Fall 2013): 29–67.
Hausman, Catherine and Ryan Kellogg. "Welfare and distributional implications of shale gas," *Brookings Papers on Economic Activity* (Spring 2015): 71–125.
Holyoke, Thomas T. "Choosing battlegrounds: Interest group lobbying over multiple venues," *Political Research Quarterly* 56(September 2003): 325–36.

"Hydraulic fracturing jobs and security letter to Obama," *Marcellus Drilling News*, September 20, 2011.

Irwin, Neil. "What the negative price of oil is telling us," *New York Times*, April 21, 2020.

International Energy Agency. "Total energy supply by source," *Data and statistics*, 2017.

Jones, B.D. and F. R. Baumgartner. *The politics of attention: How government prioritizes problems*. Chicago: University of Chicago Press, 2005.

Krupnick, Alan; Nathan Richardson; and Madeline Gottlieb. "Heterogeneity of state shale gas regulations," *Economics of Energy and Environmental Policy* 4(March 2015): 53–68.

Libecap, Gary D. *Contracting for property rights*. Cambridge: Cambridge University Press, 1989.

Libecap, Gary D. "The self-enforcing provisions of oil and gas unit operating agreements: Theory and evidence," *Journal of Law, Economics, and Organization* 15(July 1999): 526–48.

Libecap, Gary D. and Steven N. Wiggins. "The influence on private contractual failure on regulation: the case of oil field unitization," *Journal of Political Economy* 93(1985): 690–714.

Light, Alfred R. "State severance taxes fuel court action and Congressional debate," *Publius* 11 (Summer 1981): 85–111.

Lynch, Michael. "What ever happened to peak oil?" *Forbes*, June 29, 2018.

Nordhaus, William. *Climate casino*. New Haven: Yale University Press, 2013.

Rabe, Barry G. and Rachel L. Hampton. "Taxing fracking: The politics of state severance taxes in the shale era," *Review of Policy Research* 32(2005): 389–412.

Richardson, Nathan; Madeline Gottlieb; Alan Krupnick; and Hannah Wiseman. *The state of state shale gas regulation*. Washington: Resources for the Future, 2013.

Seismological Society of America. "Studies link earthquakes to fracking in the Central and Eastern U.S.," *ScienceDaily*, April 26, 2019.

"Sparks fly over 'Gasland' drilling documentary," *NPRNews*, February 24, 2011.

Spence, David B. "Responsible shale gas production: Moral outrage vs. cool analysis," *Fordham Environmental Law Review* 25(2013): 141–90.

Statista. *Shale gas production worldwide in 2015 and 2040, by selected country*, 2020. https://www.statista .com/statistics/653200/shale-gas-production-forecast-worldwide-by-country/, accessed 12/28/20.

United States. Energy Information Administration. *Today in Energy*, August 20, 2019.

United States. Energy Information Administration, Form EIA-23L, *Annual Report of Domestic Oil and Gas Reserves*, 1978–2018, December 13, 2019.

United States. Energy Information Administration. *Annual Report of Domestic Oil and Gas Reserves*, 2011–18. Form EIA-23L. December 3, 2019.

Warner, Barbara and Jennifer Shapiro. "Fractured, fragmented federalism: A study in fracking regulatory policy," *Publius* 43(Summer 2013): 474–96.

Wiggins, Steven N. and Gary D. Libecap. "Oil field unitization: Contractual failure in the presence of imperfect information," *American Economic Review* 75(June 1985): 365–85.

World Energy Council. *World Energy Resources 2016*.

# 9    Climate change

The little town of Verkhoyansk has the reputation of being one of the coldest towns on earth. The town of 1,300 residents is located in Russian Siberia, lying north of the Arctic Circle. In the winter, locals keep their cars running all day, afraid that if they turn them off, they may not start up again until spring ["Growing up in -60C"]. This is why it was so striking when, on June 20, 2020, the local temperature topped 100 degrees Fahrenheit (37.8 Celsius). This was not only the highest temperature ever recorded in the town since records began to be kept in 1885 [Borunda (2020)]. It may have been the hottest temperature ever recorded north of the Arctic Circle [Aleem (2020)]. But, to climate scientists, this event is merely one of the latest pieces of evidence, in a growing mountain of evidence, that the earth is getting warmer.

Ongoing climate change, as seen in rising global temperatures, is one of the great environmental challenges of our time. It is difficult to think of an environmental problem that is more important while presenting such intractable policy challenges. A warming earth means generally hotter temperatures, warmer oceans, and more frequent extreme climate events like typhoons, hurricanes, tropical cyclones, heatwaves, flooding, wildfires, and drought. Furthermore, extreme climate events may be becoming more intense, as evidenced by recent extended heatwaves like in Yerkhoyansk; severe droughts in East Africa, South Africa, and Australia; the California wildfires of 2018 and 2020; and recent powerful annual hurricanes that originate in the southern Atlantic. Finally, climate change is likely causing the collapse of the Antarctic ice sheet and the melting of glaciers in Greenland, which contribute to rising sea levels, threatening low-lying coastal areas around the globe and posing a threat to the existence of island countries.

One thing that scientists know is that weather, including air and water temperatures, is subject to natural trends and fluctuations. Some years are hotter than others, and the Ice Age was an extended period of relatively cool global temperatures. However, almost all scientists are now convinced that much of our recent weather experience over the past 40 years or so is attributable to human activities and, in particular, the continued burning of massive amounts of fossil fuels. The problem is that we rely heavily on fossil fuels for heating our homes, fueling our cars, producing a wide variety of goods, and, in general, maintaining our standard of living. This all means that humankind is presented with an enormous challenge: How to address the massive, worldwide problem of climate change?

## Scientific causes of climate change: The greenhouse effect

Let us begin by examining the evidence on global temperatures. Figure 9.1 graphs global temperatures over the last 140 years, courtesy of the National Aeronautic and

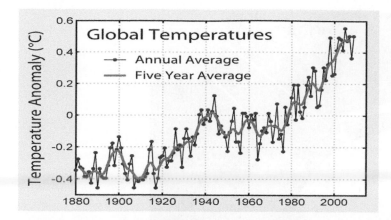

*Figure 9.1* Global temperatures, 1880 to present. *Source*: NASA Goddard Institute for Space Studies (Wikimedia Commons)

Space Administration [NASA (2006)]. This figure suggests that there has been a noticeable uptick in global temperatures over time. The black dots indicate actual global mean temperatures in each year, while the solid line shows the general implied trend, while eliminating ("smoothing") the year-to-year fluctuations.[1] These data reflect two things. First, there is indeed short-term variation in global temperatures from year to year. Second, global temperatures are trending noticeably upwards. Notice in particular that since 1980 or so, global temperatures have risen by about 0.5 degree Celsius and they are on track to rise by nearly 1 degree Celsius by 2020. This is nearly 2 degrees Fahrenheit!

This recent trend coincides with a steady build-up in the atmosphere of certain types of gases that scientists call greenhouse gases (GHGs). These GHGs include carbon dioxide ($CO_2$), methane, nitrous oxides (NOx), and a few other gases. This correlation is no accident. Scientists believe that rising GHGs are responsible for rising global temperatures. The mechanism for this connection, shown in Figure 9.2, is known as the *greenhouse effect*.

To explain: if you have ever been in a greenhouse when the sun is shining, you may have noticed how warm it is. This is because the sunlight enters the greenhouse through the transparent roof, which traps some of the sun's energy, heating the interior of the greenhouse. Similarly, the atmosphere of the earth traps the sun's energy, heating the earth's surface. Without this greenhouse effect, the earth's surface would be frigid, like the surface of the moon.

Here's the thing: the higher the concentration of GHGs in the atmosphere, the greater is the greenhouse effect; that is, the more heat gets trapped in the atmosphere. This is why global temperatures are rising: GHGs are accumulating in the atmosphere. Figure 9.3 shows the time trend for $CO_2$, the most common greenhouse gas.[2] This figure shows that from 1958 to 2020, atmospheric $CO_2$ has increased by nearly 100 parts per million, or over 31%. Where are these GHGs coming from? Though the overall picture is complicated, scientists almost entirely agree that a key cause is the burning of fossil fuels.

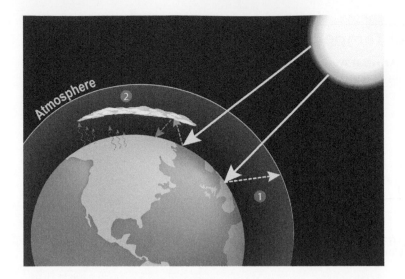

*Figure 9.2* The greenhouse effect. *Source*: Shutterstock

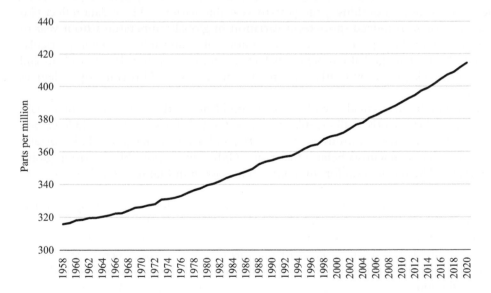

*Figure 9.3* Atmospheric $CO_2$, 1958 to 2020. *Source*: Keeling et al. (2001) Scripps CO2 Program.

## Environmental impacts of climate change

Rising global temperatures are unleashing a highly complex set of changes in the global environment. As temperatures rise, heatwaves become more common and intense in certain regions, which can result in heat-related stress to land, forests, bodies of water, and ecosystems, not to mention humans. Higher temperatures can also alter rainfall patterns, increasing the risk of drought and wildfires in certain regions. In recent days, this includes widespread wildfires in California in 2018 and 2020 and bushfires in Australia in 2019–20.

The 2020 California wildfires took place in a record-setting heatwave. On August 16, the temperature in Death Valley reportedly hit 130 degrees Fahrenheit (54.4 degrees Celsius), which, if confirmed, would be the hottest temperature ever reliably recorded on earth. Similarly, the 2019–2020 bushfires in Australia occurred in the midst of perhaps the hottest summer in the recorded history of the country, with national average temperatures reaching a record high of 107.4 degrees Fahrenheit (41.9 degrees Celsius) on December 18, 2019. These fires burned an area of roughly 72,000 square miles, an area over twice the size of Portugal. The fires caused the deaths of at least 34 people and over *one billion* animals in Australia, and smoke from the fires drifted as far away as South America, where it caused the skies in Santiago, Chile to turn grey ["Australian bushfire smoke" (2020)]. According to the *Global Carbon Project*, hundreds of millions of tons of CO2 were spewed into the atmosphere.[3]

Rising temperatures are also causing extreme weather events to occur more frequently and with greater intensity. This includes not just heatwaves and drought but also tornadoes, heavy precipitation, flooding, and tropical cyclones such as hurricanes and typhoons. Hurricanes, the most costly individual natural weather disasters, are likely becoming more severe as ocean temperatures rise. One recent study, for example, estimates that from 1986 to 2015, hurricane wind speeds in the Atlantic Ocean increased by about 4.4 miles per hour per decade [Balaguru et al. (2018)]. In addition, warmer hurricanes will tend to cause more rain because warmer air can hold more moisture. This helps explain what happened in 2017, when Hurricane Harvey dumped 60 inches of rain in southeast Texas, an amount that researchers say should occur only once every 9,000 years [Berardelli (2019)].

Rising temperatures are also likely to have major effects on the world's oceans. Rising air temperatures cause ocean temperatures to rise, and greater concentrations of CO2 in the atmosphere make ocean waters more acidic, because the oceans absorb CO2. This is a deadly combination for tropical coral reefs, which may be completely destroyed by the year 2100 [Young (2020)]. In addition, warmer, more acidic oceans can harm fish populations, both by stressing the fish populations themselves and by reducing their food supplies. A recent well-regarded study that appeared in 2019 in the journal *Science* found that warming oceans have already resulted in significant declines in fish populations over the past 90 years[4] [Free et al. (2019)]. With ongoing climate change, this trend is likely to continue in the future.

Another important impact of climate change on oceans is rising sea levels. The main reason is that rising temperatures are causing the two largest land ice sheets in the world – one in Antarctica and the other in Greenland – to shrink and deposit water into the ocean. To see why this matters, consider this: the continent of Antarctica is covered in a thick layer of ice. On average, this ice layer is over a mile-and-a-half deep. At its thickest point, it is almost three miles deep [Cockburn (2020)]. That is almost five kilometers. Recall the last time you ran a 5K road race and how long it took you to complete it. Ignoring gravity, that is how long it would take you to run from the bottom to the top of that ice sheet at that point! This massive ice sheet covers virtually the entire continent of Antarctica and may contain over 60% of all of the earth's fresh water. Rising temperatures are causing parts of the Antarctic ice sheet to break off into the ocean, and, because there is so much ice, this is having a measurable effect on global sea levels.

Rising temperatures are also likely causing the Greenland ice sheet to melt at unprecedented rates. The Greenland ice sheet is the second largest mass of ice in the world after

Antarctica, and it is currently melting at its fastest rate in over 400 years. Between 1992 and 2018, it is estimated to have lost 3.8 trillion (*that's trillions with a t!*) tons of ice. The largest contributor to this melting is likely an increase in summer temperatures of 1.2 degrees Celsius over the past 150 years. At present rates, it will contribute to raising global sea levels by anywhere from three to five inches.[5]

## Economics of climate change

You will not be surprised to hear that much of the economics of climate change is about the weighing of costs and benefits, with a few wrinkles that I will discuss shortly. The costs of climate change have largely to do with the physical impacts of climate change, some of which we have just seen, and figuring out how these translate into various kinds of costs, like the cost of sea level rise or the cost of reduced crop production.

But what do I mean by the *benefits* of climate change? Here, I am actually not referring to a few ways in which climate change itself may actually benefit (some of) us, such as lengthening growing seasons or reducing fuel costs at high latitudes. Rather, I am refer-ring to the fact that doing something about climate change – like reducing fossil fuel use or building sea walls – will require resources. And so, in a real sense, we reap benefits from *not* having to do these things. So this argument is similar to the benefits enjoyed by the rancher in chapter four when he did not have to expend money to pen up his cattle.

In any case, thinking about costs and benefits in this way, it may be apparent to you that the bigger challenge is figuring out the costs of climate change. The reason it is such a challenge to calculate the cost of climate change, of course, is that the world is a huge, incredibly complex place. The effect of ongoing emissions and growing concentrations in the atmosphere can have all sorts of interrelated effects on rainfall patterns, humidity, the world's oceans, ecosystems, human health, and so forth that scientists are still trying to understand. Furthermore, the effects of climate occur over long periods of time, requiring the evaluation of costs for decades, perhaps centuries, into the future. And the further into the future we have to project, the less certain we can be of what the impacts will be, let alone the costs associated with those impacts.

In order to assess the costs of GHG emissions, the standard practice is to use *integrated assessment models* (IAMs), which combine climate models and models of the world econ-omy. These IAMs permit calculation of economic costs by feeding economic scenarios into the climate models and then calculating the costs using so-called *damage functions*, which translate climate impacts into economic damages. These scenarios can generate dif-ferent amounts of GHG emissions that correspond to different amounts of global warm-ing. Economists use these models to estimate the magnitude of economic costs associated with rising global temperatures. These costs are typically expressed as percentages of world gross domestic product (GDP), to more clearly convey their magnitude.

One of the most well-known of these models is the Dynamic Integrated Climate-Economy (DICE) model, which was created by the Yale economist William Nordhaus. Nordhaus won the Nobel Prize in economics for his research on the environment, espe-cially the problem of climate change. Using the DICE model, Nordhaus has estimated the cost impact for a range of increases in global temperature. For example, if the global mean temperature increases by 3 degrees Celsius (5.4 degrees Fahrenheit), Nordhaus estimates this will mean additional costs equivalent to roughly 2.1% of world GDP. With global GDP at $142 trillion in 2019, this estimated cost is nearly $3 trillion [Nordhaus (2017), p. 1519; Statista (2020)].

I should emphasize that many economists have undertaken similar exercises using IAMs and calculated their own cost estimates. Overall, the estimates vary widely, with some estimates being much higher than those of Nordhaus. One of the higher estimates is found in the famous *Stern Review*, a 2006 study by a team of specialists headed by the British economist Nicholas Stern, which was commissioned by the government of the United Kingdom. In sharp contrast to the estimates of Nordhaus, Stern estimated costs of at least 5% of world GDP, going possibly as high as 20% or more [Stern (2006), pp. 186–7]. The Stern findings have generated a sharp divide among economists [Cole (2008), pp. 54–5]. Just about every economist agrees, however, that at the projected rises in global temperature that seem increasingly likely with each passing day, climate change is likely to impose large costs both on the environment and on the human race.

There are a number of reasons for vastly divergent estimates of the costs of rising temperatures. One thing to keep in mind is that there is much uncertainty regarding the connection between temperature increases and environmental impacts, and between environmental impacts and economic damages. And, again, keep in mind that these studies are trying to estimate costs over a very long time frame, which just adds to the uncertainty. Studies that also estimate measures of variation like confidence intervals or standard deviations of their estimates often come up with a quite wide range of values.[6]

Another important reason why you observe different estimates gets at the heart of the economic approach, which, as we have seen, is all about people responding to incentives. You see, some studies assume that people adapt to climate change, and others do not. In the latter category are studies that take scientific evidence regarding the physical impacts of climate change, assign each one a price, and then add them up.[7] So, for example, scientific papers may contain predictions for the effect of climate change on crop yields, and then changes in production are evaluated at market prices. So, very roughly speaking, if increasing temperatures were to decrease production of wheat by one bushel per acre and the price of wheat was $20 per bushel, the cost of climate change would be $20 per acre planted in wheat.

For many economists, this approach is problematic because it ignores the fact that farmers may have other options besides continuing to plant wheat as temperatures rise. For example, they could switch to planting corn or other crops whose yield is less susceptible to rising temperatures. Or they could get out of raising crops entirely. And, to the extent that they do these things, they may be able to reduce the cost they incur from rising temperatures.

Figure 9.4 illustrates the argument. Here, there are three possible uses of a farmer's land. She could either produce wheat or corn, or turn the land into grazing pasture. Each of these hills represents the amount of value the farmer could receive, as a function of the temperature. So, for example, the optimal temperature for growing wheat is $T_0$, because that is where the value of growing wheat is maximized. A farmer facing that temperature would plant her fields in wheat. As the temperature rises past $T_0$, wheat yields start to go down. But, as long as the temperature is less than $T_1$, it still makes sense to grow wheat. However, past $T_1$, the value of growing corn exceeds that of growing wheat, so the farmer would switch to growing corn. And, then, when the temperature rises above $T_3$, the farmer would switch to grazing pasture. So this analysis embodies the economic assumption that people respond to incentives in order to make themselves as well-off as possible.

Notice what ignoring this adaptive behavior does to our estimates of the cost of rising temperatures. Suppose the temperature were to increase from $T_0$ to $T_2$. Ignoring the possibility that the farmer will switch to corn will yield a loss estimate of $(a - c)$. Allowing for

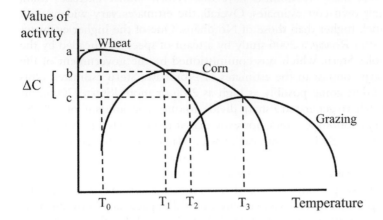

*Figure 9.4* Adaptive behavior [adapted from Mendelsohn et al. (1994)]

the possibility that the farmer will switch to corn will yield a loss estimate of only (a − b). The difference, denoted $\Delta C$, represents how much you overestimate the cost of rising temperature when you ignore the possibility of switching crops. Evidence suggests that estimated costs are indeed lower, perhaps significantly lower, when you allow for adaptation.[8]

I should mention that the economic approach to valuing climate change costs has been evolving over time, for a number of reasons. First, the approach that assumes that people adapt to climate change is being increasingly adopted, driven by evidence that adaptation, in things like crop choice and forest management practices, matters. Second, advances in our understanding of (a) physical systems such as the carbon cycle and (b) the climate-economy connection are allowing us to refine our specifications of the damage functions used in IAM models. Our improved understanding of the connection between climate and the economy has been driven in part by advances in methodologies used by economists to quantify the effect of climate change on economic variables [Dell et al. (2014)]. Finally, economists have begun to investigate and quantify the effects of climate change on a broader range of environmental outcomes [Krupnick and McLaughlin (2012); Diaz and Keller (2016); Anthoff et al. (2016)]. The overall result of these changes is that estimates of costs were trending downward in the 2000s but, more recently, these cost estimates have been increasing.[9]

But perhaps the most important source of differences in cost estimates has to do with the choice of discount rate. Differences in choice of discount rate account for a huge portion of the difference between Nordhaus's and Stern's estimates. Nordhaus uses discount rates that are quite a bit higher than those used by Stern [Stern (2008), pp. 12–17]. You can see that this might make a huge difference in their cost estimates, given the extremely long time frame for evaluating the impacts of climate change. Because this means that Stern is implicitly weighing a long stream of future costs of climate change much more heavily than Nordhaus.

The question of what discount rate to use in assessing climate change damages may well be the single most intractable issue in the economics of climate change. It is a gross understatement to say that the discount rate is important in determining the relative magnitude of the costs and benefits of climate change. Because of the extremely long time frame of climate change, choice of discount rate may literally swamp all other cost-benefit considerations [Weitzman (2013), p. 873]. Making matters worse, economists violently disagree on what discount rate to use in these models. Stern and others have argued that ethical considerations call for a small or even zero discount rate. Others, like Nordhaus and Weitzman, argue that the discount rate we use should be moored to market evidence on investment returns and consumption behavior, which implies much higher discount rates than the ones Stern favors. In chapter three, we saw that a compromise position would be to use hyperbolic discount rates, such as Weitzman has favored.

I should also mention that the costs of climate change are unlikely to be borne equally by different countries, raising further important questions of equity. Lower-income countries tend to be more heavily agricultural, an economic sector that is likely to suffer damages from climate change, especially in countries located at lower latitudes nearer the equator. Furthermore, low-lying countries with substantial coastal area such as Bangladesh, and island nations such as the Maldives and the Marshall Islands, are likely to be heavily affected by rises in sea level. In thinking about the costs of climate change, it is important to keep in mind the large divide between low-income and high-income countries.

## What to do?

The various consequences of ongoing and projected climate change have provoked spirited public debate regarding what to do about it. The types of policies being considered fall into three general categories: mitigation, geoengineering, and adaptation. Let us discuss each of these options in turn.

### I. Mitigation

The most commonly discussed policy option is *mitigation*: simply reducing the amount of GHG's emitted into the atmosphere. The idea is simple enough: reducing GHG emissions will, over time, lower ambient GHGs in the atmosphere, alleviating the greenhouse effect. The practical question is more challenging: what sorts of policies will get us there? And how well will they work? Two mitigation policies have garnered quite a bit of interest among economists. These are the so-called *carbon tax*, and a system of *cap-and-trade*.

### A. Carbon taxes

A carbon tax is a tax on the burning of carbon-based fuels, primarily coal, oil, and natural gas – the fossil fuels. Carbon taxes are levied on manufactured products depending upon how much carbon is combusted in their production. The more carbon is burned, the higher the tax.[10] If companies have to pay a tax for every unit of carbon they use in production, they will be discouraged from burning fossil fuels and therefore, from producing GHGs.

#### I. THE BASIC ECONOMICS OF CARBON TAXES

The basic economic argument for carbon taxes can be seen using our externalities framework. Figure 9.5 depicts a competitive market for steel, an industry that uses much fossil fuel energy. In this figure, demand reflects the social marginal benefits ($MB^{Soc}$) of steel production and supply captures the private marginal costs ($MC^{Priv}$) incurred by steel producers

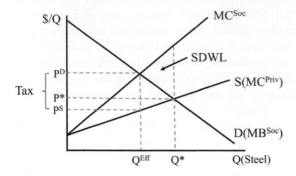

*Figure 9.5* Effect of a carbon tax, market

(see chapter two). Since steel production requires much energy, there are external costs associated with GHG emissions. Adding (marginal) external costs to private marginal costs yields social marginal costs ($MC^{Soc}$). The result is a market equilibrium quantity $Q^\star$ that exceeds the efficient quantity $Q^{Eff}$, which generates social deadweight loss equal to the area of the triangle labeled SDWL.

The idea of a carbon tax is to increase production costs in order to discourage production. So imagine requiring steel firms to pay a tax of so many dollars for every ton of steel they sell. This would have the effect of driving a wedge between the price consumers pay and the price firms receive for a ton of steel. Imposing the tax would then move output from $Q^\star$ toward $Q^{Eff}$. Figure 9.5 shows the efficient tax, which results in production of the efficient quantity. At that point, the price paid by consumers is $P^D$, the price being received by steel firms is $P^S$, and the difference is the magnitude of the tax.

Carbon taxes are very popular among economists. It is important to understand why, because it might have occurred to you that there may be other ways to reduce GHGs. For example, an alternative strategy might be to impose regulations that reduce GHG emissions directly. These could include setting emission limits (standards) for power plants or mandating that manufacturers produce more energy-efficient homes, appliances, trucks, and automobiles. These are examples of what economists call *command and control* strategies. These strategies were, for example, components of President Obama's climate change strategy, as contained in his *Clean Power Plan*.

However, carbon taxes have certain advantages over these types of strategies. Perhaps the most common argument you hear for carbon taxes is that they permit emissions reductions to be achieved at lower cost than command and control policies. To see why, let us continue to consider our steel industry example but, now, let us focus on the fact that there are a number of different steel firms. Each of these firms use energy and each one emits GHGs. The question is: How do you achieve any desired reduction in overall GHG emissions?

To see the argument graphically, let us focus on two of these firms, each of which have access to a particular technology for reducing GHG emissions. Given the technologies they use, each one incurs certain costs of reducing (abating) their respective emissions. In Figure 9.6, these are shown as their *marginal abatement cost* (MAC) schedules, which show the marginal cost of abating emissions as a function of the total amount abated. Notice that I have drawn the MAC schedules at different heights. This reflects the likely possibility

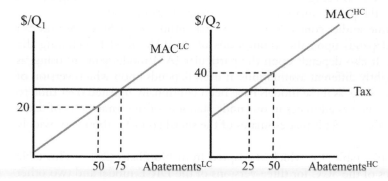

*Figure 9.6* Taxes vs. standards

that some firms may be more efficient at abating emissions than others and therefore, would incur lower costs. In Figure 9.6, I denote the more efficient firm's curve as MAC$^{LC}$ (MAC lost-cost) and the less efficient firm's curve as MAC$^{HC}$ (MAC high-cost).

Assume that initially they are not abating any emissions at all. Now suppose you are a regulator and you want to reduce total emissions by 100 tons. Figure 9.6 shows two ways to accomplish this. One is to mandate that each firm abate 50 tons. The other is to levy the tax at the level shown. This tax would encourage the low-cost firm to abate 75 tons and the high-cost firm to abate 25 tons. It should be obvious that total costs of abatement are lower under the tax. Intuitively, the same tax levied on both firms causes the low-cost firm to abate more, which as a policymaker is exactly what you want. More generally, as long as the marginal cost of abatement differs *on the margin*, you can reduce overall costs by reallocating emissions to whichever firm has the lower (*marginal*) marginal abatement costs. Carbon taxes result in equal marginal abatement costs for different firms on the margin, which serves to minimize overall abatement costs.

## II. CALCULATING CARBON TAXES

OK, now that we have seen the nice properties of a carbon tax in theory, let us consider the more difficult question of exactly how to determine how large it should be. As we have seen, the magnitude of the carbon tax should be based on the damages caused by the GHGs emitted by steel firms and other polluters. So imagine a ton of GHGs, say $CO_2$, being emitted into the atmosphere at a particular point in time. The damages of this one ton of $CO_2$ would consist of the *present discounted value of ALL of the damages inflicted on the world henceforth indefinitely into the future*. Holy smokes. How in the world do we calculate this value?

The answer is that economists use the integrated assessment models (IAMs) we encountered in the previous section, such as Nordhaus's DICE model. As we have seen, these models integrate climate models and economic scenarios and use damage functions in order to calculate the costs of particular changes to climate. A model like DICE will first set a baseline climate/economy scenario that generates a relationship between $CO_2$ emissions and an entire stream of economic costs. It then adds one more ton of $CO_2$ emissions to this baseline scenario and calculates the impact on the stream of costs. The discounted value of this impact on the stream of costs is the (marginal) social cost of carbon (SCC).

You can think of this SCC as being the real–world marginal external cost in Figure 9.5: the difference between the private marginal cost and social marginal cost.

So, what do economic studies conclude about the magnitude of the SCC? As you can probably imagine, it depends upon the assumptions of the IAM model, including the assumed discount rate. It also depends upon the particular IAM model you are using, as each one embodies slightly different assumptions. It even depends upon which version of an IAM model you are using, as these models have been continually updated over time to reflect new knowledge, such as advances in our understanding of the parameters of damage functions. The practical effect is that estimates of the social cost of carbon vary widely across studies.

Consider, for example, Table 9.1, which is adapted from Nordhaus (2017). This table shows various estimates of the SCC for three versions of the DICE model and two other widely used IAM models, PAGE and FUND. For example, the DICE-2010 model yields an estimate of the SCC of $12/ton of $CO_2$, assuming a discount rate of 5%. The first three rows are all from one study that applies the same model assumptions in order to draw meaningful comparisons of the results of the three models [Nordhaus (2017), p. 1521]. The DICE suffix indicates the year the new DICE version was introduced, the most recent being 2016.

Aside from the estimates ranging all over the map, three things are noteworthy here. First, as expected, the estimates of the SCC get larger at lower discount rates. This, of course, reflects the fact that lower discount rates translate into higher total costs of climate change, by weighting the future costs more heavily. Not to belabor the obvious, but one thing that leaps out here is the large quantitative impact of changes in the discount rate on the SCC estimates.

Second, the different IAM models yield noticeably different estimates, with the PAGE estimates being consistently higher, and the FUND estimates being consistently lower, than the earlier DICE estimates. The differences derive in part from different assumptions of the models. For example, one important reason the PAGE estimates are higher is that this model incorporates the possibility of catastrophic events [Cole (2008), pp. 57–8]. One reason the FUND estimates are lower is that the FUND model assumes robust adaptation to rising sea levels.

Finally, notice that the cost estimates are increasing over later versions of the DICE model. These increases reflect various tweaks to the DICE model, based on better knowledge of earth systems, and changes in the parameters of the damage functions and the cost of emissions abatement, based on better information on the relationship among the

*Table 9.1* Estimates of the social cost of carbon, various studies

|  | 5% | 4% | 3% | 2.5% |
|---|---|---|---|---|
|  | | *(2010 $)* | | |
| DICE-2010 | 12 | NA | 40 | 59 |
| PAGE | 23 | NA | 74 | 105 |
| FUND | 3 | NA | 22 | 37 |
| DICE-2013 | 15 | 26 | 50 | 74 |
| DICE-2016 | 23 | 41 | 87 | 140 |

variables. But the most important factor is economics: the newer version is based on larger measured output and methodological changes in the valuation of goods across different countries [Nordhaus (2017), p. 1522].

### B. Cap-and-trade

At this point, let us return to Coase's argument that we encountered in chapter four. According to Coase, it makes no sense to think of steel firms as imposing costs on everyone else. As we have seen, Coase would argue that steel firms and other parties mutually impose costs on each other. So he would not necessarily support a carbon tax, without knowing more about the relative magnitudes of the costs: to the steel firms on the one hand, and to the rest of the world on the other. As Coase famously put it: "the problem is to avoid the more serious harm" [Coase (1960)].

Now, I think it is arguable that the costs to the rest of the world are greater than the costs to the steel firms (and other generators of GHGs). If so, this might argue for imposing the tax, which awards a right, to the rest of the world, to be free of damages. However, this raises another issue. As we have just seen, there is huge uncertainty regarding the correct carbon tax to impose. So if it is quite easy to get it wrong, as seems likely, it is doubtful that imposing a tax will allow us to achieve anything like an efficient outcome. Under these conditions, Coase might argue for a property rights solution, which would entail placing entitlements into people's hands and letting them work things out among themselves.

With all this in mind, let us now consider a second policy that is commonly proposed for mitigation of GHGs: the so-called *cap-and-trade* system. Under cap-and-trade, the government decides how much total emissions it wants to allow. It then issues legal rights to emit GHGs, which are called *allowances*. So, for example, possession of an allowance entitles a firm to emit a ton of $CO_2$. By allowing firms to trade in allowances, the government is in effect creating a market for the right to emit GHGs.

To be clear, in a cap-and-trade system, individual firms holding allowances are not trading with everyone else in the world who are experiencing the effects of the GHG emissions. That would be a tall order, by which I mean there would be enormous transaction costs. Rather, cap-and-trade systems identify specific firms who are eligible to trade with each other, like electric utilities in a particular geographic region. So, in effect, cap-and-trade systems set up regional markets, allocate allowances somehow, and then stand back and allow firms to trade among themselves.

An important feature of cap-and-trade is that, in principle, it allows us to achieve cost efficient reductions in GHG emissions. Specifically, high-cost abaters will be able to buy allowances from low-cost abaters because they would be willing to pay more for allowances than they are worth to the low-cost abaters. This has the practical effect of placing abatements in the hands of low-cost abaters.

For example, in Figure 9.7, firm 2 would be willing to pay up to the amount b for an allowance, because this would allow it to avoid paying that amount in extra abatement costs. By the same token, firm 1 would be willing to sell an allowance for the amount a. Notice that when this allowance changes hands, the marginal cost to firm 2 of abating the last unit goes down, while the marginal cost to firm 1 goes up. In principle, such trades would occur until enough allowances have changed hands to equate the MC of abatement for the two firms on the margin. As with carbon taxes, total abatement costs would be minimized.

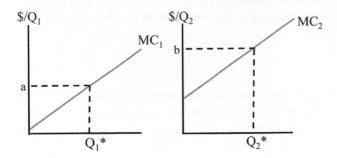

*Figure 9.7* Incentives to trade allowances under cap-and-trade

## C. Carbon taxes vs. cap-and-trade

Both carbon taxes and cap-and-trade enjoy cost advantages over the traditional policy of setting emissions standards (that is, without the possibility of trade). So, aside from the issue of where to set the tax, does this mean that we should consider carbon taxes and cap-and-trade identical as policies for achieving GHG mitigation? The answer is not quite, because there are some subtle but important differences having to do with their distributional consequences and the likelihood of actually achieving the emissions targets.

First, you may have noticed that carbon taxes set a price for GHG emissions, whereas cap-and-trade sets a quantity; namely, whatever total number of allowances the government decides to issue. Under a system of carbon taxes, emitters respond to the tax level in deciding how much to emit. Under cap-and-trade, the amount of emissions is set at the outset. This means that cap-and-trade affords more certainty regarding the quantity of abatements that will be achieved. If the government sets the carbon tax at the "wrong" level, it may end up with more or less emissions than expected.

Second, by design, cap-and-trade permits the government to target specific firms to receive the benefits of the program. In particular, whoever receives the initial allocation of allowances can benefit either by avoiding abatement costs or selling them to other emitters. Under a system of carbon taxes, all emitters share the tax burden, in proportion to how much GHGs they emit.

Third, carbon taxes can provide significant tax revenues to the government, depending on the size of the tax and the total emissions of GHGs. A cap-and-trade system can provide revenues as well, but how much depends on how the government decides to allocate the allowances initially. If it chooses to auction them off, it may be able to garner significant auction revenues. However, another possibility is to simply issue allowances for free, in which case there will obviously be no revenues. To give you an idea of how real-world cap-and-trade systems handle this issue, consider the state of California. California has had a cap-and-trade system in place since 2013, and it relies heavily on auctioning allowances while also issuing some free to utilities, suppliers of natural gas, and industries vulnerable to competition from companies in other states [Cart (2018)].

Fourth, given that cap-and-trade is a market-like system based on free trading of allowances, there may be transaction costs associated with its operation. For example, if you hold an allowance, there may be costs involved in identifying interested potential buyers. If these costs are large enough, allowance brokers may be able to step in to bring buyers and sellers together. But even if so, there may be brokers' fees to pay. There is no consensus on how large these transaction costs are, but they are unlikely to be zero, which may make it difficult to achieve the efficient allocation of allowances.

Finally, the two systems may be different in terms of sheer political feasibility. Some recent events suggest that carbon taxes are hugely unpalatable politically. In 2018, the state of Washington tried to adopt a statewide carbon tax by putting it to a vote in a popular referendum. It lost by a landslide. In that same year, the government of France tried to impose an energy tax on gasoline and diesel fuel. The result was rioting in the streets, and the French government had to back off. These events have led some knowledgeable observers to conclude that the carbon tax is simply politically infeasible, at least under current political conditions.[11]

On the other hand, the state of California managed to launch an ambitious cap-and-trade program in 2013 and as of late 2018, several other states, including Connecticut, Delaware, Maryland, Oregon, and Pennsylvania are eyeing them cautiously [Plaven (2018); McKenna (2018)]. The European Union has had one in place since 2005, when it set up the first international emissions trading program. China is currently in the process of launching a countrywide cap-and-trade program that, if it lives up to expectations, will be nearly twice as large [Temple (2018); Schmalensee and Stavins (2019)]. And as we shall see shortly, cap-and-trade systems have been built into the structure of international climate change agreements, suggesting that in principle, they may have broad political acceptability. Overall, market mechanisms may well be an easier political sell than taxes.[12]

## II. Geoengineering

A second policy option is to reverse the effects of rising GHG levels by using scientific/engineering solutions, sometimes referred to as *geoengineering* [Nordhaus (2013)]. There are two broad categories of geoengineering solutions: *solar radiation management* and *negative emissions technologies*.[13]

### A. Solar radiation management

As we have seen, the greenhouse effect occurs because, as GHG levels rise, more of the sun's solar radiation is trapped in the earth's atmosphere. The idea behind solar radiation management is simple: either reflect more solar radiation back out into space or keep it from entering in the first place.

There are various ways to reflect more solar radiation away from the earth, including spraying seawater into the atmosphere (to increase the reflectivity of clouds), reforestation of tropical areas, covering desert areas with plastic sheets (really!), and even painting rooftops white. But the idea that has probably received the most serious attention by scientists and policymakers is to inject aerosols containing sulfates into the atmosphere. By essentially making parts of the atmosphere more opaque, less sunlight would penetrate into the lower atmosphere.

The 1991 eruption of Mount Pinatubo in the Philippines provides a good natural illustration of how this could work and how effective it might be. The eruption spewed a cubic mile of material into the atmosphere, creating an ash cloud some 22 miles high and over 100 miles across. The result was a measurably cooler atmosphere, with average global temperatures falling by about 0.5 degrees Celsius for the next year or two [USGS (1997)].

Studies suggest that sulfate aerosol injection could be both effective and relatively inexpensive, costing anywhere from one-tenth to one-hundredth as much as mitigation for an equivalent amount of global cooling [Nordhaus (2013), p. 153]. However, there are two main concerns. First, there could be unpredictable and potentially negative side effects on global climate. Climate models suggest that precipitation patterns could be affected in potentially unpredictable ways, possibly increasing the frequency of droughts. In addition, there is a risk of possible ozone depletion. Finally, aerosol injection would not address the problem of ongoing ocean acidification because it would not reduce existing levels of ambient $CO_2$ in the atmosphere [Blackstock and Long (2010)].

The second main concern with sulfate aerosol injection is the need for coordination among countries who may not always agree on how, when, and to what extent it should be used. There may be winners and losers from climate modification, which could heighten tensions among countries. Thus, international agreements may need to specify policies and mechanisms for losers to be compensated. These may be difficult to agree upon. There is even the possibility that aerosol injection might be intentionally abused. It places a considerable amount of power in the hands of the countries or other entities that would be administering the injection program. This power could be used to gain political or even military advantage.

## B. Negative emissions technologies

The second type of geoengineering solution is what are known as negative emissions technologies (NETs), which involve physical removal of $CO_2$ from the atmosphere and storing it. The rationale should be obvious: if there is too much $CO_2$ in the atmosphere, let us take some of it out.

### I. FOREST MANAGEMENT

As with solar radiation management, there are various ways this could be done. Some involve natural processes, such as planting more trees and changing farming practices to increase carbon storage in soils. It is well known that trees absorb $CO_2$ in the natural process of photosynthesis. So planting trees, or keeping them from getting cut down, are ways to reduce $CO_2$ in the atmosphere.[14] Studies have shown that either strategy is probably cost-competitive with other ways of mitigating carbon, such as switching to alternative fuels or energy conservation [Richardson and MaCauley (2012), p. 5]. A possible downside is that using large amounts of land in these ways could affect food production and possibly harm biodiversity. Furthermore, it is unclear whether these natural methods could remove large enough amounts of $CO_2$ at a reasonable cost [NASEM (2018)].

### II. DIRECT AIR CAPTURE

From an engineering standpoint, the NET process you probably hear about most often is *direct air capture* (DAC). DAC involves using chemical processes to remove $CO_2$ from air, concentrate it, and inject it into a storage medium, such as an underground reservoir or a saline groundwater aquifer. While scientists have been studying DAC for some time, it has generally been considered to present onerous technical challenges and to simply be too expensive to really help. At present, it is probably still not cost-competitive but experts believe it could be in the foreseeable future, and some believe it could make a significant impact. A recent scientific panel commissioned by the National Academies of Sciences, Engineering, and Medicine (NASEM) believes DAC has sufficient potential to merit a serious multi-billion dollar research effort and should be considered a "component of the mitigation portfolio" [NASEM (2018); Kolbert (2018)].

### III. Adaptation

The final option is to adapt to changing climate conditions. As we have seen, this is already occurring to some extent. Future adaptation measures would include measures like: erecting sea barriers; retiring farmlands; building larger and more extensive irrigation systems; changing forest management practices; moving populations away from coastal areas; installing more extensive air conditioning systems; and, in general, doing our best to adjust to changing environmental and economic conditions. Basically, the strategy here is to do nothing to mitigate GHG emissions and simply live with the consequences of ongoing climate change.

Adapting to changing climate conditions means, of course, accepting all of the physical consequences of climate change, including flooded coastal areas, more powerful storm systems, greater incidence of wildfires and drought, ocean acidification, and potential damages to ecosystems and biodiversity. We may consider these costs to be too large to bear. And we may be concerned about equity as well: the effect on heavily agricultural low-income countries and future generations. The economic approach is to weigh, as best we can, all these costs of ongoing climate change against the benefits that we derive from everything associated with burning fossil fuels, including not having to install potentially expensive abatement technologies.

I hope it is obvious that we should not consider adaptation to be something we do instead of mitigation or geoengineering. Rather, adapting to changing climate conditions is likely to be something we do even at the same time as we think about pursuing other strategies. Indeed, humans are already engaging in adaptive behavior, to the extent we are relocating our residences away from coastal areas and building seawalls against rising sea levels. More generally, you can think about all of the strategies we have been discussing over the last several pages as components of a climate change portfolio of policies. One big question moving forward is: Which, and how much of each, of these policies should we pursue?

## The challenges of international cooperation

There is, however, a larger context in which we should consider strategies for dealing with climate change. The unfortunate reality is that no individual country, even the very largest carbon emitters like the United States and China, is in a position to address climate change effectively on its own. Climate change is occurring because of the uncoordinated decisions of millions of emitters in countries all over the globe. Let's say we wanted mitigation of GHGs to be the main component of our climate change strategy. In order for mitigation to make a real difference, we need a coordinated effort by all countries to reduce emissions. For numerous compelling reasons, this is likely to be extremely difficult to pull off.

### I. A brief history of international climate agreements

Some evidence for the challenges of international cooperation to deal with climate change can be gleaned by examining the history of international climate agreements over the past 30 years.

In the early 1990s, in response to growing concerns about rising GHG levels in the atmosphere, the United Nations convened an international meeting of countries all over the world. The result was the passage of a treaty for addressing climate change, the *United Nations Framework Convention on Climate Change* (UNFCCC), in 1992, and its official enactment in 1994. The UNFCCC called for stabilizing "greenhouse gas concentrations in the atmosphere at a level that would prevent dangerous anthropogenic interference with the climate system."

Under the provisions of the UNFCCC, parties to the Convention meet annually, in so-called *Conferences of the Parties* (COPs), in order to gauge progress in meeting climate change goals. The UNFCCC called for all signatory countries to take steps to address climate change, specifically mentioning mitigation and adaptation [Article 4]. In addition, it specified that developed countries should "take the lead," recognizing the central importance of large emitters in contributing to a solution. As of late 2015, 197 signatory countries were party to the UNFCCC.

The provisions of the UNFCCC have given rise to occasional international agreements where countries pledge to take specific actions to address climate change. Let me briefly discuss two of the more famous agreements.

## A. The Kyoto Protocol (1997)

The first agreement occurred in 1997, when countries signatory to the UNFCCC negotiated the *Kyoto Protocol*, a step toward attempting to meet the climate change goals of the UNFCCC. The Kyoto Protocol called for high-income (so-called Annex 1) signatory countries to set specific limits on GHG emissions, with the goal of reducing overall emissions by 5.2% below 1990 levels (Article 3). It encouraged countries to achieve reductions in various ways, including increased energy efficiency, sustainable forest and agricultural management practices, and research into renewable energy and carbon sequestration technologies.

It also called for a market-based approach to achieving emissions reductions. For example, it called for phasing out fiscal incentives, taxes, and subsidies that promote GHG emitting sectors. It also called for the use of market instruments to achieve reductions. This included the establishment, in a subsequent COP, of principles, rules, and guidelines for emissions trading (Article 17), to support a cap-and-trade system. The Kyoto Protocol went into effect in 2005 after being ratified by 55 countries.

Ultimately, the Kyoto Protocol has apparently been largely unsuccessful in attaining its emissions reduction goals. Figure 9.8 shows total worldwide $CO_2$ emissions from 1990 through 2018. There is absolutely no evidence in these data that the Kyoto Protocol had any effect in slowing down $CO_2$ emissions. Whereas the Kyoto Protocol called for a 5.2% decrease in emissions below the 1990 level, Figure 9.8 shows that, currently, total emissions are way in excess of that level.

## B. The Paris Agreement (2015)

In response to the failure of the Kyoto Protocol, as well as increasing urgency regarding climate change, the twenty-first COP convened in 2015 and enacted the *Paris Agreement*. Under the Paris Agreement, the countries set more ambitious targets for emissions reductions and instituted a series of measures to support efforts by countries to reduce emissions. Emissions reductions were to be pursued that would limit global temperature increases to no more than 1.5 degrees Celsius. Each country was charged with setting an emissions

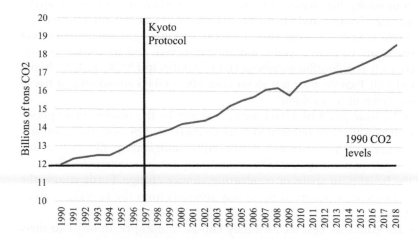

*Figure 9.8* Worldwide CO2 emissions over time, 1990 to 2018. *Source*: International Energy Agency. *Global energy and CO2 status report 2019.*

reduction target for itself, its so-called "nationally determined contribution" (NDC), which was to become increasingly stringent over time. In addition, the Agreement continued to support cap–and–trade arrangements by allowing countries to bank emissions reductions and trade with each other [Article 6].

In addition to mitigation, the Agreement went beyond the Kyoto Protocol in how much it stressed and supported adaptation to climate change. It mandated that all countries should engage in adaptation planning: to enhance adaptive capacity and reduce vulnerability to climate change. Furthermore, it specified that low-income countries should be especially targeted for support in their adaptation efforts [Article 7]. Finally, it called for the development of ways to help particularly vulnerable countries, such as low-lying ones, to deal with the effects of climate change.

The Paris Agreement represents a more flexible approach to addressing climate change than the Kyoto Protocol. The Kyoto Protocol drew a sharp distinction between high-income countries and low-income countries, in terms of what they were expected to do to address climate change. The Paris Agreement, on the other hand, is specially designed to take into account the capacities of low-income countries to address climate change without threatening their economic development.

The Paris Agreement is, of course, still fairly recent so we have little evidence so far on how effective it will be in addressing climate change. Much depends upon the measures designed by subsequent COPs to implement its provisions. However, even in best-case scenarios it faces major challenges. The decision by the Trump administration in 2017 to withdraw the United States from the Agreement will almost surely make it more challenging to meet climate goals. And that factor aside, as of late 2018 there are simply insufficient commitments for emissions reductions to even come close to the Paris target of 1.5 to 2 degrees Celsius.

## II. The free-rider problem and transaction costs

Let us now don our economist hats to understand exactly why international cooperation on controlling GHG emissions is so challenging.[15] I should mention that, over the years, the international community has managed to hammer out many multinational agreements to protect natural resources and environmental quality. Over the last century, our best guess is that there have been more than 140 international agreements to protect globally valued ecosystems, to protect plant and animal species, and to control certain damaging emissions, such as chlorofluorocarbons (CFCs) [Libecap (2014), p. 425]. Many of these have worked well. Exactly what is it about controlling GHG emissions that has so far defeated the best efforts of humankind?

Let us start with the basic problem: GHG mitigation may be the mother of all public goods. Recall from chapter five that public goods are characterized by non-exclusion and non-rivalry in consumption. This seems to describe GHG mitigation perfectly. If country A engages in mitigation, it is impossible to exclude any other country, like Country B, from enjoying the benefits, in terms of combatting climate change. Furthermore, the amount of benefit enjoyed by Country B in no way subtracts from the benefits enjoyed by Countries, C, D, and E.

The problem is that countries bear a cost if they choose to mitigate. In order to meet their mitigation goals under an international agreement, maybe they have to cut back on their fossil fuel consumption. So I can see each country reasoning to itself: *Why should I engage in costly mitigation just to benefit everyone else when I will get virtually the same benefits even if I do not?* In other words, every country has incentive to engage in *free-riding* on the efforts of other countries and not engage in mitigation efforts itself.

But it is not simply a matter of looking out for number one. Climate change is not any old run-of-the-mill environmental issue. It is a serious, in some cases existential, problem for every country on earth. Wouldn't you think that highly motivated countries engaging in repeated face-to-face negotiation in annual Conferences of the Parties would be able to come up with some effective solution? After all, the international community managed to protect the stratospheric ozone layer in 1987 when it agreed to phase out worldwide production of chlorofluorocarbons (CFCs).[16]

I think the answer is a perfect storm of factors that produce enormous transaction costs in negotiations over mitigating GHGs. First, despite the best efforts of climate scientists and economists, there remains much scientific uncertainty regarding the economic and environmental impact of rising GHG emissions. General circulation models can predict worldwide patterns in the relationship between emissions and climate change with reasonable accuracy, but they perform much more poorly in predicting changes at the regional level. Furthermore, long-term estimates of damages are highly uncertain because of the limitations of climate models and because future GHG emissions are related to future economic conditions, which are difficult to predict. All of this makes it difficult for different countries to assess the magnitude of their stakes in the outcomes of different policies, which increases the transaction costs of negotiation.

Second, various factors make the internal politics of mitigating GHGs contentious within each country. Studies have shown that support for action on climate change is affected by the following factors: educational level, political party, gender, age, and income level. To put it starkly: in the United States, if you are a younger, female, well-educated, higher-income Democrat, you are much more likely to favor climate action than if you are an older, male, less well-educated, lower-income Republican. Large demographic disparities among voters along these dimensions will mean sizable voter heterogeneity. This is likely to increase the political transaction costs for a country to arrive at a position regarding GHG mitigation.

Exacerbating this political issue is the fact that, for many people, climate change is a pollution problem that is largely invisible. When people get sick from lead in their drinking water, or when an oil spill befouls ocean beaches, it is clear what is going on and the adverse impacts are easy to observe. GHG emissions into the atmosphere are generally unseen, and the connection to climate-related environmental damages is often not obvious. Even when droughts occur more frequently or hurricanes increase in intensity, it is not obvious to many people that climate change is the culprit. These factors raise transaction costs within countries, both by dampening demand for policy action and by generating stronger differences of opinion over what position the country should take.

Third, countries sharply differ from each other in their preferences for what policies to implement. A key issue is a major divide between high-income and low-income countries regarding who should bear the mitigation costs. Low-income countries argue that high-income countries should bear greater responsibility for mitigation because they have historically contributed most to causing the problem of climate change in the first place. The transaction costs argument here is not merely different parties haggling over who should receive a larger piece of the pie. Concerns over fairness may hold up reaching a negotiating solution and thus be viewed as transaction costs [Husted and Folger (2004)].

Finally, there is the issue of enforcement. No international agreement will meet its climate objectives if participant countries do not live up to their obligations under the agreement. The problem is that parties to an agreement may have strong incentive to defect from the terms of the agreement. Whether they in fact will depends upon two factors: *ease of monitoring*, and *sanctions*.

## A. Monitoring

Monitoring refers to the ability of the international community to observe the actions taken by individual countries to actually reduce GHG emissions. If the actions of countries cannot be observed, of course, there is no way to tell whether they are living up to their obligations under the agreement. Technologies exist to detect emissions levels, and these are in use. However, it is challenging to come up with accurate figures when there are so many sources of GHG emissions. Furthermore, it is difficult to know whether a country is complying because there are other factors, such as business conditions, that affect emission levels. For example, if a country goes into an economic downturn, it may appear that it is in compliance, when its reduced emissions are simply due to declining production. And its lack of compliance may only be revealed when economic conditions improve.

Another reason monitoring may be ineffective is the issue of *leakage*. Leakage refers to situations where parties may try to circumvent the limits on their emissions under the terms of the agreement. For example, if the agreement mandates that less stringent reductions in emissions for low-income countries, firms may relocate some of their operations from high-income countries. This practice is difficult to monitor and results in increases in emissions levels.

## B. Sanctions

To the extent that emissions levels can be monitored, enforcement of the agreement requires that it be possible to levy sanctions on countries that do not live up to their obligations. There are at least a couple of ways in which sanctions could be imposed on countries that are not in compliance: *carbon tariffs*, and other sorts of trade sanctions.

Carbon tariffs are tariffs on imports or exports based on their carbon content. So they are basically the same thing as carbon taxes except that they are imposed on goods that are traded internationally. As part of the enforcement mechanism of an agreement, countries could agree that higher tariffs be imposed on internationally traded goods with higher carbon content. To the extent this can be effectively enforced, it increases the prices of goods containing more carbon, making them less attractive in world markets.

Enforcement is, however, a key issue. Importantly, it is no trivial matter to determine the exact carbon content of every single good that is internationally traded. For example, it would not work to base carbon estimates on the production techniques used in any one country, say, the United States. This is because the cost of factor inputs varies widely around the world, encouraging the use of varying production techniques, depending upon which inputs are less expensive. Practically speaking, this means that the carbon content of a good might vary greatly depending on which country produced it [see Frankel (2010)]. And tailoring carbon content estimates to country-specific conditions would likely be quite costly.

In principle, other sorts of trade sanctions could be used to penalize traded goods containing greater carbon. For example, rather than a general system of carbon tariffs that apply broadly to all countries, specific countries could be targeted if they are found to not be in compliance with their obligations under the agreement, with, say, an *anti-dumping duty*. These are generally used by individual governments when they believe that imports from another country are being priced below market value (that is, they are being "dumped" into the country's markets). Since they typically take the form of fees, when they are applied to goods with high carbon content, they are similar to carbon tariffs in that they penalize "undesirable" goods.

One major challenge with targeted actions like anti-dumping duties is that they can be misused when applied against individual countries. In particular, a country may be targeted for an anti-dumping duty for reasons having nothing to do with carbon content. For example, a country might claim it is targeting carbon content when, in reality, it is simply trying to protect its industries from competition. The difficulty of measuring carbon content, and the internal political pressures for protectionism that a country's government may be under, may make it difficult to use these measures to effectively reduce GHG emissions.

### III. Nordhaus climate clubs

Given the tremendous challenges of negotiating international agreements on climate change, Nordhaus has recently proposed the idea of using what he calls *climate clubs* to get the international community to up their mitigation game. The basic idea here is similar to the use of trade sanctions based on carbon content. However, Nordhaus argues that if you do it right, you do not need to base sanctions on carbon content at all. This should come as a relief to you, given what we have just discussed about the challenges of implementing carbon tariffs and associated policies. Let us now consider how these climate clubs would work.

To see the idea, let us go back to the notion of a *club good* that we saw earlier in chapter five. To refresh your memory, club goods are goods characterized by non-rivalry in consumption and *ease* of exclusion. Nordhaus's idea is to invite every country to participate in a climate club where every member agrees to adopt a carbon tax. If a country does not join, then its goods are subject to a tariff if it wants to trade with any country in the club. Thus, each country is penalized in terms of lost trade if it does not join the club. So climate clubs take advantage of the fact that tariffs can be used to exclude countries from the club, but, once they are in, they enjoy all the (non-rivalrous) benefits of being a member. So this climate club is kind of like a non-congested movie theater that requires you to buy a ticket in order to get in, another example of a club good.

One big challenge of establishing a climate club is that the carbon tax cannot be set too high. The higher it is set, the more countries will decide that joining is not worth it. This is, of course, because each country is weighing the cost of joining (having to adopt the carbon tax) against the benefits (not having tariffs levied on its goods). If the carbon tax is set too high, few countries will join and the tariff imposed by the club starts to be meaningless. In the extreme case, if the carbon tax is set so high that no country agrees to join, then the tariff loses all bite and, of course, no country would end up adopting a carbon tax.

Unfortunately, the maximum carbon tax to make this idea feasible is probably way too low to slow global warming down to the Paris Agreement target of 1.5 degrees Celsius. We know this based on simulations that Nordhaus has done using a version of his DICE model. So to accompany the climate club arrangement, Nordhaus also proposes a massive program of research and development (R&D) on low-carbon energy technologies; that is, renewable energy. This combined policy could permit achievement of the Paris Agreement targets, depending upon the success of the R&D in lowering the costs of renewable energy. The larger question is, of course, political feasibility: How receptive will the international community be to the climate club idea? On this question, only time will tell.

## Key takeaways

- Climate change is contributing to rising global temperatures, more frequent extreme climate events, warmer, more acidic oceans, and rising sea levels.
- Economists typically use integrated assessment models in order to assess the costs of climate change. These IAMs combine climate models with models of the world economy.
- An important element of the economist's approach to valuing the costs of climate change is to allow for adaptive behavior, which tends to lower the cost estimates.
- Different assumptions about the discount rate make a huge difference in estimates of the costs of climate change.
- The impacts of climate change are more likely to be borne by lower-income countries.
- The three basic strategies for dealing with climate change are mitigation, geoengineering, and adaptation.
- Two mitigation strategies are carbon taxes and cap-and-trade.
- International climate agreements are subject to the free-rider problem, transaction costs, and enforcement costs. Enforcement costs occur because of difficulties in effective monitoring and imposing sanctions.

## Key concepts

- Greenhouse effect
- Integrated assessment models (IAMs)

- Mitigation
- Social cost of carbon
- Carbon taxes
- Cap-and-trade
- Geoengineering
- Solar radiation management
- Negative emissions technologies
- Adaptation
- Climate clubs

## Exercises/discussion questions

9.1.  Suppose you have two firms, Lo-Cost and Hi-Cost. Lo-Cost has a marginal abatement cost schedule that may be written as: $MAC^{LC} = 0.01A$, where MAC is marginal abatement costs in dollars per ton of CO2 abated and A is abatements of CO2, in tons. Hi-Cost has a marginal abatement cost schedule that may be written as: $MAC^{HC} = 40 + 0.01A$. Suppose neither firm is currently abating emissions and a regulator wants to achieve total abatements of 2,000 tons. (a) What carbon tax will allow the regulator to achieve this amount of total abatements? (b) How much more costly would it be if instead of imposing a carbon tax, the regulator simply had each firm abate 1000 tons?

9.2.  Suppose that, currently, 1,000 tons of carbon are being emitted, which inflict costs in each of three years of: $100 in the current year (year 0), $120 next year (year 1), and $200 the following year (year 2). Now suppose you increase carbon emissions to 1,001 tons, and costs become: $105 in the current year, $125 next year, and $500 the following year. Assuming a discount rate of 10%, calculate the implied social cost of carbon.

9.3.  Consider the three main types of policies for dealing with climate change: mitigation, geoengineering, and adaptation. Which policy do you think would be subject to the lowest political transaction costs to get implemented, and why?

9.4.  The discussion argues that a combination of these policies moving forward may be called for. Where would you start in thinking about *how much* we should rely on each policy as part of our overall climate change strategy?

9.5.  Can you see any ways to make the carbon tax more politically acceptable?

9.6.  According to the discussion, the international community managed to agree to phase out worldwide production of chlorofluorocarbons in 1987. Why would you think that the transaction costs of that negotiation would be lower than the transaction costs of negotiating over reductions in greenhouse gases? What evidence would you look for to support your argument?

9.7.  When evaluating the potential effectiveness of the climate club idea, consider the fact that countries vary a great deal in terms of reliance on international trade. For some countries, importing from other countries forms a large component of their total economy. Other countries rely much less heavily on imported goods, and they would therefore not be affected as much by the tariffs imposed by climate club on other countries. Consider these countries negotiating to set up the climate club in the first place. Would this difference affect the transaction costs of negotiation among the countries? How? Would this difference make it easier or harder to set up the climate club? Or would it make no difference?

# Notes

1 On the vertical axis, a value of zero represents the global mean temperature over the period 1961 to 1990.
2 The series shown is atmospheric CO2 concentrations, derived from local air measurements at Mauna Loa Observatory, Hawaii, in July of each year. See Keeling et al. (2001).
3 For the various facts in this paragraph, see de Leon and Schwartz (2020); Patrick and Freedman (2019); Rott (2020); Freedman (2020).
4 The study found some variation in the effect of rising temperatures across different fish populations, with a few populations actually benefiting from warming ocean waters. But many more populations were adversely affected by warming waters.
5 For the facts in this paragraph, see "Greenland ice sheet melting" (2018); NASA (2019).
6 See, for example, Tol (2009), p. 31.
7 The economist Richard Tol calls this approach the *enumerative* method. See Tol (2009), pp. 31–32.
8 See, for example, Mendelsohn et al. (1994); Sohngen et al. (1998); Mendelsohn and Dinar (1999); Weber and Hauer (2003)].
9 See the striking differences in the conclusions of the studies by Tol (2009), p. 36 and Heal (2017), p. 1051.
10 You often read that carbon taxes are based on the carbon content of the products, but this is a bit misleading. If a product such as chemical solvents or feedstocks contains petroleum products, these are not taxed. It is combustion that matters [Horowitz et al. (2017)].
11 See, for example, Cowen (2018); Gillis (2018).
12 But see Libecap (2014) for a more pessimistic view of cap-and-trade markets.
13 To clarify, some experts consider negative emissions technologies to be a different type of solution than geoengineering. Technically, negative emissions technologies do not involve engineering the earth. I discuss them together here because they are both engineering approaches.
14 Other methods include forest management practices such as delaying harvests and thinning the forest by removing smaller saplings (understory) [Sedjo and Sohngren, pp. 132–3].
15 This discussion draws heavily on Libecap (2014), pp. 451–66.
16 This example of effective international agreement was the Montreal Protocol.

# Further readings

## I. Behavioral economics and climate change

Brekke, Kjell Arne and Olof Johansson-Stenman. "The behavioral economics of climate change," *Oxford Review of Economic Policy* 24 (2008): 280–97.
*Applies various insights of behavioral economics to address a number of issues relating to climate change, including climate negotiations and choice of optimal discount rate.*

## II. Discounting

Weisbach, David and Cass R. Sunstein. "Climate change and discounting the future: A guide for the perplexed," *Yale Law and Policy Review* 27(Spring 2009): 433–57.
*Clear, non-technical discussion of the debate over the appropriate choice of discount rate for guiding climate change policy.*

## III. Public choice and climate change

Cazals, Antoine and Alexandre Sauquet. "How do elections affect international cooperation? Evidence from environmental treaty participation," *Public Choice* 162(2015): 263–85.
*This article offers evidence that the election cycle within individual countries may affect the ability of the international community to negotiate successfully over international environmental agreements.*

# References

Aleem, Zeeshan. "A Siberian town near the Arctic Circle just recorded a 100-degree temperature," *Vox*, June 21, 2020.

Anthoff, David; Francisco Estrade; and Richard S. J. Tol. "Shutting down the thermohaline circulation," *American Economic Review* 106(May 2016): 602–6.

"Australian bushfire smoke affecting South America, UN reports," *Reuters*, January 7, 2020.

Balaguru, Karthik; Gregory R. Foltz; and L Ruby Leung. "Increasing magnitude of hurricane rapid intensification in the central and eastern tropical Atlantic," *Geophysical Research Letters* 45(2018): 4238–47.

Berardelli, Jeff. "How climate change is making hurricanes more dangerous," *Yale Climate Connections*, July 8, 2019.

Blackstock, Jason J. and Jane C.S. Long. "The politics of geoengineering," *Science* 327(January 29, 2010).

Borunda, Alejandra. "What a 100-degree day in Siberia really means," *National Geographic*, June 23, 2020.

Cart, Julie. "Checking the math on cap and trade, some experts say it's not adding up," *KQED Science*, May 23, 2018.

Cockburn, Harry. "'Teetering at the edge': Scientists warn of rapid melting of Antarctica's 'Doomsday glacier,'" *Independent*, July 13, 2020.

Coase, Ronald H. "The problem of social cost," *Journal of Law and Economics* (1960): 1–44.

Cole, Daniel H. "The 'Stern Review' and its critics: Implications for the theory and practice of benefit-cost analysis," *Natural Resources Journal* 48(Winter 2008): 53–90.

Cowen, Tyler. "The carbon tax is dead, long live the carbon tax," *Bloomberg Opinion*, November 8, 2018.

de Leon, Concepcion and John Schwartz. "Death Valley just recorded the hottest temperature on Earth," *New York Times*, August 17, 2020.

Dell, Melissa; Benjamin F. Jones; and Benjamin A. Olken. "What do we learn from the weather? The new climate-economy literature," *Journal of Economic Literature* 52(September 2014): 740–98.

Diaz, Delavane and Klaus Keller. "A potential disintegration of the West Antarctic ice sheet: Implications for economic analyses of climate policy," *American Economic Review* 106(2016): 607–11.

Frankel, Jeffrey A. "Global environment and trade policy," in *Post-Kyoto International Climate Policy*. Joseph E. Aldy and Robert N Stavins(eds.). Cambridge: Cambridge University Press, 2010.

Free, Christopher M. et al. "Impacts of historical warming on marine fisheries production," *Science* 363(March 1, 2019): 979–83.

Freedman, Andrew. "Australia's greenhouse gas emissions effectively double as a result of unprecedented bush fires." *Washington Post*, January 24, 2020.

Gillis, Justin. "Forget the carbon tax for now," *New York Times*, December 27, 2018.

"Greenland ice sheet melting doubled over the last century," *Yale Environment 360*, March 29, 2018.

"Growing up in -60C," *BBCNews*, November 14, 2017.

Heal, Geoffrey. "The economics of the climate," *Journal of Economic Literature* 55(September 2017): 1046–63.

Horowitz, John et al. "Methodology for analyzing a carbon tax," Office of Tax Analysis. Working paper 115(January 2017).

Husted, Bryan W. and Robert Folger. "Fairness and transaction costs: The contribution of organizational justice theory to an integrative model of economic organization," *Organization Science* (2004): 719–29.

Keeling, C.D.; S. C. Piper; R. B. Bacastow; M. Wahlen; T. P. Whorf; M. Heimann; and H. A. Meijer. *Exchanges of atmospheric CO2 and 13CO2 with the terrestrial biosphere and oceans from 1978 to 2000. I. Global aspects*, SIO Reference Series, No. 01-06, Scripps Institution of Oceanography, San Diego, 2001.

Kolbert, Elizabeth. "Climate solutions: Is it feasible to remove enough CO2 from the air?" *YaleEnvironment360*, November 15, 2018.

Krupnick, Alan and David McLaughlin. "Valuing the impacts of climate change on terrestrial ecosystem services," *Climate Change Economics* 3(2012): 1–11.

Libecap, Gary D. "Addressing global environmental externalities: Transaction costs considerations," *Journal of Economic Literature* 52(June 2014): 424–79.

McKenna, Phil. "9 states target transportation emissions with new cap-and-trade plan," *Inside Climate News*, December 20, 2018.

Mendelsohn, Robert and Ariel Dinar. "Climate change, agriculture, and developing countries: Does adaptation matter?" *World Bank Research Observer* 14(August 1999): 277–93.

Mendelsohn, Robert; William D. Nordhaus; and Daigee Shaw. "The impact of global warming on agriculture: A Ricardian analysis," *American Economic Review* 84(September 1994): 753–71.

National Academies of Sciences, Engineering and Medicine (NASEM). "Technologies that remove carbon dioxide from air and sequester it need to play a large role in mitigating climate change, says new report," *News release*, October 24, 2018.

National Aeronautic and Space Administration (NASA). Goddard Institute for Space Studies. "Global temperature change, " *Proceedings of the National Academy of Sciences* 103 (2006): 14288–93.

National Aeronautic and Space Administration (NASA). "Greenland's rapid melt will mean more flooding," *Sea Level News*, December 11, 2019.

Nordhaus, William. "Critical assumptions in the Stern Review on Climate Change," *Science* n.s. 317 (July 13, 2007): 201–02.

Nordhaus, William. *The Climate Casino*. New Haven: Yale University Press, 2013.

Nordhaus, William. "Revisiting the social cost of carbon," *PNAS* 114(February 14, 2017): 1518–23.

Patrick, A. Odysseus and Andrew Freedman. "Australia has its hottest day for a second straight day as areas face 'catastrophic' fire conditions," *Washington Post*, December 19, 2019.

Plaven, George. "Oregon lawmaker: Cap and trade coming," *Capital Press*, December 3, 2018.

Richardson, Nathan and Molly MacCauley. "Forest carbon economics: What we know, what we do not and whether it matters," *Climate Change Economics* 3(2012): 1–18.

Rott, Nathan. "Australia's wildfire are releasing vast amounts of carbon," *NPR*, January 12, 2020.

Schmalensee, Richard and Robert Stavins. "Learning from thirty years of cap and trade," *Resources*, May 16, 2019.

Sedjo, Roger and Brent Sohngen. "Carbon sequestration in forests and soils," *Annual Review of Resource Economics* 4(2012): 127–44.

Sohngen, Brent; Robert Mendelsohn; and Ronald Neilson. "Predicting CO2 emissions from forests during climatic change: A comparison of natural and human response models," *Ambio* 27(November 1998): 509–13.

Statista. *Global gross domestic product (GDP) at current prices from 2009 to 2021*, 2020. https://www.statista.com/statistics/268750/global-gross-domestic-product-gdp/, accessed 12/28/2020.

Stern, Nicholas et al. *Stern Review: The Economics of Climate Change*. London: HM Treasury, 2006.

Stern, Nicholas. "The economics of climate change," *American Economic Review* 98(May 2008): 1–37.

Temple, James. "China is creating a huge carbon market – but not a particularly aggressive one," *MIT Technology Review*, June 18, 2018.

Tol, Richard S. J. "The economic effects of climate change," *Journal of Economic Perspectives* 23(Spring 2009): 29–51.

United States. Geological Survey (USGS). "The cataclysmic 1991 eruption of Mount Pinatubo, Philippines." Fact Sheet 113-97.

Weber, Marian and Grant Hauer. "A regional analysis of climate change impacts on Canadian agriculture," *Canadian Public Policy* 29(June 2003): 163–80.

Weitzman, Martin L. "Tail-hedge discounting and the social cost of carbon," *Journal of Economic Literature* 51(September 2013): 873–82.

Young, Jessie. "Climate change could kill all of earth's coral reefs by 2100, scientists warn," *CNN*, February 20, 2020.

# 10 Air and water quality

New Delhi, the capital of India, may have the dubious honor of being the most-polluted capital city in the world. In the winter, farmers in nearby rural areas burn their fields to prepare them for the next crop, and the smoke and ash blows into the city, mixing with vehicle emissions and dust from construction sites to create a thick gray haze of polluted air [Jain (2019)]. On bad days, the air quality index tops 400, or roughly eight times the recommended maximum. The result? Stinging eyes and throats, respiratory ailments, hypertension, and heart problems for the unfortunate residents of the city. The gray haze is so thick that during the 2020 COVID-19 crisis – that halted much economic activity and briefly cleared the skies above the city – solar installations in the city registered a roughly 8% increase in power output ["COVID-19 shutdown" (2020)].

Climate change is, of course, only one of many environmental problems faced by humankind. Air pollution, water pollution, groundwater contamination, toxic waste disposal, overfishing, deforestation, species extinction, aquifer depletion, topsoil depletion, and many more environmental problems present important challenges to economies all over the world. In this chapter, we focus on air and water quality issues resulting from pollutants of various kinds finding their way into our air and water resources, which can:

- Degrade air quality;
- Pollute drinking water;
- Damage forests, ecosystems, and fisheries;
- Harm wildlife; and
- Pose serious threats to human health and welfare.

The economist's approach to understanding and modeling air and water quality is, as always, to focus on tradeoffs, costs and benefits, and incentives. In this chapter we will develop the economic rationale for policies designed to address threats to air and water quality. In addition, we will explore the institutional approach to understanding the genesis of air and water quality policies, whose interests they serve, and their likely consequences for the environment.

## Causes and consequences

### I. Air quality

Outdoor air pollution comes from a number of different anthropogenic (human-made) sources, especially the burning of fossil fuels in industrial and power production in industrialized countries, fuel combustion in motor vehicles, residential heating, and waste incineration in cities and farms. In addition, certain parts of the world are particularly prone to air pollution from natural sources such as forest fires and dust storms, which can be exacerbated by climate change, the desiccation (drying-up) of bodies of water, and soil erosion. Especially in low-income countries, indoor air pollution is also a serious problem, commonly caused by combustion of biofuels in fireplaces and cook stoves.

Of highest concern to public health officials are the effects on human health, resulting in increased mortality (death) and morbidity (illness). The World Health Organization (WHO) has estimated that, currently, about 4.2 million people worldwide die prematurely every year from exposure to outdoor air pollution, while another 3.8 million die from indoor air pollution [World Health Organization(2020)]. One recent study, using a new methodology, has put the mortality numbers considerably higher, with annual premature deaths from outdoor air pollution estimated to be around 8.8 million [Lelieveld et al. (2019)]. Regardless of the exact numbers, the effects of air pollution on human health continue to be a major societal concern.

The human health impacts of air pollution operate mainly through their effects on the cardiovascular and respiratory systems. The leading categories of death from air pollution are: heart disease, stroke, pulmonary disease, lung cancer, and acute respiratory infections. In addition, air pollution is responsible for various health issues such as respiratory distress, neurological disorders, decreased lung function, asthma, and negative birth outcomes such as low birth weight and preterm birth, all afflicting millions more.

For a better sense for the magnitude of the health effects of air pollution, consider a new pollution index created by researchers at the Energy Policy Institute of the University of Chicago, the *Air Quality Life Index*. This index measures the effect of one particular pollutant – airborne particulate matter – on average life expectancy around the world. According to the research that went into making this new index, the average human being can expect to live nearly two years less due to exposure to particulate pollution. This is larger than the effect of smoking and *30* times the effect of conflict and terrorism (see Figure 10.1).[1]

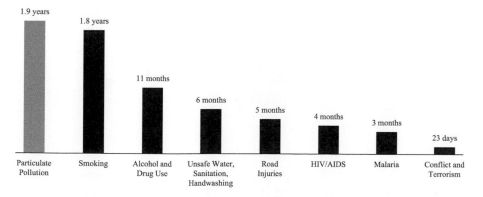

*Figure 10.1* Average life expectancy lost per person, various causes. *Source*: Air Quality Life Index®, Annual Update July 2020, aqli.epic.uchicago.edu

The incidence of adverse health effects of air pollution is spread very unevenly around the world and within countries, falling disproportionately on poorer populations. According to WHO, nearly 90% of the 4.2 million people who die prematurely from outdoor air pollution are from lower-income countries.

For a more detailed breakdown, Figure 10.2 shows a map of the world in 2017, with the figures for death rate from particulate air pollution broken down by country. Keeping in mind that the darker areas represent higher mortality rates, you can see that most of North America and Europe, along with selected other industrialized countries like Japan, Korea, and Australia, fare relatively well by this metric. On the other hand, mortality rates tend to be relatively high in central Africa and central Asia. Certain demographic groups are especially vulnerable, including children and the elderly, and people with underlying medical conditions like lung or heart disease.

At the same time, there is evidence that, in many ways, environmental conditions have been improving in recent years. Consider the period from 1990 to 2017. During this 28-year period, deaths from indoor pollution fell by nearly 65% worldwide. Recalling that indoor pollution is primarily a problem that plagues low-income countries, this represents significant progress for that set of countries. Similarly, deaths from outdoor particulate matter and ozone pollution fell by nearly 13% and 43%, respectively.[2]

### *II. Water quality*

Threats to water quality also come from a variety of sources, both natural and human-made. In many parts of the world, surface rivers and streams receive runoff from farming operations, as well as municipal, commercial, industrial, and mining wastes. Depending

## Death rates from air pollution, 2017

Death rates are measured as the number of deaths per 100,000 population from both outdoor and indoor air pollution. Rates are age-standardized, meaning they assume a constant age structure of the population to allow for comparisons between countries and over time.

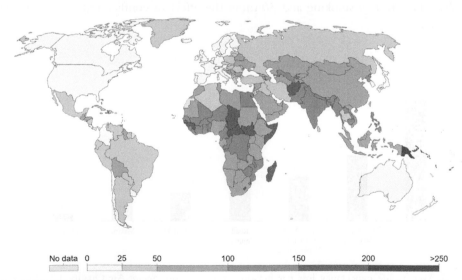

No data   0      25      50          100         150         200        >250

Source: IHME, Global Burden of Disease                    OurWorldInData.org/air-pollution • CC BY

*Figure 10.2* Death rate from air pollution by country, 2017. *Source*: Institute for Health Metrics and Evaluation, *Global Burden of Disease*, (Our World in Data)

upon the source, wastewaters can contain a variety of contaminants with varying effects on surface water quality. Heavy metals from industrial manufacturing can cause cancer, anemia, and, potentially, acute liver, kidney, intestinal, and neurological damage. Organic chemicals found in many household products and used in industrial processes can cause damage to kidneys, liver, and the nervous and reproductive systems.

In the residential sector, lack of, or incomplete, treatment of residential wastes can result in contaminants entering our rivers and streams, including microorganisms and nitrates that can cause gastrointestinal and other illnesses. Another issue is the widespread presence of lead, a highly toxic substance, in water mains and pipes in residential housing. Corrosive chemicals in delivered water can mobilize the lead, which causes it to enter residential drinking water supplies. This issue came to the fore in the United States in the early 2010s with an infamous episode involving Flint, Michigan (see Box 10.1). However, Flint may be only the tip of the iceberg, as other cities have experienced elevated lead levels in their drinking water since then, including Newark, NJ, Virginia Beach, VA, and a number of cities across Canada.[3]

---

**Box 10.1: Flint water crisis**

Beginning in early 2015, residents of Flint, Michigan began to complain to city officials about feeling ill after drinking water supplied by the city. Studies showed elevated levels of lead in children that began after the city switched drinking water supplies. It turns out that water from the new sources contained chemicals that corroded the lead in water service lines, which entered the water supplies. Thousands of children were exposed to drinking water containing high levels of lead. In addition, the change in drinking water supplies was also likely responsible for an outbreak of legionnaires' disease that killed 12 people [Zahran et al. (2018)]

---

Finally, runoff from farming operations often contains nitrates, sulfates, and fertilizer and pesticide residues. All of these can pollute wildlife habitat and local ecosystems, and they can also cause adverse effects on health if they make their way into drinking water. Furthermore, elevated nitrogen and phosphorus in rivers and streams can contribute to oxygen depletion, which, if severe enough, can result in so-called *dead zones* with insufficient dissolved oxygen to support marine life.

Pollutants can also make their way into groundwater aquifers, contaminating drinking water. These pollutants include pesticides and fertilizers, industrial wastes, processing materials in mining operations, and untreated discharge from leaky septic tanks. And recall that groundwater contamination may be associated with the new fracking technologies, as we saw in chapter eight.

Groundwater contamination presents some different, and in some ways more intractable, challenges to polluted surface water because once an aquifer is contaminated, it is more difficult to clean. Free-flowing rivers and streams have natural self-cleansing tendencies, once a pollution source is removed. However, contaminants tend to remain in the confined layers of a groundwater aquifer, requiring special efforts for groundwater cleanup (*remediation*).

## Clean air and water policies

### I. Air quality

Air and water pollution, of course, have been with us for a long time. But, over time, we have become increasingly aware of their adverse effects on humans and the environment.

In October 1948, the town of Donora, PA experienced a temperature inversion that trapped dirty air from local steel and zinc smelters within the town. Over a period of five days, nearly half of the town's 14,000 residents experienced severe cardiovascular or respiratory problems. Twenty people died. The Donora event has been called "the worst air pollution disaster in U.S. history" [Boissoneault (2018)]. In 1952, thick smog enveloped the city of London, the result of another inversion. This event, which became known as the *Great Smog of 1952,* is believed to have been responsible for thousands of deaths, both in the immediate aftermath of the event and in following months[Stone (2002)].

Events like these dramatized the problem of air pollution and its potentially deadly consequences, resulting in legislative action. In 1955, Congress passed the *Air Pollution Control Act,* which supported research and technical assistance for pollution control, but left it to the states to take steps to control air pollution. The following year, in response to the Great Smog and its aftermath, the British Parliament passed the *Clean Air Act of 1956,* which was the first serious attempt in Britain to regulate air quality on a national scale. Among other things, the Clean Air Act regulated smoke from chimneys and established smoke control areas in municipal areas, where only smokeless fuel could be burned. These laws did not solve the problem of air pollution[4], but they were important first steps toward national environmental policy in these countries.

In the 1960s, Congress enacted additional environmental legislation to fight air pollution. In 1963, it passed the first *U.S. Clean Air Act,* which provided support for the development of state pollution control agencies, established a permanent program for federal aid for environmental research, and enabled federal involvement in pollution cases crossing state lines. Then, a major step occurred seven years later, when Congress passed the *Clean Air Act Amendments of 1970.* This important law, which is commonly known as the modern *Clean Air Act* (CAA), established a major role for the federal government in fighting air pollution and has been called "the most important and far-reaching environmental statute enacted in the United States" [Aldy et al. (2020)]. Its provisions included:

- National standards for air quality [*National Ambient Air Quality Standards* (NAAQS)] for six major pollutants: carbon monoxide, nitrogen oxides, ozone, sulfur dioxide, and solid particulate matter. Lead was added in 1977. These are known as the *criteria pollutants.*
- Pollution standards for new manufacturing plants [*New Source Performance Standards* (NSPS)];
- Emissions standards for trucks and automobiles; and
- Requirements for states to come up with a detailed plan [a *state implementation plan*, or SIP] that specified how the state would meet federal air quality standards.

That same year, the *Environmental Protection Agency* (EPA) was created through a reorganization of the powers of various other federal agencies [Ahlers (2015), p. 117]. The EPA is the primary federal agency responsible for protecting air and water quality.

The 1970 Clean Air Act has since been revised and amended a number of times. In 1977, Congress passed the 1977 *Clean Air Act Amendments* (CAAA77), which called for prevention of the deterioration of air quality in areas of the country that were relatively clean [*prevention of significant deterioration*, or PSD]. Thirteen years later, Congress passed the 1990 *Clean Air Act Amendments* (CAAA90). This law addressed ozone depletion, calling for the phasing out of chlorofluorocarbons. It also established standards for regulating a variety of other air pollutants. Finally, it contained a novel approach to combatting acid rain, calling for the creation of a program of emissions trading for coal-burning electric utilities. Additional laws enacted since then have called for, among other things: banning the use of leaded gasoline in new vehicles, mitigating nitrogen oxides (NOx), and regulating air pollution that crosses state borders [Aldy et al. (2020)]

## II. Water quality

In the United States, the dumping of wastes into surface waters was virtually unregulated for a long time. Prior to 1948, the only federal law that dealt with dumping of wastes into rivers was the 1899 *Refuse Act*, but this law was more concerned with wastes impeding navigation than with water pollution. When information came to light in the 1940s regarding the magnitude of the water pollution problem in the United States, Congress responded by passing the *Federal Pollution Control Act* in 1948. This law authorized the Surgeon General of the Public Health Service to prepare a comprehensive program for "preventing, abating, and controlling water pollution." However, it left implementation of water pollution policy to the states, and it did not set standards for water quality or limit new sources of water pollution.

The Federal Pollution Control Act turned out to be largely ineffective in controlling water pollution. As a result, amid growing public awareness of the problems associated with water pollution, the law was significantly amended in 1972 when Congress passed the *Water Pollution Control Act Amendments*, which became commonly known as the *Clean Water Act*. Among other things, the Clean Water Act:

- Authorized the EPA to regulate water pollution in various forms;
- Made it unlawful for anyone to discharge point-source pollutants into navigable waters without a permit; and
- Funded the construction of sewage treatment plants throughout the U.S through a construction grants program.

Since 1972, the Clean Water Act has been amended several times. In 1977, Congress expanded regulation of discharge of untreated wastewaters into rivers, lakes, and coastal waters. In 1987, it began phasing out the construction grants program and replaced it with a revolving fund program that supported partnerships between states and the EPA. The EPA now enforces a variety of laws regulating water quality, including ones governing safe drinking water, wetlands, water quality in the Great Lakes, tailings from uranium mills, ocean dumping, and pesticide use. The Clean Water Act and subsequent amendments provide the basic federal regulatory structure for water quality policy in the United States today.

## Economic analysis of air and water quality

Standard economic analysis of air and water quality begins with the basic externalities model we have encountered a couple of times so far in this book. The basic analysis is similar to the analysis of GHGs in chapter nine. Like GHGs, air and water pollutants impose costs on others that are not taken into account by the emitter. These are modeled as negative externalities, resulting in social marginal costs that exceed private marginal costs. In equilibrium, firms that generate pollutants overproduce from the viewpoint of efficiency, resulting in social deadweight loss.

In response, standard economic analysis proposes two types of mechanisms to deal with air and water pollution: *emissions taxes*, and *tradable emissions permits* (TEPs). Emissions taxes are sometimes called *Pigouvian taxes* after the famous economist Arthur Pigou who originally proposed the use of taxes to correct market failures. Analytically, emissions taxes are very similar to the carbon taxes we encountered in chapter nine. In fact, I think the best way to think of carbon taxes is as a particular type of emissions tax. The only real distinction is that the magnitude of carbon taxes is based specifically on the carbon content of the goods being produced, whereas emissions taxes are based on pollutant content more generally.

Similarly, you can think of the allowances we encountered in the earlier discussion of cap-and-trade as being a particular form of tradable emissions permit. Like allowances, possession of a TEP allows the holder of the permit to emit a unit of pollution (say, a ton) and like allowances, TEPs can be traded among eligible companies. The administrative authority decides how much pollution it wants to allow, which determines how many permits to issue. And, finally, the administrator must decide on a mechanism to get the TEPs initially into the hands of polluters, such as an auction, random distribution, or distribution on the basis of some other criterion. All of which should sound very familiar after our discussion of cap-and-trade.

In terms of how well they perform, again, there really are not significant differences between the two sets of instruments. In principle, emissions taxes and TEPs have similar efficiency implications in that both are predicted to be cost-efficient in addressing air and water pollution, as long as the transaction costs in the emissions market are low. Finally, we observe the same issue of price certainty vs. quantity certainty, with emissions taxes giving us price certainty, and TEPs giving us quantity certainty, in terms of the amount of pollution generated. Since the arguments are extremely similar for these basic results, let me refer you to the discussion in chapter nine on carbon taxes and the cap-and-trade system.

There are, however, a couple of additional things worth thinking about, which often come up in discussion of air and water quality regulation.

## I. Market power

One issue that may call the emissions tax into question concerns the market structure of the polluting industry. If the industry is characterized by *market power*, standard economic analysis predicts that producers will exploit their market power by lowering output in order to raise prices.

For example, Figure 10.3 shows the standard analysis for a monopoly firm. Here, the firm produces $Q^M$, which is less than the efficient quantity $Q^{Eff}$. In this case, notice that the efficient quantity $Q^{Eff}$ is less than $Q^\star$, the quantity that would be efficient in a monopoly analysis without the externality factored in. And, correspondingly, the SDWL is smaller as well.

The point is that negative externalities and the exercise of market power may have offsetting effects. While negative externalities tend to lead to overproduction from the viewpoint of efficiency, market power tends to lead to underproduction. In this case, imposing an emissions tax may very well be counterproductive, pushing production even further below the efficient level [Buchanan (1969)]. This analysis suggests that we might want to think twice before imposing an emissions tax, one of the factors to consider being the market structure of the polluting industry.

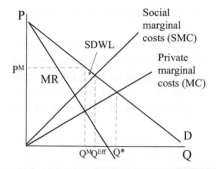

*Figure 10.3* Externalities in a monopoly model

In markets for tradable emissions permits, the market power issue is slightly different. Here, an important question is whether one or a small number of firms may be able to adversely affect the operation of the allowances market through their trading behavior. In principle, the answer appears to be yes: in general, when a monopoly firm exercises market power in the allowances market, the overall cost of abatement will not be minimized, as it would be if the market were competitive [Hahn (1984)]. However, even if costs are not minimized, it is believed that the existence of an allowances market is probably far better than not having a market. In addition, problems could arise if firms are successful in deterring entry by new firms into an industry. This is why it is important to design these markets carefully [Hahn and Hester (1989)].

None of this is to argue that market power issues will never occur when considering carbon taxes or cap–and–trade systems for GHGs. But these issues generally tend to come up less in that context.

### II. Double dividend

Another issue regarding the emissions tax concerns the use of the tax revenues raised from imposing the tax. If you think about it, if you were actually to administer an emissions tax, you would be collecting tax revenues. We did not consider this issue before, but now that I have you thinking about it, you might ask yourself: What should be done with these tax revenues?

Some economists have argued that we might want to use it to correct other distortions in the economy caused by the tax system. For example, in many economies, labor is taxed pretty heavily, through policies such as the income tax and social security tax. These tax policies likely distort decisions to supply labor, specifically by making wage labor less attractive than alternatives such as home production or leisure activities. The resulting undersupply of labor could lead to social deadweight loss in labor markets. Perhaps revenues from emissions taxes could be used to offset labor market inefficiencies. For example, they could be used to lower personal income taxes or fund job retraining programs. So maybe we could kill two birds with one stone: correcting distortions in two different markets, enabling us to reap a so-called *double dividend*.

It turns out that the whole issue is a good deal more complicated. The problem is that the emissions tax itself exacerbates labor market inefficiency. It does this by raising production costs, which tends to exert upward pressure on output prices. This in turn lowers the real (inflation-adjusted) wage, further discouraging labor supply. So if labor markets are already distorted, emissions taxes may make matters worse.

Furthermore, the effect is likely not to be small. You might be thinking to yourself: surely the effect on output prices of imposing an emissions tax is likely to be so small as not to warrant bothering to worry about. However, economic analyses have consistently concluded that the effect is likely to be considerable. The problem is that labor income is a large component of total national income. This means that even small reductions in the real wage can result in large welfare losses.

How large is the effect in absolute terms? According to economic analyses, this inflation-induced reduction in the real wage is probably greater than labor market benefits from using the tax revenues to encourage labor supply. This means that the double dividend likely does not exist [Parry and Oates (2000), p. 607]. It also means that even in the best-case scenario, where we are using emissions tax revenues as efficiently as possible, it is unlikely that we will be able to attain the efficient pollution outcome using emissions taxes. And achieving the second-best outcome requires imposing a tax that is less than the marginal external costs. How much less depends on a number of factors, including the level of environmental damages and the elasticity of labor supply.

## The property rights approach to air and water quality protection

OK, so much for the standard economic treatment of air and water quality policy. Let us now take a step back and consider air and water quality from a transaction costs perspective. Again returning to the basic Coase argument, consider a factory dumping sludge into an adjacent river. Coase would argue that there are two ways to think about what is going on here. One is the perhaps more intuitive interpretation that the factory is imposing costs on people living downstream on the river. The other interpretation is that the people downstream are potentially inflicting costs on the factory, which would have to incur costs to keep from dumping sludge in the river. Again, we have the idea of mutually inflicted costs where, from society's perspective, the best (that is, most efficient) solution would be to minimize the cost.

As the Coase Theorem states, a key issue is the level of transaction costs in negotiations between the factor and the people downstream. Let us consider the two polar extreme cases in turn.

### I. Transaction costs low

If transaction costs are low, the solution is straightforward. Just specify clearly the relative rights of the factory and the people downstream by either making the factory liable for damages, or not. Then, if the costs to the factory of keeping its sludge out of the river exceed the costs of sludgy water to the people downstream, we would expect negotiations to result in the sludge being deposited in the river, which is the least-cost outcome. This outcome may not sit well with the fishers or with many environmentalists, but we can address their concerns to some extent by making the factory liable for the damages. This would have the effect of compensating the people downstream for the negative environmental outcome.

If, on the other hand, the costs to the people downstream are greater, negotiations should result in a clean river, again the least-cost outcome. Again, the relative wealth position of the people downstream would be improved by making the factory liable for damages. Here, however, the reason is slightly different as, in this case, they would not have to negotiate to pay the factory not to pollute the river. If any of this argument is unclear, you may want to reread the discussion of the Coase Theorem in chapter four.

## II. Transaction costs high

The more likely case is that transaction costs are not low, for the reasons given at the end of chapter four. In this case, we can no longer count on private negotiations between the factory and people downstream to arrive at the low-cost outcome. Here, whether we obtain the low-cost outcome depends upon assigning rights to the low-cost party. So, if the costs of keeping sludge out of the river exceed the costs to the people downstream, we would not make the factory liable for damages. In this case, we get the low-cost outcome; namely, that the factory does not have to invest in expensive abatement equipment. If the costs to the people downstream are greater, we would want to make the factory liable for damages, which would provide it with incentive to abate its discharges.

The bottom line of this discussion of the property rights approach is that we may not need to rely on government policies like emissions taxes and tradable emissions permits in order to effectively address air and water pollution. By appropriately assigning liability for damages (or not), the property rights approach leaves the onus on the affected parties to work things out between them (or not!). Some economists strenuously argue the virtues of this approach for two reasons. First, it enables us to capitalize on the information available to the affected parties regarding things like the costs to them of abating pollution or the costs to them of polluted water.[5] This information may not be available to government regulators when they are setting emissions taxes or the number of emissions permits to issue.

Second, the property rights approach steers disputes over the environment into the court system, as opposed to leaving their resolution up to government regulators. This is viewed as good for several reasons. For one thing, courts tend to be impartial arbiters of disputes, kind of like baseball umpires calling balls and strikes. They tend to resolve disputes based on actual harms done, awarding remedies based on the magnitudes of those harms. Over time, they create a body of rules (*the common law*) that clearly state what the law is and that clearly distinguishes permissible from impermissible actions. These rules provide certainty to the competing parties regarding exactly what their rights are, which encourages them to take correct actions like keeping costs down and perhaps avoiding disputes in the first place. All this is kind of like only swinging at pitches that are actually over the plate, which batters tend to do more when the umpire is doing his job right.

All of this is in sharp contrast to many government environmental regulations, which tend to be complex, opaque, and take insufficient account of particular circumstances. This might be the direct result of the hurly-burly of politics, which commonly results in policies that reflect political exigencies more than being scientifically and economically informed solutions. To some economists, the government regulator umpire has little incentive to call balls and strikes accurately, while being yelled at by people from all sides, some of whom are surreptitiously slipping money into his pocket.[6] All of this is a perfect segue into the next thing I wanted to discuss: public choice in environmental policy.

## Public choice in environmental policy: Clean air policy in the 1970s

In the standard economic model, government policies are created to improve economic outcomes; say, by imposing emission taxes or using new mechanisms like tradable emissions permits in order to correct market failures. That is, under this model, government decision makers act in the public interest. However, under the public choice approach, policy emerges largely from the supply and demand for policy by largely self-interested interest groups, politicians, and agency administrators. Clean air policy in the United States provides a complicated picture of how these different actors have interacted with each other to produce the actual policies we observe.

In this section, we will examine two major pieces of U.S. air quality legislation that we encountered earlier: the 1970 Clean Air Act (CAA) and the 1977 Clean Air Act Amendments (CAAA77). Many observers agree that, even though plenty more remains to be done, these laws contributed in a major way to a decades-long trend of gradually improving air quality in the United States. On the surface, this result is consistent with clean air policy in the public interest. However, for a number of reasons related to politics, here is the bottom line in advance: we could probably have achieved progress toward cleaner air at lower, perhaps considerably lower, cost to the U.S. economy. Let us see all this in the context of clean air policy with regard to sulfur dioxide (SO2) emissions, a key aspect of clean air policy during this 20-year period.

As we have seen, the 1970 CAA set national ambient air quality standards (NAAQS) and new source performance standards (NSPS). The NAAQS set ceilings on permissible concentrations of each criteria pollutant in the air, with states being mostly responsible for making sure that localities met these standards. NSPS set caps on emission rates by new plants; that is, how much pollutants new manufacturing plants could spew into the air (as a fraction of total energy burned). Regarding SO2 specifically, we are mainly talking about emissions by electric utilities operating coal-fired power plants.

So: two key elements of the CAA were ambient standards and emissions ceilings. These are examples of what economists call *performance standards*, which establish limits or goals for things – like emissions or air quality – that regulated firms need to achieve. I should emphasize that, by themselves, performance standards do not prescribe exactly how firms are supposed to achieve these targets. Sometimes performance standards afford firms a good deal of flexibility in how they attempt to meet these standards. As you might expect, how much flexibility firms are given is an important factor in assessing the efficiency of the standards.

I am mentioning this issue of flexibility because it came to play an important role in the implementation of the CAA and the follow-up legislation CAAA77. Under the CAA, utilities were required to limit SO2 emissions to 1.2 pounds per million British thermal units (BTUs) of energy produced. To achieve this standard, they were directed to adopt:

> The best system of emission reduction which (taking into account the cost of achieving such reduction) the Administrator determines has been adequately demonstrated.
> [Section 111]

Sound prescriptive to you? Well, it was meant to be, as environmentalists really pushed for strong technological solutions to air pollution.[7]

However, ironically, it turned out that this provision afforded utilities a good deal of flexibility in meeting the standard. At the time, there were basically three options available to them. First, they could apply water to pulverized coal before burning it, which would leach out a substantial amount of sulfur oxides ("*washing*"). Second, they could apply a high-tech process known as *flue gas desulfurization* to emissions from plant smoke stacks ("*smoke scrubbing*"). Finally, they could substitute low-sulfur coal for high-sulfur coal in the production process. This low-sulfur coal was sometimes referred to as "*compliance coal,*" for obvious reasons [Hercher (1980), p. 749].

Important to the political story was the fact that, at the time, high-sulfur coal was largely mined in the eastern and Midwestern United States, while low-sulfur coal came from mines out West.[8] When utilities weighed the different options, it became apparent to many that the lowest-cost option was to switch to using low-sulfur coal, despite extra transportation costs. For many, it did not make economic sense to use the high-tech smoke-scrubbing technology. And these decisions to use low-sulfur coal displeased two key interest groups: environmentalists and coal-producing states in the East and Midwest. Eastern and Midwestern states were not pleased about losing business, and environmentalists were not pleased that utilities were apparently able to "get away with" not using the best technologies available to combat air pollution.

As a result, in 1977 Congress amended Section 111 of the CAA in two important ways. First, it amended the standard to read that limitation of emissions was to be achieved through "the best technological system of continuous emission reduction which ... the Administrator determines has been adequately demonstrated." To emphasize the change, Box 10.2 lists the new wording side by side with the original wording of the CAA. You can almost see the authors of this clause wagging their fingers at all those utilities who had switched to compliance coal.

---

**Box 10.2: 1970 CAA vs. 1977 CAAA**

| *1970 Clean Air Act* | *1977 Clean Air Act Amendments* |
| --- | --- |
| "the best system of emission reduction which ... the Administrator determines has been adequately demonstrated." | "the best **technological** system of **continuous** emission reduction which ... the Administrator determines has been adequately demonstrated." |

---

Second, recall that the original performance standard in the CAA capped total SO2 emissions at 1.2 pounds per million BTUs of energy produced. The 1977 CAAA amended this standard to also require that new plants reduce SO2 emissions by anywhere from 70% to 90%. The practical result of these two amendments was to force utilities to use the higher-cost scrubbing technology. They had to use a technology that reduced emissions massively and continuously. Smoke scrubbers were the only option that could do both.

I hope it is clear how this benefited eastern and Midwestern coal producers. With utilities forced to use scrubbers, all incentive for them to use low-sulfur western coal suddenly vanished. Environmentalists were of course delighted with the extremely stringent standards that now had to be achieved using the "best" technology possible. So CAAA77 emerged as the result of an unholy alliance between environmentalists and dirty coal states. But cleaner air came at a much higher cost than necessary. Estimates at the time suggest the extra costs of mandating scrubbers ran into the tens of billions of dollars [Ackerman and Hassler (1980), pp. 528–9]. Furthermore, the fact that the standards only applied to new plants built after 1977 had the perverse result of encouraging utilities to continue to use older, dirtier plants for longer than they would have otherwise. Overall, the air in many places may well have become dirtier as a result.

There are a couple of important takeaways from the story of these two important pieces of clean air legislation. Perhaps the more obvious one is that politics can play an extremely important role in shaping environmental policy. And this can occur both to the detriment of the efficiency of those policies and to the environmental goals of those policies.

But there is a more subtle takeaway, one that is emphasized by the two scholars who have contributed more to our understanding of the politics of these two clean air policies than anyone: Bruce Ackerman and William Hassler. They point out that forcing utilities to adopt scrubbers mostly to benefit one small interest group – eastern coal producers – was never a slam dunk. A provision of the bill crucial to enforcement of forced scrubbing barely passed the Senate when it came to be understood that the bill was engaging in regional protectionism [Ackerman and Hassler (1980), pp. 1504–5]. And, despite everything we know and understand about concentrated benefits and dispersed costs, doesn't it seem strange that rent-seeking by one relatively small interest group could inflict literally tens of billions of dollars of costs on the rest of society?

Ackerman and Hassler make the additional point that committees probably played an important hand in making smoke scrubbing a reality. In both the House and the Senate, bills mandating scrubbers originated in subcommittees, which possessed important gatekeeper and agenda-setting powers in the congressional legislative process. However, the constant press of business can make it challenging for senators and representatives to give adequate attention to all pieces of business. So they focus on salient issues, allowing low-visibility ones to slip by with less attention than they merit. This situation gives congressional aides, who play a key role in drafting legislation, a lot of the agenda-setting power. Ackerman and Hassler argue that this dynamic permitted the smoke-scrubbing mandate essentially to fly under the radar until it was too late [Ackerman and Hassler (1980), pp. 1512–14].

## Emissions trading: 1990 Clean Air Act Amendments[9]

It would not do to leave this discussion of U.S. clean air policy without examining an extremely important subsequent development: the enactment of the 1990 Clean Air Act Amendments (CAAA90). One key provision of CAAA90 was Title IV, which established a program that permitted emissions trading by electric utilities. This new program represented a sharp turnaround from the technology-forcing mandate of the earlier clean air legislation and has been called "pathbreaking" by economists [Schmalensee and Stavins (2019), p. 37].

Title IV specifically targeted emissions of SO2 from the burning of coal by electric utilities. At the time, the phenomenon of acid rain had become a major environmental issue, causing significant damage to forests, aquatic ecosystems, and buildings, especially in the eastern United States, Canada, and continental Europe. As we have seen, the burning of coal produced SO2, the primary source of atmospheric sulfur that produces acid rain.[10]

To implement the trading program, Title IV created allowances, which we encountered in chapter nine in our discussion of climate change. Recall that allowances are essentially permits to emit, denominated in terms of the target pollutant. Under Title IV, possession of an allowance entitled a utility to emit one ton of SO2. Under the terms of the program, allowances could be freely traded among utilities. If a utility did not use all of its allowances, it was allowed to bank them for use in a future year. Over time, the number of allowances was designed to fall to roughly 50% of their 1980 levels, which was to be accomplished by gradually retiring allowances over time.

One interesting thing about the CAAA90 was that it was extremely uncontroversial, passing by overwhelming, bipartisan majorities in both the House and the Senate, with almost equal levels of support among Democrats and Republicans. This bipartisan consensus may seem surprising, given the heated political battles over CAA and CAAA77 and also given the current extreme polarization in the United States around many issues, including the environment. How did this come to pass? The answer seems to be a perfect storm of public interest and special interest considerations that is unlikely to happen again, at least in the foreseeable future.

First, it occurred in response to a serious environmental issue with both domestic and international implications. Acid rain started to be seen as a serious issue in the early 1980s, and many bills were proposed in Congress to address the problem during that decade. However, throughout the 1980s acid rain legislation was stymied by strong opposition from eastern and Midwestern coal-producing states. In addition, key congressmen who opposed acid rain legislation enjoyed positions of legislative power in both the House and the Senate. These included Robert Byrd, senator from West Virginia – a big high-sulfur coal-producing state – who was Senate majority leader.[11] Finally, Ronald Reagan, who can by no stretch of the imagination be considered a supporter of environmental regulation, was president until 1988.

By 1990, however, much of those political realities had changed. Population shifts and a declining high-sulfur coal industry were gradually reducing the political power of the high-sulfur coal states. Environmental interest groups were growing in size and political influence as environmental protection was becoming increasingly important to the public. Robert Byrd had been replaced as Senate majority leader by George Mitchell from Maine, who strongly supported some sort of policy to limit acid rain. Ronald Reagan had been replaced by George H.W. Bush, a moderate Republican who ran as "the Environmental President." By 1990, the political stars were aligned for Congress to do something about acid rain.

But why the market-based approach of tradable permits? A couple of factors seemed to help. First, in 1988 the Environmental Defense Fund, a moderate environmental group, had developed an emissions-trading proposal, which provided some political cover for congressmen concerned about attacks from more progressive environmental groups. Second, true to his Republican roots, President Bush was interested in market-based innovative solutions to environmental problems. When the Bush administration sent a market-based proposal to the Senate, it found a receptive audience in committee. To get the eastern coal states on board, the final proposal did two things. First, it gave extra allowances to utilities that used scrubbers instead of switching to low-sulfur coal over an initial implementation period.

Second, it fiddled with the approach for allocating the allowances more generally. A key issue with important distributional consequences was the initial allocation of allowances into the hands of utilities. At least two approaches were possible: to simply distribute them to utilities for free, or to auction them off. The advantage of auctioning was that revenues could be generated, which could be used to offset other distortions. However, the EPA chose to distribute them for free, many of the initial allowances going to eastern utilities that were generating much of the SO2 that was causing the acid rain problems. This had the advantage of defusing a lot of political opposition to the trading program [Joskow and Schmalensee (1998); Schmalensee and Stavins (2019)].

By virtually all accounts, the emissions trading program turned out to be a huge success. By the time it was completely phased in, in 2000, it covered virtually all electric utilities operating in the continental United States. By imposing a penalty of $2,000 per ton of emissions in excess of the number of allowances, it managed to achieve almost 100% compliance by utilities. Between 1990 and 2004, SO2 emissions from all electric power plants *decreased* by 36%, while total power generation from coal-fired plants *increased* by 25%. And it achieved this at significantly lower cost than traditional command-and-control policies, with estimates by economists of cost savings of anywhere between 15% and 90%.[12] Furthermore, the program likely gave utilities extra incentive to engage in innovation, which shifted MC of abatement curves downward and probably brought total abatement costs down even further.

Finally, it is likely that the societal benefits of the program have hugely outweighed the costs. The consensus of economic studies is that, for an average annual cost of about $0.5 – $2 billion, we have been reaping annual benefits of anywhere from $59 to $116 billion. Interestingly enough, only a small portion of the benefits of the program is due to its original intent of combating acid rain. Instead, virtually all of the quantifiable benefits have come from improvements in human health. Discouraging the use of sulfur-laden coal led to a significant reduction in airborne particulate matter in many communities. Economists have estimated the magnitude of these health benefits to be in the tens of billions of dollars.

Despite this impressive track record, subsequent developments have rendered this emissions trading program largely inoperative. In response to legal challenges by states and utilities, the courts have ruled that any changes in the administration of the trading program require new congressional legislation and cannot be done administratively by the EPA. This created problems when the Bush administration tried to direct the EPA to tighten the overall emissions cap in 2005. When the courts said no, the EPA was forced to rely on regulation that focused on individual sources in order to meet new local air quality standards. This had the effect of crippling the emissions markets by driving down demand for the tradable allowances. As a result, allowances prices plummeted [Schmalensee and Stavins (2013), pp. 115–16]. And new legislation to revive the emissions trading program has been stymied in the recent political climate, in which emissions trading programs have been demonized as "cap-and-tax" policies.

The bottom line of this story is that there is no simple political interpretation of the enactment of the emissions trading program in the CAAA90. Special interest groups certainly mattered, but they did not determine the policy outcome, which was also a response to public demand for improved environmental quality. Committees played an important agenda-setting role, but the final policy did not completely reflect their interests either. And the enormous economic benefits of the program turned out to have little to do with combatting acid rain but instead hinged on improving public health, which was not part of the political discussion over enactment of the act. All of this points to both the explanatory power but also the limitations of the public choice approach.

### Key takeaways

- Degraded air quality has major negative effects on human health, especially in lower-income countries.
- Dramatic incidents of degraded air quality resulted in the enactment of air quality legislation in the United States and the United Kingdom beginning in the 1950s.
- Major legislation to protect air and water quality in the United States occurred in the early 1970s with the enactment of the Clean Air Act in 1970 and the Clean Water Act in 1972.

- Two key policies for addressing air and water pollution with similar economic impacts are emissions taxes and tradable emissions permits.
- Emissions taxes may enable us to reap a double dividend, by using the tax revenues to correct inefficiencies in other markets, like labor markets. However, there is some question regarding how large the double dividend is empirically, or if it even exists.
- There may be a role for the property rights approach in addressing air and water pollution, but, in many cases, the transaction costs of private negotiations may be quite high.
- Air quality policy in the United States has likely been shaped in part at the behest of special interests, including eastern coal interests, northern regions concerned with gaining a competitive advantage over other regions, and eastern electric utilities.
- Emissions trading among electric utilities has likely resulted in huge societal benefits, almost all from improvements in human health.

## Key concepts:

- Emissions taxes
- Tradable emissions permits
- Double dividend
- Performance standards
- Technology forcing

## Exercises/discussion questions

10.1. Which of the following policy instruments are command-and-control? If you cannot tell without further information, explain why.
   - Emissions tax
   - Ambient air quality standards
   - Mandating the best available control technology (BACT)
   - Ceilings on emissions rates
   - Cap-and-trade
10.2. Why might command-and-control environmental policy result from the interaction of special interests, as the public choice framework would predict?
10.3. *[Exercise to see the effect of externalities on the efficiency of a monopolized industry]* Suppose that the market demand for steel can be written as: $P = 120 - 0.1Q$, where $P$ is the price of steel in dollars per ton, and $Q$ is the quantity of steel, in thousands of tons. Suppose also that the market for steel has been monopolized by the firm Bigsteel, which can produce steel according to the following marginal cost schedule: $MC = 0.1Q$, where $MC$ is private marginal costs, in dollars per ton. However, suppose that production of steel also imposes external costs on nearby communities, which can be written as $MC^{Ext} = 0.1Q$, where $MC^{Ext}$ is external marginal costs, in dollars per ton. Calculate the social deadweight loss resulting from Bigsteel's exercise of monopoly power.
10.4. Comment on the economic desirability of banning smoking in public places. Would such a ban be efficient?
10.5. Use the public choice framework to explain why tradable emissions permits might be more politically feasible than emissions taxes.

## Notes

1 More information on the methodology used to perform this calculation can be found at https://aqli .epic.uchicago.edu/about/methodology/.
2 These numbers represent age-standardized death rates, per 100,000 population. See *Our World in Data*, https://ourworldindata.org/air-pollution for more details.
3 Iati (2019); Martin (2019); "Lead in Canada's drinking water" (2019).
4 In 1957 and 1962, London was hit again by similar smog events, each of which is thought to be responsible for hundreds more deaths.
5 See, for example, Anderson and Leal (2001).
6 For more on the case for relying on the courts rather than government regulation, see the readings in Meiners and Morris (2000). See also Yandle (2013).
7 To see how environmental demands actually made it into the CAA, see, for example, Sunstein (2002); Yandle (2013).
8 This discussion draws heavily on Ackerman and Hassler (1980; 1981).
9 This discussion draws heavily on Joskow and Schmalensee (1998); Schmalensee and Stavins (2013); and Schmalensee and Stavins (2019).
10 The more correct term for acid rain is acid precipitation, but for simplicity I will call it acid rain in this discussion.
11 Another influential congressman was John Dingell of Michigan, who was chair of the House Energy and Commerce Committee and used his position to block any acid rain legislation.
12 These estimates may be found in Schmalensee and Stavins (2019), p. 37; and Goulder (2013), p. 91.

## Further readings

### I. More on the effects of the 1990 CAAA

Popp, David. "Pollution control innovations and the Clean Air Act of 1990," *Journal of Policy Analysis and Management* 22(Autumn 2003): 641–60.
*Provides evidence that the 1990 CAAA promoted innovation in smoke-scrubbing technology.*

### II. Public choice in environmental policy

Hussain, Anwar and David N. Laband. "The tragedy of the political commons: Evidence from U.S. Senate roll call votes on environmental legislation," *Public Choice* 124(September 2005): 353–64.
*This paper provides evidence from voting records that U.S. senators tend to vote against environmental regulation that imposes costs in constituents in their own states, and in favor when such regulation imposes costs on other states.*
Maloney, Michael T. and Robert E. McCormick. "A positive theory of environmental quality regulation," *Journal of Law and Economics* 25(April 1982): 99–123.
*Argues that environmental policies can increase incomes of the producers of goods that generate externalities, providing incentive for environmentalists and regulated industries to form political coalitions.*
Potoski, Matthew. "Clean air federalism: Do states race to the bottom?" *Public Administration Review* 61(May/June 2001): 335–42.
*Examines the effect of U.S. federalism on state environmental policy, and whether it encourages a "race to the bottom" where states relax their environmental standards in order to compete with other states to attract industry.*
Stanton, Timothy J. and John C. Whitehead. "Special interests and comparative state policy: An analysis of environmental quality expenditures," *Eastern Economic Journal* 20(Fall 1994): 441–52.
*An attempt to separate out and evaluate the explanatory power of competing political and economic explanations for state environmental policy expenditures. Some, not terribly challenging, econometric analysis.*

## References

Ackerman, Bruce A. and William T. Hassler. "Beyond the New Deal: Coal and the Clean Air Act," *Yale Law Journal* 89(July 1980): 1466–71.

Ackerman, Bruce A. and William T. Hassler. *Clean coal, dirty air*. New Haven: Yale University Press, 1981.

Ahlers, Christopher D. "Origins of the Clean Air Act: A new interpretation," *Environmental Law* 45(Winter 2015): 75–127.

Aldy, Joseph et al. "Looking back at 50 years of the Clean Air Act of 1970," *Resources*, June 15, 2020.

Anderson, Terry L. and Donald R. Leal. *Free market environmentalism*. New York: Palgrave, 2001.

Boissoneault, Lorraine. "The deadly Donora smog of 1948 spurred environmental protection – But have we forgotten the lesson?" *Smithsonian*, October 26, 2018.

Buchanan, James M. "External diseconomies, corrective taxes, and market structure," *American Economic Review* 59(1969): 174–7.

"COVID-19 shutdown led to increased solar power output," *Science News*, July 22, 2020.

Goulder, Lawrence H. "Markets for pollution allowances: What are the (new) lessons?" *Journal of Economic Perspectives* 27(Winter 2013): 87–102.

Hahn, Robert W. "Market power and transferable property rights," *Quarterly Journal of Economics* 99(November 1984): 753–65.

Hahn, Robert W. and Gordon L. Hester. "Marketable permits: Lessons for theory and practice," *Ecology Law Quarterly* 16(1989): 361–406.

Hercher, David W. "New source performance standards for coal-fired electric power plants," *Ecology Law Quarterly* 8(1980): 748–61.

Iati, Marisa. "Toxic lead, scared parents and simmering anger: A month inside a city without clean water, " *Washington Post*, October 3, 2019.

Jain, Rishab R. "Air quality sinks to 'severe' in haze-shrouded New Delhi," *ABCNews*, November 12, 2019.

Joskow, Paul L. and Richard Schmalensee. "The political economy of market-based environmental policy: The U.S. acid rain program," *Journal of Law and Economics* 41(April 1998): 37–84.

"Lead in Canada's drinking water worse than Flint crisis, investigation says," *CBSNews*, November 4, 2019.

Lelieveld, Jos; Klaus Klingmüller; Andrea Pozzer; Ulrich Pöschl; Mohammed Fnais; Andreas Daiber; and Thomas Münzel "Cardiovascular disease burden from ambient air pollution in Europe reassessed using novel hazard ratio functions," *European Heart Journal* 40(May 21, 2019): 1590–6. doi: 10.1093/eurheart/ehz135.

Martin, Jeffery. "Virginia Beach latest community affected by high levels of lead in drinking water," *Newsweek*, November 6, 2019.

Meiners, Roger E. and Andrew P. Morriss (eds.). *The common law and the environment*. New York: Rowman and Littlefield, 2000.

Parry, Ian W.H. and Wallace E. Oates. "Policy analysis in the presence of distorting taxes," *Journal of Policy Analysis and Management* 19(Autumn 2000): 603–13.

Schmalensee, Richard and Robert N. Stavins. "The SO2 allowance trading system: The ironic history of a grand policy experiment," *Journal of Economic Perspectives* 27(Winter 2013): 103–21.

Schmalensee, Richard and Robert N. Stavins. "Policy evolution under the Clean Air Act," *Journal of Economic Perspectives* 33(Fall 2019): 27–50.

Stone, R. "Counting the cost of London's Killer Smog," *Science* 298(2002): 2106–07.

Sunstein, Cass R. *Risk and reason: Safety, law, and the environment*. Cambridge: Cambridge University Press, 2002.

World Health Organization. *Ambient air pollution – a major threat to health and climate*, 2020. https://www.who.int/airpollution/ambient/en/, accessed 12/28/2020.

Yandle, Bruce. "How Earth Day triggered environmental rent seeking," *Independent Review* 18(Summer 2013): 35–47.

Zahran, Sammy et al. "Assessment of the Legionnaires' disease outbreak in Flint, Michigan," *PNAS* 115(February 20, 2018): E1730–9.

# 11 Energy conservation and energy efficiency

The environmental issues that accompany the burning of fossil fuels have prompted searches for alternative ways to meet our energy needs. One possibility is simply to conserve energy. That is, to use less energy in satisfying all of our various needs. If, for example, we want to maintain a comfortable temperature in our homes in the wintertime, there are two basic ways to go about it. The first is to add more heat to our living spaces by, say, burning more natural gas in our furnaces. The second is to keep heat from escaping from our living spaces. We could do this by, say, increasing the amount of insulation in our walls, weatherstripping external doors, or installing energy-efficient double-paned windows.

Similarly, if we want to commute to work every day, we can put a lot of gasoline into a gas guzzler, or we can buy a fuel-efficient hybrid car that uses less gasoline. From biking to buying an energy-efficient furnace to lowering the thermostat setting in your house to adjusting the screensaver setting on your computer, many options are available to reduce energy use and still obtain the things we want.

When we use less fossil fuel energy, we produce fewer greenhouse gases as well as other pollutants that reduce environmental quality. Less coal burnt in power plants means less coal ash residue that requires disposal, less particulate matter in the air we breathe, and less sulfur dioxide emitted into the atmosphere. All of this confers benefits on society.

At the same time, there are costs associated with energy conservation. These include cooler-than-you-would-like temperatures in your house, higher ambient radon levels in houses that are better insulated, and hybrid cars that are less safe (because they are lighter). In addition, more energy-efficient processes often mean greater use of other factor inputs (materials, capital), to make up for the less energy used. Greater use of other inputs implies real resource costs, since it forecloses those inputs from being used in other ways to produce other things we value.

## The origins of energy conservation efforts

Energy conservation as deliberate government policy originally stems from major geopolitical events in the 1970s. At the time, the vast majority of oil sold on world oil markets came from a handful of countries, mostly located in the Middle East. In 1960, these countries formed an organization – the Organization of the Petroleum Exporting Countries (OPEC) – in order to coordinate production, pricing, and export decisions for international sales of crude oil.

In 1973, the OPEC countries declared an embargo on exports to the United States and other industrialized countries who supported Israel in the Yom Kippur War. Then, in 1979, the government of Iran was overthrown and replaced by an Islamic fundamentalist government. This event fueled tensions between Iran and Iraq and ultimately triggered the Iran-Iraq War, which began in September 1980. These two sets of events disrupted the flow of oil from the Middle East to world oil markets and resulted in a dramatic increase in world oil prices. In 1973–1974, the world price of oil quadrupled from $3 per barrel in October 1973 to $12 by March 1974. Then, in 1979–1980, the price doubled again.

These developments caused great consternation among western countries accustomed to uninterrupted flows of cheap oil to fuel their economies. In the United States, concern was heightened by severe shortages of gasoline caused by rising oil prices and wage-and-price controls set in place by the Nixon administration. These wage-and-price controls limited the ability of gasoline markets to adjust to the oil shocks by keeping gas prices artificially low. The result was widespread shortages of gasoline.

The short-term government response was to enact various policies designed to reduce energy consumption. Beginning in the 1970s, Canada, France, the United Kingdom, Germany, the Netherlands, Japan, Italy, and Sweden all embarked on major efforts to reduce energy use. Several European countries banned non-essential driving on Sundays. The UK temporarily imposed a three-day workweek in order to limit electricity use, and launched programs to upgrade insulation in residential housing and public buildings, and fund demonstration projects for industrial energy conservation.[1]

In the United States, there was a flurry of federal legislation to promote energy conservation beginning in 1974. In that year, Congress enacted the *Emergency Highway Energy Conservation Act*, which reduced the speed limit to 55mph. This was followed, in a few short years, by a series of federal laws aimed at, among other things: regulating fuel efficiency requirements for automobiles, providing assistance to low-income families to increase home energy efficiency, and setting energy efficiency standards for household appliances. Table 11.1 lists additional major energy conservation legislation enacted in the United States since 1974.

Beginning in the early 1970s, we began to observe a steady secular decline in the amount of energy used to produce goods and services, the so-called *energy intensity* of production. Figure 11.1 shows the U.S. time trend of one measure of energy intensity – primary energy consumption per dollar of GDP – since 1950, plus projections into the future. Beginning in the early 1970s, there is a noticeable decline: by 2013, energy intensity had fallen by nearly 40%. Furthermore, this trend is expected to continue into the future, with energy intensity projected to roughly halve between 2013 and 2040 [U.S. Energy Information Administration (2013)].

## Economics of energy conservation

If you think about it, it may not be clear to you why energy conservation policies are even warranted, at least on grounds of economic efficiency. After all, price increases such as occurred during the oil shocks of the 1970s should induce reductions in energy use on their own. When faced with higher energy prices, you might very well be interested in buying a more fuel-efficient car or dishwasher, or biking more, or adding insulation to your house. Why do we need government policies to encourage you to do these things?

There are actually a number of reasons why energy conservation programs of various kinds may permit us to achieve more efficient energy use. These reasons fall into two basic categories: *market failures*, and *behavioral factors*.[2] All of these reasons explain why consumers, even informed ones, may not make efficient energy use decisions. This is the so-called *energy-efficiency gap*, to which we shall return shortly.

*Table 11.1* Major congressional legislation on energy efficiency, 1974 to 2011

| Policy | Year | Subject |
|---|---|---|
| Emergency Highway Energy Conservation Act | 1974 | National speed limits |
| Energy Policy and Conservation Act | 1975 | Established Strategic Petroleum Reserve; Energy conservation: consumer products; Fuel efficiency requirements (CAFÉ): vehicles |
| Department of Energy Organization Act | 1977 | Established the Department of Energy; Established Federal Energy Regulatory Commission |
| National Energy Act | 1978 | Income tax credit, loans for home solar and wind installations; "Gas guzzler" tax |
| Public Utility Regulatory Policies Act | 1978 | Energy conservation/efficiency: utilities; Subsidized use of renewable energies |
| Energy Security Act | 1980 | Established Synthetic Fuels Corporation; Established acid precipitation rain task force |
| Low-income Home Energy Assistance Program (LIHEAP) | 1981 | Home weatherization, low-income households |
| Energy Policy Act | 1992 | Energy efficiency: buildings; Energy efficiency standards: appliances; Deregulated power industry |
| Energy Policy Act | 2005 | Tax incentives for energy conservation; Use of alternative fuels |
| Energy Independence and Security Act | 2007 | Increased fuel efficiency standards: vehicles; Phased out incandescent light bulbs |
| American Recovery and Reinvestment Act | 2009 | Short term tax credits for home energy efficiency improvements |

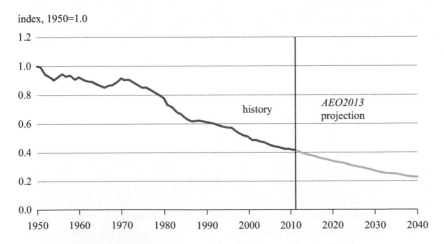

index, 1950=1.0

*Figure 11.1* Energy consumption per real dollar of GDP index, U.S. *Source*: Energy Information Administration. "U.S. energy intensity projected to continue its steady decline through 2040," *Today in Energy*, March 1, 2013.

# I. Market failures

## A. Externalities

There are several forms that market failures could take. First, keep in mind that the market price of energy may not be the price that encourages efficient energy consumption when externalities are taken into account. As we have seen, the world price of oil does not reflect the unpriced externalities of greenhouse gases and other pollutants. This implies that the price of energy should be higher and, if it is too low, firms and consumers are given insufficient incentive to adopt energy conservation measures. Furthermore, if the price of oil is too low, this translates directly into prices of products that oil is used to produce, such as gasoline, home heating oil, and chemicals, that may also be too low.

Some evidence suggests that these externality costs are substantial. One study, for example, estimated that the costs associated with various externalities associated with driving – pollution, getting into accidents, and increased congestion on the roads – are roughly $2–3 per gallon of gasoline [Parry et al. (2007)]. So, according to this study, if gasoline costs $3 per gallon at the pump, this implies that the efficient price of gasoline is anywhere from 67% to 100% higher.

Similarly, in 2010 the National Research Council estimated that the externality costs of coal-powered electricity, *not including impact on climate*, were about 3–4 cents per kilowatt-hour. At the time, the average price of power in the United States was about 12 cents per kilowatt-hour. So the efficient price of power may have been about 25–33% higher than the price consumers actually had to pay.[3] And again, this doesn't even include climate impact costs.

## B. Imperfect information

Another type of market failure arises when consumers do not have good information on new, more energy-efficient technologies such as hybrid cars and energy-saving appliances. When new products are first introduced, it may take time for many consumers to even be aware they exist. And, even when they find out about them, they may lack good information on whether it would be profitable to switch to the new product. This problem is exacerbated by the fact that new products lack a proven track record on which to base an informed buying decision. I should add that information regarding energy-saving products behaves in many ways like a public good. Use of information is non-rivalrous and available to everyone once it is supplied. This implies, of course, that markets are likely to undersupply this information.

## C. Split incentives

A final type of market failure occurs when the individuals who make decisions on energy use do not bear the costs of their decision-making. This can happen, for example, when people rent apartments from landlords. If renters pay for utilities, landlords lack sufficient incentive to make energy-saving investments such as energy-saving appliances and insulation. On the other hand, if landlords pay for utilities, renters may have little incentive to keep their thermostats set low. One empirical study found, for example, that renters who do not pay for their heating expenses are less likely to change their thermostat settings in order to reduce energy consumption [Gillingham et al. (2012)].

## II. Behavioral factors

You can think of the market failure factors we just discussed as being traditional arguments based on standard economic theory. These arguments assume rational, individually optimizing behavior subject to constraints of various kinds (technology, income, time). In addition, various behaviors do not fall easily into this category of rational, optimizing behavior, at least as traditionally defined by economists. The result may be decisions not to purchase new energy-saving products even when it makes economic sense to do so. Let us think about a few of these factors, what economists have called behavioral factors.

### A. High discount rates

Behavioral explanations for decisions by consumers not to adopt energy-saving technologies fall into several categories. One has to do with the question of how much consumers appear to discount future energy savings. Notice that the decision to invest in an energy-saving technology involves a tradeoff between upfront costs in the present and a stream of future energy savings. For example, you insulate your house now in order to reduce your heating bills into the future. Or you buy a hybrid car now in anticipation of fuel savings. For a consumer assessing this tradeoff, a key issue is the rate at which future energy savings are discounted. The higher the discount rate, the less attractive are those future energy savings and, thus, the less likely that a new energy-saving product will be purchased.

Some evidence suggests that consumers apply extremely high discount rates when purchasing energy-saving appliances. This includes the results of several studies, which found that consumers apply discount rates of anywhere from 20% to 800% (that is not a typo!).[4] The fact that these implied discount rates are so high and, in particular, well above market interest rates suggests that consumers may not be making rational energy conservation choices. And this would explain why (some) valuable energy-saving investments are not being made.

Some economists have argued, however, that these apparent high discount rates may not be real. For example, consumers may be reluctant to invest in energy-saving appliances if they are unsure how much energy they will save [Sutherland (1991)]. Or there may be hidden costs not taken into account in these discount rate studies. This would include things like the time it takes to research product alternatives. More generally, there may be various costs experienced by consumers that are not accounted for in these studies. In these cases, decisions not to invest may be perfectly rational. The research challenge is that it is difficult to quantify these costs, or the effect of uncertainty in energy savings. At this point, the jury is still out on whether subjective high discount rates can explain decisions not to invest in energy-saving appliances.

### B. Limited attention

A related issue is the fact that making an energy-saving purchase can be a complex problem for many consumers. For many types of purchases, consumers may have to consider a wide range of factors that are relevant to energy savings. Think, for example, about the energy efficiency of a house that you are considering purchasing. Its energy efficiency depends upon how much insulation are in its walls; the thickness and energy rating of that insulation; the number of windows; whether or not those windows are double-paned; the thermal efficiency of its furnace and air conditioning system; and so forth.

The fact that there are so many factors to consider makes it harder for consumers to effectively assess the energy savings component of their decisions. As a result, they may downplay or dismiss energy costs and focus on the house's more obvious characteristics. Like its beautiful landscaping, wood-burning fireplace, or the quaint gazebo in the backyard. That is, they are less attentive to the factors they need to pay attention to in order to make an informed energy-saving decision. This is the issue of *limited attention*.

A recent study has found that limited attention may have a significant impact on whether or not consumers undertake an energy audit of their homes [Palmer and Walls (2015)]. Energy audits assess the energy efficiency of a house, including ways a house may be losing energy and recommendations for investments that more than recoup their costs in energy savings. Furthermore, energy audits do not cost very much. My local utility company charges $30 for one, which includes a comprehensive inspection of a house's windows and exterior doors, amount of insulation, and the efficiency and general condition of its heating and cooling systems. One would think many consumers would take advantage of such a good deal.

However, surprisingly few homeowners have one done for their house. The study constructed a measure of inattentiveness based on the result of a survey of homeowners and found that consumers who were more inattentive were significantly less likely to have an energy audit done. The complexity of the energy use features of houses may make people gloss over valuable steps they could take to save energy.

## The energy-efficiency gap

When you take all of these factors – both market failure and behavioral – into account, they add up to a notion that has occupied many economists and policymakers: the possible existence of a so-called *energy-efficiency gap*. By this, we mean a widespread failure to undertake energy-saving investments that yield more social benefits than they cost. By some accounts, the energy-efficiency gap is quite large and responsible for considerable economic losses. For example, a 2009 report by a highly respected consulting firm found that in the United States, many households and businesses fail to undertake simple investments that in the aggregate could dramatically reduce energy consumption and save hundreds of billions of dollars.[5]

Other economists have argued, however, that the evidence for the existence of a significant energy-efficiency gap is somewhat mixed. An important reason is that it is not clear how trustworthy many estimates of energy cost savings in fact are. The problem is that many of these estimates come from engineering or observational studies that use methodologies that are likely to yield biased estimates. For example, these types of studies commonly do not take into account behavioral responses to price changes. Thus, these studies are likely to overstate the potential energy cost savings. Other studies that have investigated energy use inefficiencies using economic methods have consistently found them to be considerably smaller than those estimated in engineering studies.

I should add that energy-saving investments have other effects that are difficult to observe. For example, one factor that is difficult for studies to assess is the effect of energy-saving investments on the health of homeowners. A health factor that might be important is the potential for increased exposure to radon gases, which can be higher in more well-insulated houses. There may also be impacts on other industries such as suppliers of related and complementary products. These and other factors can make it difficult to construct reliable estimates of costs and benefits. All of this has made it difficult to document conclusively the existence of an energy-efficiency gap.

## Two energy conservation policies

In this section, we take a closer look at two policies that have received a lot of attention from economists. These policies – *fuel taxes* and *fuel economy standards* – are designed to encourage energy conservation in the transportation sector. Describing and comparing them is instructive because they are two very different kinds of policies with different implications for economic efficiency. At the same time, they present very different public choice issues.

### I. Fuel taxes

The first option, and one favored by many economists, is to raise taxes on fuel used in automobiles and other motor vehicles. In the United States, these are typically referred to as *gasoline taxes*, though taxes are also applied to other motor fuels such as diesel fuel. The idea, of course, is to raise fuel prices and discourage consumption, while providing a source of tax revenue for the government. Figure 11.2 illustrates the effect of imposing a constant (excise) tax of t★ per unit of fuel, which drives the price up from P★ to P', resulting in a reduction in consumption from Q★ to Q'. The government raises tax revenues equal to the tax multiplied by Q', the final amount of fuel sold. Thus, total tax revenues may be calculated as the area of the shaded box.

It is useful to notice that Figure 11.2 illustrates a tension between the goals of discouraging consumption and raising tax revenues. In general, the more consumption is discouraged, the less tax revenues will be raised. Which effect dominates depends centrally upon the price elasticity of demand. If demand is inelastic (demand curve is *steep*), consumption will not decline as much, but more tax revenues will be raised. If demand is elastic (demand curve is *flat*), consumption will decline more, but less revenues will be raised.

Economic studies tend to show that the demand for gasoline, while price-inelastic in the short run, exhibits significant elasticity in the long run (see Table 11.2). So, for example, Dahl and Sterner estimate that a 1% increase in the price of gasoline will result in a 0.26% decline in the quantity of gasoline demanded in the short run. Overall, the evidence in this table suggests that raising gasoline taxes should result in significant reductions in gasoline

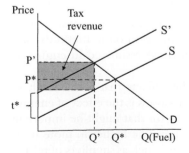

*Figure 11.2* Imposing a fuel tax

*Table 11.2* Estimates of price elasticities of demand for gasoline [adapted from Dahl and Sterner (1991)]

| Study | Elasticity | |
|---|---|---|
| | Short run | Long run |
| Dahl and Sterner (1991) | −0.26 | −0.86 |
| Dahl (1986) | −0.29 | −1.02 |
| Bohi and Zimmerman (1984) | −0.26 | −0.70 |
| Bohi (1981) | −0.22 | −0.58 |

consumption, especially in the long run. Economists point to evidence like this to support recommendations to impose gas taxes in order to save energy in the transportation sector.

Despite evidence that gas taxes could make a big difference in helping us to conserve energy, it turns out that they are used surprisingly little as a policy instrument. In the United States, the current federal gasoline tax of 18.4 cents per gallon was enacted in 1993 and has not changed since then. In the meantime, consumer prices have increased by nearly 70%.[6]

Perhaps the most important reason the United States does not use gasoline taxes more is that they are extremely unpopular politically. A recent Gallup poll found, for example, that 66% of Americans oppose increasing gas taxes, even if the proceeds are used to fund mass transit and repairs to roads and bridges. And such opposition spans the political spectrum, though Democrats are somewhat more likely to support such a tax.[7] And this opposition is not only a recent phenomenon. Ever since the first oil crisis in 1973, attempts to increase gas taxes have mostly failed (see Box 11.1).

---

**Box 11.1: The historical unpopularity of gas taxes in the United States**

Over the past 50 years, there have been periodic calls to increase gasoline taxes. At the time of the 1973–4 embargo, President Nixon proposed a ten cent increase in the gas tax, in order to reduce excess demand. However, it ran into opposition in Congress and was never **implemented**, and the idea went nowhere under Nixon's successors, Gerald Ford and Jimmy Carter. At the time, the tax was criticized as being regressive: that is, burdening lower-income consumers. It was also opposed by representatives of rural areas, which were also depicted as being unduly burdened. Various polls taken at the time all indicated strong popular opposition to gas taxes [Knittel (2013)].

---

Why is there such deep-seated opposition to gas taxes in the United States? A basic public choice story might go something like this: for a variety of reasons, Americans consume a lot of gasoline. These reasons include historically low gas prices and relatively undeveloped mass transit systems (compared to much of Europe). Furthermore, the United States has land use laws that promote low-density urban development.[8] All of this means that Americans are more invested (have greater stakes) in low gas prices than people in other countries, which explains why they would more broadly oppose gas taxes [Hammar et al. (2004)].

This story is certainly consistent with the facts: Americans do drive a lot and enjoy some of the lowest gas prices in the world. However, it seems incomplete. For one thing, in dollar terms the actual size of the stakes do not seem very high. To see what I mean, let us go through a simple numerical example. Suppose you are considering imposing a gas tax of ten cents per gallon. In 2020, the average price of gasoline in the United States was roughly $2.50 per gallon. So let us suppose that imposing this tax would increase the price of gas by 4% (10/250).[9] Now picking a state at random, suppose you are a resident of, say, New York. If so, you spent $623 on gas in 2018 [Shaw (2019)]. If your gas bill increased by 4%, imposing the tax would only add about $25 to your annual gas bill.[10] This hardly seems worth getting all hot and bothered about.

Yet people do, so what is going on? Here is one factor: people drive a lot and, as a result, they are very aware of the price of gas and how much they are having to pay. These facts make gas taxes highly visible and conspicuous. Studies have shown that taxes that are highly visible are likely to encounter the most public resistance. So politicians tend to experience large political costs if they impose taxes like gas taxes, sales taxes in general, and income taxes, as opposed to other ways they could raise revenues, such as user fees, corporate taxes, and royalties on natural resources.[11] There are taxes and there are taxes, and some you do not want to touch with a ten-foot pole if you are a government trying to raise revenues.

However, the answer probably goes deeper than this. Some argue that *path dependence* helps explain how an economy like the United States can develop an entrenched reliance on automobile travel. Once we have started down a path of reliance on the automobile, staying on that path has self-reinforcing tendencies. Figure 11.3 shows this dynamic, where the development of a road network encourages people to drive more, which results in added tax revenues, which encourages further expansion of the road network, which encourages people to drive more, and so forth. It also shows that driving can itself be self-reinforcing, as driving becomes a habit, which perpetuates our tendencies to drive. The entire process resembles the process we encountered in chapter four in which the QWERTY keyboard came to dominate typewriting and later word processing, leaving DVORAK in the dust.

The emergence of driving dominance through this path-dependent process may help explain some of the resistance in the U.S. to gasoline taxes. It is not simply that Americans drive a lot. According to this argument, Americans have become accustomed to driving, and this may make attempts to impose a gasoline tax that much more objectionable.

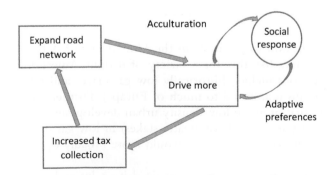

*Figure 11.3* Path dependence in driving [adapted from Unruh (2000)]

---

### Box 11.2: Path dependence and the demise of the steam car

Did you know that we could be driving cars powered by steam? In the early 20th century, cars powered by steam engines were in serious contention with cars powered by internal combustion engines. The Stanley Steamer car was designed by two inventor brothers, Francis and Freelan Stanley. Its engine was much lighter and less complicated (and therefore less likely to break down) than the finicky internal combustion engine. However, just as it was getting off the ground, there was an outbreak of hoof-and-mouth disease in the northeastern United States, which led authorities to close all of the horse troughs, thereby depriving drivers of water to refill the boilers of their steam cars. The Stanley Steamer never recovered [McLaughlin (1954)].

---

All of this is consistent with the fact that Europeans drive considerably less than Americans and also happen to have much higher fuel taxes. Consider countries in the European Union, where gasoline taxes range from a high of 0.79 euros per liter in the Netherlands, down to a low of 0.36 in Bulgaria [see Asen (2019)]. Compare those figures to the federal gasoline tax in the United States of 18.4 cents per gallon, which is around 0.04 euros per liter. That's right: on the country level, gasoline taxes are higher in Europe than in the United States by roughly an order of magnitude or more. And, even when you account for state taxes imposed in the United States, the gulf between U.S. and European gasoline taxes remains vast. Gasoline taxes are apparently nowhere near as objectionable or unpopular in Europe as they are in the United States.

### II. Fuel economy standards

Setting standards for automobile fuel efficiency levels is another policy designed to conserve energy. The European Union has in place the Automotive Fuel Economy Policy, which includes fuel economy targets for auto manufacturers. In the United States, Congress passed the *Energy Policy and Conservation Act* in 1975, which created the Corporate Average Fuel Economy (CAFE) program. Both programs require auto manufacturers to achieve minimum gas mileage standards for their entire fleet of new passenger cars. For example, under CAFÉ each manufacturer's fleet of cars produced in 1978 had to achieve average gas mileage of at least 18 miles per gallon (MPG), a number that was gradually increased over time.

As opposed to gasoline taxes, both the fuel economy targets of the Automotive Fuel Economy Policy and CAFE program are performance standards. At the same time, they provide flexibility in terms of how manufacturers may meet the standards; for example, through engineering new car designs or shifting production to more fuel-efficient models.

Let us now examine the CAFE program more closely. First, evidence suggests that it has indeed been effective in increasing auto fuel efficiency for cars produced in the United States. Figure 11.4 shows, for example, that fuel economy for new cars increased dramatically in the first ten years of the program, increasing from roughly 13MPG in 1975 to 23MPG in 1985. While part of this may have been manufacturers responding to rising oil prices in the late 1970s, notice that fuel efficiency continued to improve even after oil prices began to fall after 1982.

At the same time, despite its apparent effectiveness, it is not entirely clear that the regulatory approach taken by the CAFE program is the best way to reduce gasoline consumption. There are several related reasons for this.

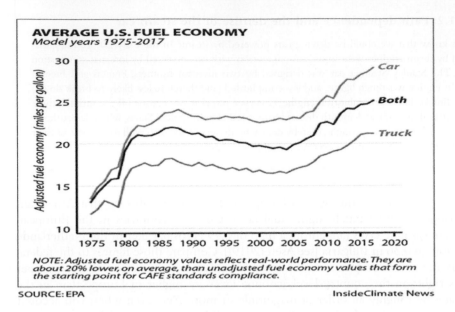

*Figure 11.4* Average fleet gas mileage, U.S. *Source*: Environmental Protection Agency, *InsideClimateNews*.

One of the biggest concerns of economists about CAFE standards is that they may have the unintended effect of actually encouraging people to drive more. If you think about it, more fuel-efficient cars makes driving cheaper. So people may compensate by driving more often or for longer distances. After they buy that new fuel-efficient Prius, maybe they become more likely to consider that long road trip to Yellowstone. This phenomenon is known as the *rebound effect*. And if the rebound effect occurs, it will tend to offset the hoped-for fuel efficiency gains. Some studies have estimated that the rebound effect could reduce fuel efficiency gains under the CAFE program by 10–20%, though the actual magnitude has been subject to debate.[12]

In addition to offsetting fuel efficiency gains, the rebound effect could have other undesirable impacts. When people drive more, there are more cars on the roads, resulting in more traffic congestion. This imposes costs on other drivers, in terms of longer commute times, wasted fuel from idling in traffic, and so forth. Furthermore, more cars on the roads means the potential for more traffic accidents, resulting in not only more property damage, but increased risk of serious injuries and even fatalities. And the potential for fatalities and serious injuries is greater if auto manufacturers respond to the fuel efficiency standards by doing things that make cars lighter or, more generally, less safe.

For all these reasons, many economists have concluded that on pure economic grounds, CAFE standards are a less attractive policy option than gasoline taxes. However, as we have seen, gasoline taxes are extremely unpopular and it may be politically infeasible to increase them much, if at all, above current levels. Furthermore, CAFE standards are an example of a less visible government intervention, which means they may be less likely to encounter political resistance. All of this means that fuel efficiency standards like CAFE, for all their imperfections, may have a policy role to play.

# The importance of social norms in energy conservation

In recent days, economists have been paying increasing attention to the importance of social norms in encouraging energy conservation. As we described in chapter four, social norms are informal rules that people abide by, in part because violating these rules may result in social sanctions of various kinds. We worry about what other people think of us. We desire social acceptance. We want to fit in. When we are unsure what to do, we take cues from what other people are doing. More generally, there are norms of behavior that we tend to abide by because there are generally accepted ways of behaving. This is an important reason why most of us, for example, do not just throw trash out the car window when we are driving on the interstate, even though the chances of being spotted and fined are probably really small.

There are various ways in which our all-too-human desire to abide by social norms might be used to promote energy conservation. To help think about these, let me introduce a distinction often made by psychologists: between so-called *descriptive* norms and *injunctive* norms. A descriptive norm refers to what we perceive others are doing, while an injunctive norm refers to what we perceive others think *we* should be doing. Both can influence our behavior, but in slightly different ways.

An example of a descriptive norm is a perception that everyone else in your neighborhood is recycling their bottles, cans, and plastics. Every Wednesday morning, you see the recycling bins in front of all of the neighbors' houses. You might feel guilty if you were the only one not recycling. Here, it is easy to believe everyone else is recycling, because you can actually see the recycling bins. But this principle applies to many things that are not so easy to observe. A car company tells you, for example, that one of their models is "the most popular car in America." Even though you realize that the company making the claim is not exactly a disinterested party, many people might find it persuasive or, at least, a factor to take into account. After all, all these people must know something!

Or consider Figure 11.5, which shows a sign used in a recent water conservation campaign. Rather than simply asking people to conserve water, this sign conveys the message

*Figure 11.5* Descriptive norm. *Source*: Shutterstock

that the person reading the sign is not going it alone. Rather, (s)he is part of a group of people all working together to conserve water. The sign leads readers to think: "Well, all of these other people are doing it, so maybe I should also."

Injunctive norms are slightly different. They are not merely about what other people are doing: they involve projecting how others must feel about what *you* are doing. In the recycling example, it goes beyond the fact that everyone else in the neighborhood is setting out their recycling bins. You might believe that they are judging you harshly for not setting out yours as well. And you can probably see how that might be a powerful motivator for many people.

As an example of an injunctive norm, Figure 11.6 shows a publicly posted sign from an anti-littering campaign. Notice that this sign says nothing about how many other people are picking up after themselves: its strategy is quite different from a descriptive norm. Rather, it is trying to convey social disapproval of littering, the message that people who litter are inconsiderate and immature.

The water conservation and anti-littering campaigns are both examples of policies exploiting social norms in order to elicit better – that is, more pro-social – behavior. Many others can be found, including policies regarding energy efficiency, recycling, binge drinking, combatting youth violence, smoking, drug use, and drunk driving.

Regardless of the policy, there is always the question of how effective it will be in influencing people's behavior. After all, people are not being *required* to recycle or to refrain from littering. They are not being threatened with arrest or having to pay a fine (in most cases). In all of these situations, policymakers are relying upon a social norm to get people to do the right thing. But people can always choose to ignore it if they wish.

*Figure 11.6* Injunctive norm. *Source*: Shutterstock

In fact, the effectiveness of social norming varies quite a bit. Social norm-based campaigns against binge drinking on college campuses, for example, have been largely unsuccessful. On the other hand, government policies regarding cigarette smoking have been, by all accounts, highly effective in discouraging consumption [see Lessig (1995)]. And, similarly, as we shall see below, various policies designed to encourage more efficient energy use have also turned out to be effective.

What distinguishes the effective policies from the ineffective ones? The answer is complicated, but here are a few things we do know. First, context probably matters a lot. If a policy is aimed at exploiting a social norm, it helps if the targeted audience is receptive to the message. For example, in the United States, advances in science that made clear the health impacts of second-hand smoke helped to transform social norms regarding smoking. At the same time, the cultural context matters, as evidenced by differences in smoking behavior between the United States and many European countries that persist to this day. It is unclear exactly why, but the culture of the United States made the American public more receptive to anti-smoking policies, with the result that social norms regarding smoking changed significantly and relatively quickly. As two observers put it:

> Like surfers, legislators ... who wish to change everyday social norms must wait for signs of a rising wave of cultural support, catching it at just the right time.
>
> [Kagan and Skolnick (1993), p. 85]

Second, for norm-based policies to be effective, it helps if there is convincing evidence of public harms from existing behavior. In the smoking example, advances in science probably played an important role, at least in the United States. The converse is likely true as well: behavior is unlikely to change if there is a lack of evidence concerning public harms that sufficient numbers of people find convincing.

Third, norm-based policies will likely be more effective if it is relatively easy for people to conform to the requirements of the policies. As a simple example, one study of towel use in a hotel found that hotel patrons were much more likely to reuse hotel towels when the hotel left little information cards in their room that other hotel patrons were reusing *their* towels. And the effect was greater than when the cards had different messages stressing respect for nature or saving energy for future generations.[13]

These last two points are illustrated by a historical example regarding energy conservation. During the energy crisis in the 1970s, President Jimmy Carter gave speeches actively exhorting the American people to lower their thermostats. He was not proposing making excessive energy use illegal and subject to fines or other penalties. Rather, he was using his bully pulpit to try to change the prevailing norm that keeping your thermostat set high was socially acceptable. Carter was hoping that people would start to conserve energy if they believed that others were conserving energy, and the energy conservation bandwagon would start to roll.

As it turned out, it is not clear that Carter's appeal had any real effect on people's behavior. One study showed that, after his speech, only a fraction of households had turned their thermostats down, and that there was little difference between households that had heard his appeal and those that had not [see Luyben (1982)]. Apparently, the hoped-for changes in behavior were either too costly or inconvenient. Or perhaps people were not sufficiently convinced that the energy crisis was serious enough to warrant change.

Some recent studies, however, have found that certain types of interventions exploiting descriptive norms can be effective in significantly reducing energy use. One program that has received a lot of attention is a home energy conservation program used by utility companies across the United States. Under this program, when customers receive their monthly utility bills, they also receive additional home energy reports that compare their energy usage to that of their neighbors. Studies have found that these households tend to reduce their energy consumption more than other households who do not receive these reports.[14] So, apparently, customers are convinced to conserve energy when they are made aware of what their neighbors are doing to conserve energy.

## Key takeaways

- Energy conservation as a conscious government policy originated in the 1970s in the wake of two significant oil shocks in the Middle East. These events fed widespread concern that western economies were vulnerable to oil supply disruptions and even some concerns that the world was running out of energy. The result was a flurry of government programs in the United States and Europe to reduce energy consumption.
- With consistent political resistance to carbon or emissions taxes, western economies have resorted to second-best programs that targeted energy use in other ways. For the most part, these have turned out to be cost-ineffective.
- It is likely that many opportunities to adopt cost-beneficial energy-saving technologies – both products and processes – are being left on the table. However, the jury is still out on how widespread this so-called energy-efficiency gap in fact is.
- Exploiting both descriptive and injunctive norms has been shown to have significant impacts on energy-saving behavior. This is a promising area for future energy conservation policy.

## Key concepts

- Energy-efficiency gap
- Gasoline taxes
- Fuel economy standards
- Descriptive norms
- Injunctive norms

## Exercises/discussion questions

11.1. Behaviorally, why would the discount rate consumers apply to energy savings be so much higher than market interest rates?

11.2. *[Exercise to see the effect on price of imposing a tax under different demand conditions]* Suppose the demand for gasoline could be expressed as: $P = 10 - 0.01Q$, where P is the price of a gallon of gasoline, in dollars; and Q is quantity demanded of gasoline, in gallons. On the other hand, suppose the supply of gasoline is: $P = 2 + 0.01Q$. (a) How much would the price of gasoline increase if you imposed a tax of 25 cents per gallon? (b) Reanswer (a) if demand is: $P = \$5$ (perfect price elasticity). (c) Reanswer (a) if demand is: $Q = 50$ (perfect price inelasticity).

11.3. Assume that in your local market, one million gallons of gasoline are currently being sold at a price of $3 per gallon. Assume, in line with the estimates provided by Dahl and Sterner, that the short-run price elasticity of demand for gas is −0.25. Finally, assume that there is currently no gas tax. You are considering imposing a gas tax of 50 cents per gallon. (a) How much do you think gas consumption will decline, in gallons? (b) How much tax revenues will you be able to raise? {*Note: In answering this question, you may assume that the supply of gasoline is subject to constant marginal cost of production*}

11.4. Can you construct a public choice story based on competing special interests that might explain the political challenges to imposing CAFE standards in the United States?

11.5. Why do you think there is such a much stronger norm against smoking in the United States than there is in Europe?

11.6. Which of the following are governed by descriptive norms and which by injunctive norms?

- People clapping at the end of a concert.
- Your mom saying to you, after you proposed doing something really dumb: "If (your friend) Tracy jumped off a bridge, would you jump off too?"
- Returning your cafeteria tray to the tray return area after finishing lunch.
- Butting into line ahead of other people.

11.7. Suppose you are trying to reduce food waste in the dining halls on your campus. Describe a poster you might post in your campus dining hall that uses descriptive norms. Be specific. Similarly for injunctive norms. Which do you think would have a stronger effect, and why?

## Notes

1 For more on early European efforts at energy conservation, see Van Vactor (1978), pp. 242–4; Mallaburn and Eyre (2014).

2 See, for example, Gerarden et al. (2015). These authors also mention a third explanation – model and measurement errors – which for brevity we will not discuss here.

3 See Jiang (2011) for the estimates of the price of power. In actuality, the delivered price of electricity reflects particular arcane details regarding how electricity is regulated, and so it may or may not reflect externality costs.

4 See, for example, Hausman (1979); Gately (1980); Ruderman et al. (1987).

5 Allcott and Greenstone (2012), pp. 3–4.

6 This calculation was done using the Consumer Price Index for urban consumers between January 1993 and January 2020.

7 See Brown (2013). This poll was taken in the wake of an increase in the state gas tax in Maryland, the first such increase in 20 years. Republicans were far more likely to oppose the tax than Democrats, but no demographic group gave this gas tax majority support.

8 On this point, see Portney et al. (2003).

9 In actuality, it would only increase it by this much if demand was perfectly inelastic. In general, the price at the pump would increase by less than ten cents. See exercise 11.2 at the end of the chapter.

10 And this is assuming, of course, that paying the higher price does not discourage you from driving.

11 See, for example, Landon and Ryan (1997).

12 See, for example, Jones (1993), Portney et al. (2003), Small and Van Dender (2007).

13 See, for example, Cialdini (2005), pp. 159–60.

14 See, for example, Ayres et al. (2013); Allcott and Rogers (2014); Farrow et al. (2018).

## Further readings

### *I. Institutions and energy efficiency*

Duffield, John S. and Charles R. Hankla. "The efficiency of institutions: Political determinants of oil consumption in democracies," *Comparative Politics* 43(January 2011): 187–205.

*Argues that in democracies, legislative institutions may influence the success of energy conservation efforts. These include electoral systems with more veto players, and decentralized political parties.*

## II. Path dependence and energy policy

Lawson, Clive. "Aviation lock-in and emissions trading," *Cambridge Journal of Economics* 36(September 2012): 1221–43.
*Argues that emissions trading programs for aviation emissions can make it more difficult to achieve energy conservation under path dependent conditions where technological development gets locked-in to particular development paths.*

## III. Energy conservation programs

Alberini, Anna; Silvia Banfi; and Celine Ramseier. "Energy efficiency investments in the home: Swiss homeowners and expectations about future energy prices," *Energy Journal* 34(2013): 49–86.
*This study examines energy conservation decisions by households in Switzerland, and characterizes the factors that are effective in encouraging households to conserve energy. These include government rebates, reductions in upfront costs, changes in the time horizon of energy savings benefits, and uncertainty over future energy costs.*
Wirl, Franz. "Lessons from utility conservation programs," *Energy Journal* 21 (2000): 87–108.
*This paper examines why energy demand management programs for electric utilities have performed poorly over time. The author attributes this to strategic behavior by both consumers and utilities, both rationally responding to perverse incentives.*

# References

Allcott, Hunt and Michael Greenstone. "Is there an energy efficiency gap?" *Journal of Economic Perspectives* 26(Winter 2012): 3–28.
Allcott, Hunt and Todd Rogers. "The short-run and long-run effects of behavioral interventions: Experimental evidence from energy conservation," *American Economic Review* 104 (October 2014): 3003–37.
Asen, Elke. "Gas taxes in Europe," *Tax Foundation*, August 15, 2019. https://taxfoundation.org/gas-taxes -europe-2019/, accessed 12/28/2020.
Ayres, Ian; Sophie Raseman; and Alice Shih. "Evidence from two large field experiments that peer comparison feedback can reduce residential energy use," *Journal of Law, Economics, and Organization* 29(October 2013): 992–1022.
Bohi, Douglas R. *Analyzing demand behavior: A study of energy elasticities*. Washington DC: Resources for the Future 1981.
Bohi, Douglas R. and M. Zimmerman. "An update on econometric studies of energy demand," *Annual Review of Energy* 9(1984): 105–54.
Brown, Alyssa. "In U.S., most oppose state gas tax hike to fund repairs," Gallup, April 22, 2013.
Cialdini, Robert B. "Basic social influence is underestimated," *Psychological Inquiry* 16(2005): 158–61.
Dahl, Carol A. "Gasoline demand survey," *Energy Journal* 7(1986): 67–82.
Dahl, Carol and Thomas Sterner. "Analysing gasoline demand elasticities: A survey," *Energy Economics* (July 1991): 203–10.
Farrow, Katherine et al. "Social norm interventions as an underappreciated lever for behavior change in energy conservation," *Journal of Energy and Development* 43(August 2017/Spring 2018): 235–49.
Gately, Dermot. "Individual discount rates and the purchase and utilization of energy-using durables: Comment," *Bell Journal of Economics* 11(1980): 373.
Gerarden, Todd; Richard G. Newell; and Robert N. Stavins. "Deconstructing the energy-efficiency gap: Conceptual frameworks and evidence," *American Economic Review* 105(May 2015): 183–6.
Gillingham, Kenneth; Matthew Harding; and David Rapson. "Split incentives in residential energy consumption," *Energy Journal* 33(2012): 37–62.

Hammar, Henrik; Asa Lofgren; and Thomas Sterner. "Political economy obstacles to fuel taxation," *Energy Journal* 25(2004): 1–17.

Hausman, Jerry A. "Individual discount rates and the purchase and utilization of energy-using durables," *Bell Journal of Economics* 10(1979): 33–54.

InsideClimateNews. "Chart: How average U.S. fuel economy has changed," April 2, 2018.

Jiang, Jess. "The price of electricity in your state," *Planet Money*, National Public Radio, October 28, 2011. https://www.npr.org/sections/money/2011/10/27/141766341/the-price-of-electricity-in-your-state

Jones, Clifton T. "Another look at U.S. passenger vehicle use and the 'rebound' effect from improved fuel efficiency," *Energy Journal* 14(1993): 99–110.

Kagan, Robert A. and Jerome H. Skolnick. "Banning smoking: Compliance without enforcement," in Robert L. Rabin and Stephen D. Sugarman (eds.) *Smoking policy: Law, politics, and culture*. Oxford: Oxford University Press, 1993.

Knittel, Christopher R. "Transportation fuels policy since the OPEC embargo: Paved with good intentions," *American Economic Review* 103(May 2013): 344–9.

Landon, Stuart and David L. Ryan. "The political costs of taxes and government spending," *Canadian Journal of Economics* 30(February 1997): 81–111.

Lessig, Lawrence. "The regulation of social meaning," *University of Chicago Law Review* 62(Summer 1995): 943–1045.

Luyben, P.D. "Prompting thermostat setting behavior: Public response to a presidential appeal for conservation," *Environment and Behavior* 14(1982): 113–28.

Mallaburn, Peter S. and Nick Eyre. "Lessons from energy efficiency policy and programmes in the UK from 1973 to 2013," *Energy Efficiency* 7(2014): 23–41.

McLaughlin, Charles. "The Stanley Steamer: A study in unsuccessful innovation," *Explorations in Entrepreneurial History* 7(October 1954): 37–47.

Palmer, Karen and Margaret Walls. "Limited attention and the residential energy efficiency gap," *American Economic Review* 105(May 2015): 192–5.

Parry, Ian W.H.; Margaret Walls, and Winston Harrington. "Automobile externalities and policies," *Journal of Economic Literature* 45(2007): 373–99.

Portney, Paul et al. "Policy Watch: The economics of fuel economy standards," *Journal of Economic Perspectives* 17(Autumn 2003): 203–17.

Ruderman, Clifford S.; Mark D. Levine; and James E. McMahon. "The behavior of the market for energy efficiency in residential appliances including heating and cooling equipment," *Energy Journal* 8(1987): 101–24.

Shaw, Gabbi. "Here's how much the average person spends on gas in every state," *Business Insider*, February 15, 2019.

Small, Kenneth A. and Kurt Van Dender. "Fuel efficiency and motor vehicle travel: The declining rebound effect," *Energy Journal* 28(2007): 25–51.

Sutherland, Ronald J. "Market barriers to energy-efficiency investments," *Energy Journal* 12(1991): 15–34.

United States. Energy Information Administration. "U.S. energy intensity projected to continue its steady decline through 2040," *Today in Energy*, March 1, 2013.

Van Vactor, S.A. "Energy conservation in the OECD: Progress and results," *Journal of Energy and Development* 3(Spring 1978): 239–59.

# 12 Renewable energy

The discussion of the last few chapters has highlighted an important tradeoff between economic well-being and environmental quality. As we have seen, fossil fuels provide many economic benefits. For example, we use them to manufacture goods, produce electricity, fuel cars, and heat homes and businesses. At the same time, fossil fuel use imposes all sorts of environmental costs including air and water pollution, global climate change, and adverse effects on human health. Because of these concerns, much attention has been devoted to developing alternative sources of energy: ones that can meet our energy needs in more environmentally benign ways. These alternative sources of energy are called *renewable energy* and include solar energy, wind energy, geothermal, hydropower, and biofuels. Many people also include nuclear power in this list.

Over the past 40 years or so, we have made great progress in developing renewable sources of energy. At the same time, developing renewable energy sources has presented numerous economic and technological challenges. Aside from hydropower, renewable energy still meets a relatively small portion of total world energy needs. However, our reliance on renewable energy is definitely on the rise and all indications are that it will be an important part of the energy picture by mid-century.

## Overview of renewable energy

Renewable energy is a catch-all phrase for all types of energy that naturally renew themselves. This stands in contrast to exhaustible energy sources such as fossil fuels. When we extract oil from an oil pool in the ground, there is no tendency for the oil pool to replenish itself, so extracting now means less available in the future. However, when we power photovoltaics with energy from the sun, there is no tendency for the solar power to run out. The same is true of hydropower, wind, and geothermal, the other primary sources of renewable energy.[1] In all of these cases, the energy is essentially inexhaustible, making them attractive as energy sources as long as they can be converted to usable energy at a reasonable cost.

Currently, renewable energy is a relatively small part of the overall energy picture in the United States. In 2019 renewable energy accounted for 11% of total energy consumption and 18% of power production in the United States [U.S. EIA 2020a, 2020b]. However, there are signs that this modest role currently played by renewable energy may be changing. The U.S. Department of Energy forecasts that by 2040, renewable energy is likely to roughly triple from its 2005 levels, surpassing nuclear shortly after 2020 and surpassing coal by 2040 [U.S. EIA (2017), p. 9]. As you might expect, these specific forecasts are subject to uncertainty. However, in the broad picture there seems little doubt that renewable energy will be playing a much larger role in the United States in the foreseeable future.

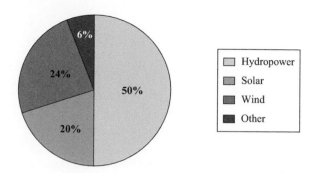

*Figure 12.1* Worldwide renewable power generation capacity, by source, 2018. *Source*: IRENA. "Renewable capacity highlights", March 31, 2019.

The worldwide picture is not very different from the U.S., though renewables have already made greater headway internationally. As of 2018, renewable energy accounted for roughly 15% of total world energy consumption and 28% of world power production [U.S. EIA (2020b)]. Figure 12.1 shows the current breakdown of renewable power production capacity by source. Hydropower comprises roughly half of all renewable capacity worldwide, while wind and solar account for 24% and 20%, respectively [IRENA (2019)]. However, hydropower expansion is increasing slowly and is being heavily dominated by expansion of wind and solar, which accounted for 84% of all new renewable capacity in 2018. In terms of growth potential, you should think of wind and solar as the most likely main sources of added power production capacity in the foreseeable future.

To be clear, renewable energy sources can be used in a variety of different ways to produce things of economic value. Windmills can be used to pump water and grind grain. Green architects design houses to use solar power and geothermal to heat houses and regulate interior temperatures. But by far the most economically important use of renewable energy is for production of electricity (power), using hydropower dams, wind turbines, solar thermal power plants, and photovoltaics. For this reason, much of the rest of the discussion in this chapter will focus on power production.

## Basic economic model of renewable energy

In the basic economic model, renewable energy is available, at some point in time, in essentially infinite amounts at a certain cost. Reflecting conditions in the real world, renewable energy is assumed to be more expensive to begin with, so that it makes economic sense to tap exhaustible sources like fossil fuels first. These so-called *conventional* energy sources provide energy services until that point in time when renewable energy can take over. The conventional sources being fossil fuels, they are exhaustible and under the conditions described in chapter eight are subject to a Hotelling price path. That is, the price of exhaustible energy increases exponentially as those energy sources are depleted over time. In this model, renewable energy is viewed as a *backstop technology*, one that provides a Plan B when conventional sources are exhausted.

Figure 12.2 shows the basic model of renewable energy as a backstop technology. It is assumed that renewable energy will be available at a certain unit cost P*, at some point in time denoted t*. Efficient energy use over time requires that the conventional energy

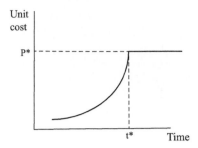

*Figure 12.2* Renewable energy as backstop technology

sources be completely exhausted right at t★, just in time for renewable sources to take over. If this occurs, no energy resources are wasted, and we will be able to meet our energy needs at minimum cost.

In the real world, of course, everything is much more complex. We have obviously not run out of fossil fuels, yet we are already developing, and using, renewable energy technologies. Like all models, this one simplifies the world in order to gain certain insights. It tells us that there is an economic logic to using less costly exhaustible energy sources first, before switching to renewable energy. At the same time, there are market forces driving up the cost of exhaustible sources as they become increasingly depleted. All of this suggests that switching to renewable energy is not a question of if but, rather, when. The key questions are: When will renewable energy be cost-competitive with fossil fuels? And how costly will it be?

## When will renewable energy be competitive with fossil fuel energy in the future? Is it now?

Let us think about the cost competitiveness of renewable energy in the context of power generation, one of the most important uses of renewable energy. Here we will focus on solar and wind energy as the renewable technologies that are likely to make the biggest inroads in electricity production in the foreseeable future.

### I. Technologies for electricity production

Nearly two-thirds of the world's electricity is produced by generators powered by steam turbines that burn fossil fuels.[2] In these systems, coal, natural gas, or oil are fed into a combustion chamber that heats water in a steam boiler. The generated steam powers a turbine connected to a generator that produces electrical power. Exhaust gases from the combustion chamber are released into the atmosphere, generally containing waste products like nitrogen oxides, sulfur dioxide, particulate matter, and mercury. Sometimes the gases are either partially recaptured or subjected to chemical processes that reduce the amount of waste. Most fossil fuel plants use either coal or natural gas, while a smaller percentage use fuel oil, shale oil, synfuels, or biomass. When coal or biomass is burned, solid waste ash is produced, which also must be disposed of.

Wind turbines work by using wind currents to power a generator that produces electricity. The typical wind turbine consists of a structure with a series of three blades that rotate when the wind blows through them. This allows them to capture the wind's kinetic energy and turn it into mechanical energy. When they rotate, they turn a shaft that causes a generator to spin, producing electricity that feeds a power grid. Wind turbines used by electric utilities affix the blade housing on the top of a tall pole structure, because air currents tend to be smoother and more regular at greater heights off the ground. Commercial applications generally consist of an entire field of turbines. Sometimes a field is situated offshore because of favorable wind patterns or to reduce conflicts with land use.

Solar-powered electricity generation relies on one of two distinct technologies: *solar thermal generation* and *photovoltaics*. In solar thermal generation, heat is applied to a thermal unit containing water (or a similar saline substance). This produces steam to power a turbine connected to a generator that produces electricity. This is all similar to the steam turbine technology. However, with solar thermal generation, the heat is generated by concentrating solar radiation using a field of mirrors. Computers control these mirrors to track the sun's movement throughout the day (see Figure 12.3).

The photovoltaic (PV) technology converts sunlight directly into electricity using solar, or photovoltaic, cells. These solar cells generate electricity when they receive sunlight. They are mounted into solar panels that can be deployed on rooftops or in large solar fields. As with solar thermal generation, the PV panels are controlled by computers and moved to track the sun in order to increase power generation efficiency. When they are connected to a power grid, the generated electricity must be processed to be compatible with the power in the grid. This is accomplished by means of an inverter, which changes the direct current (DC) voltage from the PV cells to alternating current (AC) voltage for the grid.

You should keep in mind two important distinctions between conventional fossil fuel power plants and wind- and solar-powered ones. First, both require large upfront investments in plant, machinery, and equipment. However, fossil fuel plants require regular inputs of coal, gas, or oil, resulting in ongoing fuel costs. Wind- and solar-powered plants require relatively little in the way of ongoing costs besides the costs of operating and maintaining the facilities (O&M costs). Taking all costs into account, wind and solar have historically been more expensive ways to produce power, but that may be changing [Scott (2020)].

The other important distinction is that the burning of fossil fuels produces greenhouse gases and other pollutants, contributing to ongoing climate change and other

*Figure 12.3* Solar thermal field. *Source*: Shutterstock

environmental problems. This is not to say that there are no environmental impacts associated with wind and solar energy. These impacts include:

- The use of toxic materials to make PV cells;
- The use of hazardous fluids in solar thermal plants;
- Local effects on land use from siting and building of the generation facilities;
- Possible effects on local water supplies from water used in cooling solar thermal plants; and
- Impacts on wildlife; for example, birds killed by rotating blades.

Overall, however, the environmental impacts of solar and wind are generally considered to be considerably less than those associated with conventional fossil fuel plants.

### II. An important concept: Levelized costs of electricity

Now that we understand the technologies, let us define a measure of costs so that we have a basis for comparison. Economically speaking, when you build a power plant, you incur costs in order to generate a stream of revenues over time. The costs consist of both upfront fixed costs and a stream of variable costs over time. For example, for a conventional fossil fuel-driven plant, there are the construction costs for the plant and equipment itself, and then the ongoing costs of buying coal or gas and hiring labor and other factor inputs. Analysts have defined a cost measure known as the *levelized cost of electricity*, or LCOE. Intuitively, the LCOE measures, over the lifetime of a plant, total costs per unit of power output:

$$LCOE = Total\ power\ costs\ /\ total\ output,\ in\ kilowatt-hours \qquad (12.1)$$

That is, the LCOE measures the unit, or average, cost of producing power over the entire time period when the plant is expected to be in operation.

The LCOE is useful for comparing the cost competitiveness of different technologies. Say, for example, that you wanted to compare the cost of producing electricity using a wind turbine versus a conventional coal-fired plant. You probably would not want to simply compare upfront construction costs because, generally speaking, wind turbines have lower ongoing (variable) costs, in terms of fuel costs and O&M costs. So, even if a wind turbine is more costly to construct, it might still be worthwhile on an overall basis, taking into account all costs, including future costs. The LCOE provides the basis for such a total cost comparison.

A simple example will make the point. Suppose you want to compare two technologies, which we will call *Wind* and *Coal*. To keep things simple, let us assume that each technology has an expected operating lifetime of three years: the current year (t = 0), next year (t = 1), and the following year (t = 2). Suppose the Wind plant costs $1 million to build, while the Coal plant costs $900,000. However, the Coal plant requires $100,000 per year in fuel costs, while the Wind plant requires nothing in the way of fuel. Finally, suppose they produce the same amount of electricity: 50,000 kilowatt-hours per year.

If you just compare upfront construction costs, then it looks like the Coal plant is the better option. Just counting construction costs, it looks like you are able to produce this amount of power for $100,000 less using the Coal plant. But taking into account ongoing fuel costs, it is obviously cheaper, in terms of overall costs, to build the Wind plant.

The comparison we have made is, of course, not quite fair, because we seem to be forgetting our earlier discussion of discounting. According to that discussion, we should be discounting the future fuel costs of the Coal plant, which in this simple example would make it look not quite so bad compared to the Wind plant. The nice thing about the concept of levelized costs is that it both accounts for all costs, both present and future, and discounts them appropriately. It does so by applying the following formula to the stream of costs:

$$LCOE = \sum C_t (1+r)^t / \sum X_t (1+r)^t \qquad (12.2)$$

$C_t$ is the cost incurred in time period t, $X_t$ is power output in time period t, and r is the discount rate. Comparing equation (12.2) to equation (12.1), it can be seen that they are saying the same thing: total costs divided by total output. The only real difference is that total costs and total output are occurring in specific time periods and, because of this, everything is appropriately discounted.

Let us apply this formula to our simple example so we can compare the levelized costs of wind vs. coal. For this calculation, let us assume that the discount rate is 10%. Then the LCOE for the Coal plant is:

$$LCOE^{Coal} = \frac{\left[ 900,000 + 100,000 + 100,000 / 1.1 + 100,000 / (1.1)^2 \right]}{\left[ 50,000 + 50,000 / 1.1 + 50,000 / (1.1)^2 \right]} \qquad (12.3)$$

This equals 8.58. So the average cost per unit of power – the LCOE – using the Coal plant is $8.58 per kilowatt-hour. On the other hand, the levelized cost for wind is:

$$LCOE^{Wind} = 1,000,000 / \left[ 50,000 + 50,000 / 1.1 + 50,000 / (1.1)^2 \right] \qquad (12.4)$$

This equals 7.31. So the LCOE using the Wind plant is $7.31 per kilowatt-hour. So, even when we appropriately discount in this simple example, the wind-powered plant still comes out on top.

Of course, in the real world, many additional factors complicate how we apply the LCOE formula. They include:

- Power plants have fairly long lifetimes, typically 20 or 30 years or more.
- Physical plant wears down over time, reducing the productivity of these plants.
- Tax laws permit physical plants to be depreciated in various ways, which affects taxable income.
- Future streams of costs are often subject to much uncertainty, making it difficult to know what values to use in the LCOE formula.
- The proper choice of discount rate is subject to debate.
- At various times, federal and state governments have provided tax incentives for investing in renewable energy, which reduces the upfront costs.

### *III. What are the relative costs of renewables vs. fossil fuels?*

In order to make this comparison, there is a concept that attempts to capture the relative cost of producing electricity using renewable sources instead of fossil fuels. This concept is called *grid parity*, which is defined as the point where the levelized cost of electricity generated using a renewable energy source equals the price of energy purchased from a power grid powered by conventional fuels. For example, it has become steadily cheaper over time to produce electricity using photovoltaics. When the PV unit cost falls to the same level as the fossil fuel unit price, we say that the two technologies have achieved grid parity.

Figure 12.4 illustrates this concept. Here, we are depicting electricity produced using fossil fuels as increasing over time, as one might expect under a Hotelling price path. On the other hand, the cost of production using PV is declining due to technological change. Grid parity occurs at time period $t^\star$, where the cost per kilowatt-hour (KWH) is equal using either technology.

One reason it is challenging to use the concept of grid parity is that there is large variation in the estimates of the relative costs of renewable energy and fossil fuel energy. To make matters worse, you will even observe large differences in the costs of different plants using the same type of fuel, like natural gas or solar. These variations occur in part because different studies make different assumptions about various economic variables like future fuel costs, government subsidies, and discount rates. The result is that when you read a lot of these studies, you can become uncertain about the answer to a pretty basic question: Which technology is cheaper to use?

In evaluating these studies, the first thing to keep in mind is that, for most renewables, the costs of production depend very much on where the plant site is located. For example, solar power is more cost-competitive in regions that receive more sunlight: in the United States, this especially includes the West and Southwest. In Europe, this includes much of southern Europe, especially Spain. More abundant sunlight means that solar plants can operate at closer to full capacity more of the time, which reduces the generation cost per kilowatt-hour.

Similarly, the cost competitiveness of wind power depends very much upon the speed and regularity of wind flows. Higher and more regular wind speeds increase wind turbine efficiency, thus lowering unit generation costs. Rural areas tend to work better because

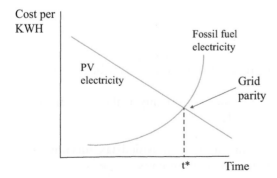

*Figure 12.4* Grid parity

there are fewer structures to cause air turbulence. Particularly attractive are hilltops or wide open plains.

The competitiveness of hydropower depends on the characteristics of the river being dammed: volume, flow speed, and the suitability of the dam site. This is why, for example, the Pacific Northwest states rely most heavily on hydropower of any region in the U.S.: the Columbia River enjoys large volumes of flow year-round, and both the Columbia and Snake Rivers contain ample suitable locations for dams. These conditions mean greater, more regular flow, permitting hydro dams to operate at close to maximum capacity.

The LCOEs associated with fossil fuel plants, on the other hand, tend to vary less with location. This is both because fossil fuel inputs tend to be more standardized and operation of fossil fuel plants are less affected by local conditions such as hours of sunshine, wind conditions, and the availability of suitable dam sites. However, LCOEs can still vary dramatically depending on the specific technology used. For example, there is a large unit cost difference between single-cycle combustion turbines and combined-cycle gas turbines [Borenstein (2012), p. 72]. You do not need to understand the technical differences between these two technologies, just that the cost differences depend upon differences in upfront construction costs and technical efficiency.

The practical upshot of all this is that achieving grid parity depends heavily upon where you happen to live. Consider solar, for example. Because of wide differences in solar availability, the PV parity price varies widely throughout the United States. In my home state of Minnesota, for example, the average PV parity price was $1.79 per watt in 2014. However, at the same time the PV parity price in blessedly sunny Hawaii was over $9 per watt. And it tends to be higher in other states with relatively abundant sunshine, like California and New Mexico. Considering the United States as a whole, in the vast majority of states the PV parity price fell in the range between $1 and $3.[3]

In places with higher conventional generating costs or lower renewable generating costs, we can expect grid parity to be achieved sooner. What all this means, of course, is that efforts to expand PV power production may want to target regions where it is more cost-competitive. In states with higher PV parity prices, substituting PV-generated electricity for fossil fuel-generated electricity will result in smaller increases in overall power production costs. Conversely, we may want to be less aggressive in expanding PV capacity in states where it is not as competitive, which includes many states in the Northwest extending into the upper Midwest and perhaps surprisingly, much of the Southeast.

## Public policies regarding renewable energy

The discussion of the last section expanded on some issues that are relevant to policies to encourage development of renewable energy. For economists, a key comparison is the unit costs of renewables versus fossil fuels and, as we saw, this comparison varies quite a bit by location. However, none of that discussion provided economic justification for policies to encourage greater use of renewable energy. Let us now consider the question of whether such policies may be warranted on economic grounds.

### I. Economic justification

Probably the most important economic justification for policies supporting renewable energy concerns the environmental costs associated with the burning of fossil fuels that, among other things, contributes to global climate change. This argument provides policy justification for government interventions of various kinds into renewable energy markets, since the transaction costs of negotiating to efficient levels of fossil fuel consumption are likely to be quite high.[4] However, if you follow public debates over reliance on fossil fuel technologies, you encounter many other arguments for reducing reliance on these technologies. Let us consider a few of these in turn.

## A. Energy security

Energy security is a bit of an elusive concept but, most commonly, it refers to uninterrupted access to affordable, meaning reasonably priced, energy [See IEA (2020)]. In the context of renewable energy, you often hear the argument that reliance on imported fossil fuels, especially oil, can leave a country vulnerable to supply disruptions and the uncertainties of price fluctuations in world markets. Both of these can have negative effects on a country's economy. Furthermore, oil revenues can prop up unfriendly regimes, support terrorism, and fund the purchase of military arms [Wirth et al. (2003), pp. 133–5]. All of these factors carry the potential to compromise a country's national security: hence the term *energy security*. For all these reasons, there may be value in developing alternative sources that are not subject to such disruptions, such as renewable energy that is developed domestically.

There are a few factors to consider when evaluating this argument for subsidizing renewable energy development. A perhaps obvious one is that the value of promoting energy security depends on how reliant a country is on imported energy. It turns out that countries vary quite a bit in terms of whether they are a net importer or net exporter of oil. The more oil they import, the more vulnerable they are to oil supply disruptions. Over the past half-century or so, the countries most vulnerable to oil disruptions have been the wealthiest countries – the United States, most European countries, Japan, and Korea – who have been importing the most oil. These countries might benefit most from diversifying more into renewable energy. Other countries that import very little oil have little to gain from renewable energy development, at least in terms of promoting energy security.

It should also be kept in mind that countries that rely heavily on imports of one fossil fuel may have plenty of domestic reserves of other fossil fuels. While historically dependent on imported oil, the United States has large reserves of coal, the main fossil fuel it uses to produce electricity. And as we have seen, in recent years it has discovered large deposits of shale oil and gas. For these reasons, economists have argued that from the viewpoint of energy security, the United States has little to gain from producing more electricity from renewable energy.[5] The other countries with the largest coal reserves include Russia, China, Australia, India, and Germany. Power production in all of these countries will be less vulnerable to oil supply disruptions.

It should be added that the sheer variability of power production using renewables makes it less attractive for many countries as a means of achieving greater energy security. This is because a system that depends on renewable energy for power production will still have the need for back-up fossil fuel capacity. And of course, the need to sometimes tap that back-up capacity to produce power [Schmalensee (2012), p. 47].

A final energy security argument that you often hear is that promoting renewable energy development, by reducing the demand for imported fossil fuels, may act to depress the price of fossil fuel energy. Doing so would thus reduce revenue flows to unfriendly countries and reduce the need for military expenditures to protect uninterrupted trade in energy. The main problem with this argument is that for most countries, even heavily oil-reliant ones, the amount of oil this would displace would probably be way too little to have much effect on world oil prices [Borenstein (2012), p. 82].

## B. Information market failures

Another justification for subsidizing renewable energy development is to remedy the market failures that often occur in markets for innovation. Because innovation typically carries societal benefits that cannot be fully captured by the innovator, there tends to be less innovation than is socially optimal. In other words, innovation may confer positive externalities on others. In theory, this creates a role for the government to step in and subsidize innovative activity.

This argument can be viewed in a dynamic setting as policies to advance the point in time at which grid parity occurs. In Figure 12.5, subsidies to photovoltaic innovation shift down the PV cost curve, resulting in grid parity that occurs earlier in time, at t'. So, overall, innovation subsidies can be viewed as means to encourage innovation so that the competitiveness of renewables with conventional technologies occurs more quickly.

Two key questions arise when considering policies of innovation subsidies for renewable energy. One is the magnitude of the private response of innovators to the subsidies. If not much innovation is forthcoming in response to subsidies, then subsidy programs may be costly for relatively little gain. There seem to be plenty of possibilities for productive innovation, things like: improvements in solar efficiency; solar power energy storage; solar airplanes and automobiles; and solar desalination. However, it is unclear how much subsidies themselves have contributed to expanded innovation efforts to date.

To take an example, since 2009 the United States has been promoting the development of energy technologies – including renewables – through the Advanced Research Projects Agency-Energy (ARPA-E), an office in the U.S. Department of Energy. As of 2018, ARPA-E has awarded over $2.6 billion to various entities (universities, private firms, and national laboratories) to conduct basic research in various areas of energy production and use, including energy efficiency, energy storage, and transportation [Hart and Cunliff (2018), p. 1]. This seemingly large figure notwithstanding, studies suggest that this agency is grossly underfunded, given the many promising opportunities that exist for improved energy technologies of various kinds [Borenstein (2012), p. 82].

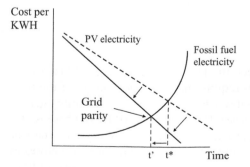

*Figure 12.5* Predicted effect of innovation subsidies for photovoltaics

*C. Green jobs*

One argument you often hear in support of subsidizing renewable energy is that it will create jobs. For example, in 2009 the *Union of Concerned Scientists* (UCS) did a study that concluded that a U.S. policy of generating 25% of its electricity by 2025 (sometimes called "25 by 2025") would create three times as many jobs as producing the same amount of electricity from fossil fuels [Union of Concerned Scientists (2009)]. The reason is that renewable energy technologies are more labor-intensive than fossil fuel technologies. This includes workers for: manufacturing parts for wind turbines; solar panel production, sales, and installation; operations and maintenance on wind farms; and the like. It is also possible that the renewable energy sector will provide "better" jobs, in the sense of requiring more skills, or by displacing low-wage jobs for the newly employed workers.

All of this may be true, but, in evaluating these claims, you should keep a few things in mind. First, it depends upon what kind of jobs you are interested in creating. If the goal is to provide jobs in general, subsidizing renewable energy is likely to succeed only when there is significant unemployment. Otherwise, subsidies will merely have the effect of pulling workers from one industry to another. So there might be more justification for subsidies, for example, during short-term downturns like the financial crisis that began in 2008. However, even here an important question to ask would be whether subsidizing one particular industry is the most efficient way to create new jobs as opposed to, say, broad-based tax incentives.

Suppose, however, that your objective is to create "green jobs." Even then, it may still be unclear that subsidizing renewable energy would be better for the economy, if renewable energy is more expensive than fossil fuel energy. If it is, this means that you might well be trading more jobs for more expensive energy, which imposes a cost on the economy. In order to evaluate this tradeoff, we would need to take into account both the cost differential and how many more jobs would be created by using renewable energy. As we have seen, the difference in cost between renewables and conventional energy varies dramatically, depending on where you happen to be located. And economists do not have a good sense for how many more jobs would be created by more heavy reliance on renewable energy technologies, the claims of the UCS notwithstanding.

## II. Types of policies

Now that we have heard the economic arguments for why subsidies for renewable energy may be warranted, let us now consider the main policies that have been used to subsidize renewable energy. These policies fall into three categories: *renewable portfolio standards* (RPSs), *feed-in tariffs* (FITs), and *investment tax credits* (ITCs).

### A. Renewable portfolio standards

To date, renewable portfolio standards have been implemented in the United States only at the state level. In most states, RPS programs require that some percentage of the total power sales of electric utilities be generated from renewable sources. When a utility generates power using renewable sources, it receives *renewable energy credits* (RECs). These RECs are used to verify that the utilities are meeting their mandated renewable energy targets. In some states, utilities can either generate the power from renewable sources themselves or purchase it from other generators. For example, they might buy renewable power from businesses or homeowners who install photovoltaic systems on their rooftops. In these cases, the purchased energy comes bundled with a certain number of RECs, which the utility can use to meet its mandated renewables target. The types of renewable energy typically covered by RPSs include: wind, solar, biomass, geothermal, and some hydropower. Some states have also included landfill gas, tidal energy, and even energy efficiency.

As of 2019, 29 states and the District of Columbia have adopted RPS programs, while another eight states have adopted voluntary renewable energy targets. The RPSs vary quite a bit across different states, with California and Hawaii being the most aggressive in terms of mandating renewable energy. Both states have set RPS targets of being 100% renewable in power production by 2045.

### B. Feed-in tariffs

Feed-in tariffs are a fairly recent phenomenon in the United States, but they have been used extensively in other countries, especially in Europe [U.S. Department of Energy (2013)]. As opposed to RPSs, which mandate that so much power be generated using renewable energy sources, FIT programs pay generators for producing power with renewable energy. That is, instead of specifying quantities, FITs specify prices. In Europe, FITs have typically worked the following way: governments mandate that utilities enter into long-term contracts with generators to buy all of their renewable generated power at specified rates. In the United States, a handful of states have adopted mandated FIT programs for all utilities, which work similarly to European-style FITs. In addition, other states allow FITs to operate on the level of an individual utility, with utilities working with generators within their service area.

Payments to generators can occur in two ways. Under a *gross metering* system, all electricity produced by a generator goes to the grid and the generator receives the FIT rate for every kilowatt-hour sent to the grid. It then pays for all electricity drawn from the grid at the normal rate per kilowatt-hour. Under a *net metering* system, the generator consumes all the electricity it generates. If it needs more electricity than it generates, it can draw it from the grid, for which it pays the normal rate per kilowatt-hour. If it needs less, it can sell the surplus to the utility at the FIT rate.

### C. Investment tax credits

A third type of subsidy system is based on investment tax credits. In ITC systems, generators of power using renewable energy are permitted to deduct a portion of their system installation costs from their taxes. In the United States, for example, the federal solar investment tax credit currently (as of 2020) permits generators to deduct 26% of the cost of installing a solar system from their federal income taxes. This tax credit is being phased down, however, to 10% by 2022. ITCs thus work differently from FITs, in that they are not based on actual production of renewable energy. This means that the magnitude of the subsidy from an ITC depends in part on technological advance. If, for example, solar panels continue to decline in cost in the future, this will reduce the value of the ITC subsidy to generators.

### D. How efficient are renewable energy subsidies?

It is important to keep in mind that subsidies for renewable energy are a policy alternative to taxing fossil fuels, which we encountered earlier in chapter eight. Both policies would be expected to increase the competitiveness of renewable energy relative to fossil fuels and, thus, both should be considered alternative ways to accomplish the same goal, at least with respect to promoting renewable energy and reducing greenhouse gas emissions. This raises the question of which policy would be better on economic grounds: targeted renewables subsidies or broad-based policies such as carbon taxes.

As we have seen, carbon taxes have the advantage of cost-effectiveness in reducing greenhouse gas emissions. By setting a common price for different emitters with different marginal costs of abatement, carbon taxes encourage more abatements by lower-cost abaters. One issue with subsidies for renewables is that, in general, they are unlikely to achieve cost-effective reductions in greenhouse gas abatement.

One study has found, for example, that instituting a federal RPS in the U.S. would tend to displace cleaner natural gas more than dirtier coal. The reason is that producing electricity with natural gas tends to be more expensive than producing it with coal. Another study has quantified the cost of instituting an RPS in the U.S. and found that it would be roughly twice as costly as instituting an emissions tax.[6] If the goal is to reduce greenhouse gas emissions, renewables subsidies should be considered a second-best policy, compared to taxing emissions. However, renewables subsidies may be an easier political sell than emissions taxes.

## Government support for nuclear power: Path dependence in action

For more insight into the economics of renewable energy, let us consider nuclear power.[7] Nuclear power has been available as a possible energy source since the 1950s, when a number of countries began intensive development programs, including the United States, Canada, France, and the United Kingdom. In the United States, nuclear power saw its heyday during the late 1960s to the mid-1970s. By 1974, there were 54 nuclear reactors operating in the United States, with plans for nearly 200 more. However, safety concerns exacerbated by two highly publicized nuclear incidents – Three Mile Island in 1979 and Chernobyl in 1986 – led to a dramatic decline in development of nuclear capacity both in the United States and Europe. Partly because of this, the estimated costs of developing nuclear have skyrocketed, to the point where it is not considered cost-competitive with either coal or natural gas.[8]

Nevertheless, nuclear power continues to intrigue many observers as a potential major source of energy. Since using it to produce electricity produces virtually no greenhouse gas emissions, it is particularly appealing to many given the threat of ongoing climate change. Famously, the Obama administration made it an important component of his Climate Action Plan. His proposed 2016 fiscal year budget included nearly one billion dollars for research and development and $12.5 billion in loan guarantees for nuclear projects [U.S. White House (2015)].

For the nuclear industry, a key feature of government subsidies has been support for research and development. This meant a good deal of government support when nuclear was just beginning to be developed as a commercially viable technology back in the 1950s and 1960s. At the time, three technologies dominated: light water reactors, heavy water reactors, and gas graphite reactors. You do not need to know all the technological details, but suffice it to say that none of these technologies was clearly superior to the others. In fact, evidence suggests that if anything, light water was considered an inferior technology: more costly, less reliable, and less safe.[9] Nevertheless, at present virtually all operating nuclear reactors in the world outside of Canada use the light water technology.

How have we come to rely on light water reactors, after such an unpromising beginning? The answer seems to be a series of events early on that just built on and reinforced each other, all of which came down in favor of the light water reactor technology.

During the formative 1950s, all three technologies were being heavily researched by various countries including the United States, Canada, France, and the United Kingdom, and there was no consensus on which was the best. However, the man in charge of the U.S. Navy's development program, Hyman Rickover, urgently wanted to build a nuclear-powered submarine, and he was partial to the light water technology. So he used his authority to ensure that light water nuclear would be used. This gave light water reactors a slight development lead over the other reactor technologies. He was then instrumental in getting the light water technology developed for use in aircraft carriers and nuclear-generating stations on land.

At the same time, concerns for national security arose as the Soviet Union was being viewed increasingly as a military threat. This caused the U.S. to accelerate its nuclear program, and to begin to share its nuclear reactor technology with other countries through its Atoms for Peace program. Because of its technological head start, the light water technology was the only one that was sufficiently developed to service the demand. And the firms General Electric and Westinghouse, the only firms constructing nuclear plants at the time, gave heavy subsidies to electric utilities to buy their plants. This was in order to secure contracts for future nuclear plants, which they expected to be profitable when they figured out how to keep costs down. So, again, we see a QWERTY-like process where small events that occur early on just snowball.[10]

The history of nuclear power provides some important lessons for government subsidies in the form of support for research and development of new energy technologies. First, it illustrates the possibility of technological lock-in that results in widespread adoption of a demonstrably inferior technology. Second, it demonstrates the potential importance of designing the early stages of the process to try to keep inferior technologies from being locked in. This might include identifying situations where technological lock-in is likely to occur and, in these cases, to delay making irreversible decisions until we have better information about the likely future implications. It also argues for greater support for experimentation with alternative approaches, especially early on in the process, and for coordinating efforts among different potential adopters.[11]

## Public choice and renewables subsidies

Let us now apply the public choice model to try to understand the politics of a policy of subsidies for renewable energy, such as an RPS or FIT. We may think of both of these policies as potentially benefiting producers of renewable energy, such as wind farms or manufacturers of components for photovoltaic systems. On the other hand, these policies might impose costs on utilities relying heavily on fossil fuels. This might especially include utilities heavily reliant on natural gas, since we saw earlier that natural gas is more likely than coal to be displaced by switching to renewables.

Quite a few studies have attempted to identify the determinants of renewable energy policies. These studies have examined a variety of economic, political, and cultural factors. Overall, studies have generally supported some of the predictions of public choice theory, with some interesting twists.

The basic methodology of many of these studies is a variant of the so-called *multiple regression* approach, which attempts to explain the enactment of a renewable energy policy with a series of variables that capture economic, political, and social factors. We saw this approach briefly in chapter three when we discussed the hedonic valuation method. The basic model here is shown in equation 12.5:

$$RE\ Policy = f\left(Economic\ factors,\ Political\ factors,\ Social\ factors\right) \tag{12.5}$$

The dependent variable *RE Policy* might be, for example, an RPS policy, the adoption of which is modeled as affected by a series of factors suggested by economic or political theory. The success of this approach depends upon having data for all of the variables that vary across different geographic units like U.S. states or countries in the European Union (EU).

The central issue concerns the relative political influence of the renewable energy and fossil fuel industries and their competing influences on policies to promote renewable energy. Studies have mostly found evidence consistent with this special interest approach. For example, studies have shown that the existence of organized lobbying groups for renewable energy tends to increase the likelihood of adoption of RPS policies in the United States and FIT policies in the EU [see Lyon and Yin (2010); Jenner et al. (2012)].

Furthermore, some evidence suggests that RPS adoption in the United States is negatively correlated with power production using natural gas but not coal [Lyon and Yin (2010); Huang et al. (2007)]. As we have seen, other studies have concluded that renewable energy is more likely to displace natural gas in power production because it is more expensive [Palmer and Burtraw (2005); Schmalensee (2012)]. So this result is consistent with the story that proponents of power production using natural gas would be more politically opposed to promotion of renewable energy. Applying the public choice model, these findings suggest that both interest group stakes and the existence of political entrepreneurs have a significant impact on renewable energy policy.

Interestingly, however, there may be more compelling evidence that the adoption of renewable energy policies is driven by ideological factors. Across the recent studies in this literature, there appears to be a strong partisan divide, both Democrats vs. Republicans and liberals vs. conservatives. In the United States, for example, studies have found that policies promoting renewable energy are significantly more likely to be found in states with more Democratic or liberal politicians.[12] Furthermore, the partisan divide may have grown in recent years.

Some studies have dug deeper to try to understand not only whether, but how, ideology matters. These studies have examined political positions taken on renewable energy policies that take different approaches: specifically, programs with regulatory mandates such as RPS vs. programs that use more market-based approaches, like tax incentives. Perhaps not surprisingly, politicians who are Republican or more conservative tend to oppose regulatory mandates in favor of more market-based approaches. Some have concluded that this divide may also stem from deep-rooted differences in political culture [Vasseur (2016); Fowler and Breen (2013)].

### Key takeaways

- Renewable energy may be thought of as a backstop technology, to which we are transitioning away from reliance on fossil fuels.
- The levelized cost of energy provides a consistent metric to compare the costs of different sources of energy.

- Grid parity occurs when the levelized cost of energy from renewable sources is equal to that of fossil fuels.
- The cost competitiveness of renewable energy is location-specific, depending upon local wind and solar conditions.
- In addition to combatting environmental externalities from burning of fossil fuels, three commonly heard justifications for promoting renewable energy are: promoting energy security, remedying information market failures, and providing green jobs.
- Taxing emissions is probably a lower-cost way of reducing GHG emissions than subsidizing renewable energy.
- Widespread adoption of potentially inferior energy technologies like light water reactors can result from a self-reinforcing, path-dependent process over time.
- Some evidence indicates that renewable energy policies are driven by special interest demands and ideology.

## Key concepts

- Backstop technology
- Levelized cost of electricity
- Grid parity
- Renewable portfolio standards
- Renewable energy credits
- Feed-in tariffs
- Investment tax credits

## Exercises/discussion questions

12.1. Consider the backstop technology model shown in Figure 12.2. What would be the predicted effect on current energy prices of a technological advance that lowered the cost of the backstop technology? A scientific setback that pushed back the date of availability of the backstop technology?

12.2. For the example in the text involving Wind vs. Coal, at what discount rate would the Coal plant be competitive with the Wind plant?

12.3. Why would you expect PV technologies to be less competitive in the U.S. South? What specific factors might cause this to be true?

12.4. Synthesize the arguments for the economic justification of renewables subsidies. What is your takeaway message regarding the conditions under which subsidies would be economically justified for the country in which you live? What information would you need to collect in order to make an informed recommendation on whether your government should subsidize renewables, and how aggressively?

12.5. A key issue regarding the competitiveness of solar energy is adequate battery storage. Suppose that there are multiple technologies that are being developed right now, and it is unclear which technology will be superior in the long-term. What policies might you recommend in order to guard against inferior technologies getting locked in a path-dependent process?

12.6. From the viewpoint of public choice theory, why would subsidizing renewable energy be less likely to encounter political resistance than imposing emissions taxes?

12.7. Taking political transaction costs into account, what would be your bottom line recommendation regarding whether to subsidize renewable energy or tax fossil fuels?

## Notes

1 Some consider nuclear energy renewable, but technically it is not because nuclear fuel – uranium used in nuclear power plants – is not renewable. However, it is considered an option to address some of the environmental issues associated with fossil fuel because it generates virtually no GHGs or other air- and water-borne pollutants.
2 Variants on this process include gas turbines with single- and combined-cycle processes and recipro-cating engines.
3 These figures come from the OPEN PV Project of the National Renewable Energy Laboratory (NREL). The numbers are based on certain assumptions, including retail electricity prices at the time the calculations were made (August of 2014). Other assumptions include the particular characteristics of the PV system including the system capacity (5KW) and the annual rate of degradation of the sys-tem (1%), and financial assumptions regarding factors such as: the discount rate, inflation rate, annual operating and maintenance costs, and various parameters for tax purposes. Importantly, the numbers include a federal tax credit of 30%, but they do not include any state or local tax incentives. So the numbers shown are only illustrative, as they will vary if we make other assumptions. However, they provide a reasonable basis for determining grid parity under current conditions.
4 The magnitude of this impact will depend on various factors. These include the relative cost of coal-fired vs. gas-fired power plants and whether the GHG emissions of country are already sub-ject to binding caps, say, because of participation on an international cap and trade program. See Schmalensee (2012), pp. 47–8. See also Palmer and Burtraw (2005) and discussion below.
5 See, for example, Borenstein (2012), pp. 81–2; Schmalensee (2012), pp. 46–7.
6 For more details of these RPS studies, see Palmer and Burtraw (2005); Fischer and Newell (2008).
7 As we have seen, nuclear energy is not technically a renewable energy source. However, many view it as important component of a strategy to combat climate change.
8 See Davis (2012), p. 50, 59, for documentation of the facts set forth in this paragraph.
9 See, for example, Cowan (1990), p. 545–7.
10 For a detailed discussion of this interesting episode, see Cowan (1990). See also Arthur (1989), p. 126.
11 For more discussion of these points, see David (1997); Kline (2001), pp. 103–4.
12 A number of recent studies have drawn this conclusion. See, for example, Coley and Hess (2012); Lyon and Yin (2010); Fowler and Breen (2013); Hess et al. (2016).

## Further readings

### I. Effects of renewable energy policy

Cullen, Joseph. "Measuring the environmental benefits of wind-generated electricity," *American Economic Journal: Economic Policy* 5 (2013): 107–33.
*A recent study that finds that renewable energy subsidies for wind power are economically justified only if the external costs of pollution are high.*
Fabrizio, Kira R. "The effect of regulatory uncertainty on investment: Evidence from renewable energy generation," *Journal of Law, Economics, and Organization* 29(August 2013): 765–98.
*Examines evidence on investments in renewable energy after enactment of renewable portfolio standards (RPSs) in the United States. Finds that investments declined under conditions of greater regulatory uncertainty.*
Hughes, Jonathan E. and Molly Podolefsky. "Getting green with solar subsidies: Evidence from the California Solar Initiative," *Journal of the Association of Environmental and Resource Economists* 2(June 2015): 235–75.
*A study of a recent initiative in California to promote solar energy through upfront rebates on residential solar installations. It finds them to be highly effective in encouraging investment in solar technologies.*

### II. Public choice in renewable energy policy

Monroe, Nathan W. "The policy impact of unified government: Evidence from 2000 to 2002," *Public Choice* 142(January 2010): 111–24.

*Uses the event study methodology to analyze the effect of party-unified government on the profitability of energy firms, as measured by changes in their stock prices. Finds that oil and gas stocks went up under unified Republican control, while renewable energy stocks took a hit. The opposite occurred after Democrats took back the Senate.*

# References

Arthur, W. Brian. "Competing technologies, increasing returns, and lock-in by historical events," *Economic Journal* 99(March 1989): 116–31.

Borenstein, Severin. "The private and public economies of renewable electricity generation," *Journal of Economic Perspectives* 26(Winter 2012): 67–92.

Coley, J.S. and D.J. Hess. "Green energy laws and Republican legislators in the United States," *Energy Policy* 48(2012): 576–83.

Cowan, Robin. "Nuclear power reactors: A study in technological lock-in," *Journal of Economic History* 50(September 1990): 541–67.

David, Paul A. "Path dependence and the quest for historical economics: One more chorus in the ballad of QWERTY," Discussion Papers in Economic and Social History, Number 20, University of Oxford, 1997.

Davis, Lucas W. "Prospects for nuclear power," *Journal of Economic Perspectives* 26(Winter 2012): 49–66.

Fischer, Carolyn and Richard G. Newell. "Environmental and technology policies for climate mitigation," *Journal of Environmental Economics and Management* 55(2008): 142–62.

Fowler, L., and J. Breen. "The impact of political factors on states' adoption of renewable portfolio standards," *Electricity Journal* 26(2013): 79–94.

Hart, David M. and Colin Cunliff. "Federal Energy R&D: ARPA-E," Information Technology and Innovation Foundation, April 2018.

Hess, David J.; Quan D. Mai; and Kate Pride Brown. "Red states, green laws: Ideology and renewable energy legislation in the United States," *Energy Research and Social Science* 11(January 2016): 19–28.

Huang, Ming-Yuan et al. "Is the choice of renewable portfolio standards random?" *Energy Policy* 35(November 2007): 5571–5.

International Energy Agency (IEA). "Energy security: Reliable, affordable access to all fuels and energy sources," 2020.

International Renewable Energy Agency (IRENA). "Renewable capacity highlights," March 31, 2019.

Jenner, Steffen; Gabriel Chan; Rolf Frankenberger; and Mathias Gabel. "What drives states to support renewable energy?" *Energy Journal* 33(2012): 1–12.

Kline, David. "Positive feedback, lock-in, and environmental policy," *Policy Sciences* 34(March 2001): 95–107.

Lyon, Thomas P. and Haitao Yin. "Why do states adopt renewable portfolio standards? An empirical investigation," *Energy Journal* 31(2010): 133–57.

Palmer, Karen and Dallas Burtraw. "Cost-effectiveness of renewable electricity policies," *Energy Economics* 27(2005): 873–94.

Schmalensee, Richard. "Evaluating policies to increase electricity generation from renewable energy," *Review of Environmental Economics and Policy* 6(Winter 2012): 45–64.

Scott, Mike. "Clean power crowds out dirty coal as costs reach tipping point," *Forbes*, March 16, 2020.

Union of Concerned Scientists. "Clean energy, green jobs," March 2009.

United States. Department of Energy. Energy Information Administration. *Annual Energy Outlook*, 2017.

U.S. Department of Energy. Energy Information Administration. "Renewable energy explained," 2020a. https://www.eia.gov/energyexplained/renewable-sources/, accessed 1/3/2021.

U.S. Department of Energy. Energy Information Administration. "Frequently asked questions," 2020b. https://www.eia.gov/tools/faqs/, accessed 1/3/2021.

United States. Department of Energy. "Feed-in tariff: A policy tool for encouraging deployment of renewable electricity technologies." *Today in Energy*, May 30, 2013. https://www.eia.gov/todayinenergy/detail.php?id=11471

United States. White House. "Fact Sheet: Obama administration announces actions to ensure that nuclear energy remains a vibrant component of the United States' Clean Energy Strategy," November 6, 2015.

Vasseur, Michael. "Incentives or mandates? Determinants of the renewable energy policies of U.S. States, 1970–2012," *Social Problems* 63(2016): 284–301.

Wirth, Timothy E.; C. Boyden Gray; and John D. Podesta. "The future of energy policy," *Foreign Affairs*, 82(Jul/Aug 2003): 132–55.

# 13 Agriculture

In July of 2020, the United Nations Food and Agriculture Organization (FAO) reported what it described as "an alarming scenario." After decades of declining world hunger due to increasing food harvests worldwide, global hunger was on the rise. In its annual report, it reported that there were 690 million hungry people worldwide, a number that had increased by nearly 60 million since 2014. It warned that if recent trends continue, the total number of hungry people will surpass 840 million by the year 2030. This would represent nearly 10% of the entire world population [FAO (2020), p. viii].

Feeding the world's population represents an enormous challenge to humankind, especially as population levels continue to rise. The FAO report speaks of increasing poverty and undernourishment due to weak or deteriorating economic conditions. Reliance on international flows of foodstuffs renders various countries and regions especially vulnerable to external shocks of various kinds. Growing income inequality and the absence of effective institutions in many regions undermine access to food, as do ethnic conflicts, violence, and civil wars. And changing climate is resulting in more frequent extreme weather events, which, along with the spread of pests and diseases, contribute to growing hunger.

In this chapter, we take a closer look at various challenges facing agriculture around the world. You can consider this chapter to be a companion chapter to our chapter seven on population. In that chapter, we emphasized the dynamics of population growth and the ways it strains our ability to feed the world's people. In this chapter, we turn to the other side and address a number of important questions regarding food production. What factors affect our ability to grow food? Will we be able to feed the world's population in a sustainable way? What policies have a good chance of being effective in improving how we grow food? And, finally, how do we understand the origins of policy in terms of the underlying institutions and political dynamics?

## Overview of world agriculture

Let us begin by defining what activities we are talking about when we say agriculture. Very generally, agriculture refers to the organized production of foodstuffs, but also other things grown and cultivated, like cotton, alfalfa, wool, and hemp. Within foodstuffs, it includes grains, beans, legumes, fruits, vegetables, poultry, dairy, aquaculture (fish farms), and livestock (cattle, oxen, llamas, hogs, sheep, goats) products such as beef, veal, and pork. And some of the foodstuffs are not for human consumption, but are used for animal feed, a major use of corn. In this chapter, we will focus mostly on foodstuffs for human consumption.

### I. Production, consumption, and trade

First, consider food production, which takes place all over the world, as every country relies to some extent on its own production to feed its people. Figure 13.1 shows, for example, a map of the world's croplands, which was created from satellite imagery.[1] Much of the light areas (not Greenland or northern Africa) are croplands, which shows heavy concentrations of croplands in Europe extending into central Asia, the United States and Canada, eastern South America, India, and selected parts of east Asia, Australia, and central-southern Africa. In total, there are roughly 1.87 billion hectares of croplands, or about 12.6% of total world land area. India is the country with the most croplands (179.8 million hectares), followed by the United States (167.8), China (165.2), and Russia (155.8).

Figure 13.1 mirrors an important fact about world food production; namely that there are certain countries that play dominant roles in the world food picture. In 2016, for example, roughly half of the total world value of food production was produced in four countries: China, India, the United States, and Brazil [FAO (2018)]. China leads the way, with nearly 24% of total production, followed by India (10.2%), the United States (10.0%), and Brazil (5.9%).

Next consider food consumption, which varies dramatically by country, for two reasons. The first is that different countries have different numbers of mouths to feed, from the 1.4 billion in China down to a few thousand in the least populated countries in the world. The second is that per capita food consumption varies considerably across different countries. Each citizen of the most well-fed countries, like the United States and many European countries, consume twice as many calories per day as citizens of the least

*Figure 13.1* The world's croplands. *Source*: U.S. Geological Survey. https://www.usgs.gov/news/new-map-worldwide-croplands-supports-food-and-water-security

well-fed countries, most of them in Africa. Figure 13.2 shows, for most countries in the world, average daily calorie consumption per capita, where the darker shaded areas represent higher caloric intake.

The disparities between food production and consumption mean that international trade in foodstuffs are an important part of the picture. A complicating factor is that the extent of trade varies quite a bit across different foodstuffs. For example, wheat, soybeans, and bananas are heavily traded internationally. On the other hand, rice is not [Alston and Pardey, p. 124]. When foodstuffs are traded internationally, there are important issues of supply and demand. The United States is, and has long been, a major exporter of grains to the rest of the world. This reflects its high agricultural productivity and relatively moderate population size. Similarly for Brazil and Argentina. On the other hand, Europe, Japan, and China are major importers of foodstuffs of various kinds.

For the least-fed countries of the world, there are two key issues. First, how can they increase their food production for their citizens? This will depend on several factors that we will discuss at length later in this chapter. For now, keep in mind that many of these very same countries are projected to experience rapid population growth until at least mid-century. This prospect only increases the urgency of finding ways to increase food production in these countries.

Second, if they are unable to boost food production sufficiently, another option is to import it. So for these countries, food trade with other countries is an important part of the picture. One challenge for these countries is that the price of food is determined in world markets, making these countries vulnerable to rising prices and price fluctuations. In some instances, another challenge is to combat disruptions in international flows of foodstuffs, for example, due to internal conflict, political strife, and, possibly, civil war. A final challenge is the recurrent problem of corruption, which keeps food aid from being distributed to where it is most needed.

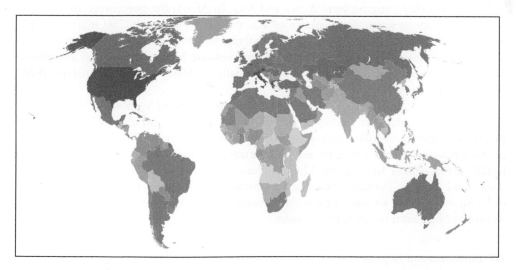

*Figure 13.2* Average per capita daily calorie consumption by country, 2006 to 2008. *Source:* Wikimedia Commons, based on 2006–2008 dietary energy consumption data from the FAO Food Consumption Nutrients spreadsheet.

## *II. Undernourishment and food security*

A key question is whether food production is sufficient to feed the world's population, both now and in the future. Currently, there is no question that hunger is a serious issue. As we have seen, the FAO has estimated that, in 2020, some 690 million people world-wide are going hungry. This is the number that are undernourished, in that they do not regularly consume sufficient calories to lead an active, healthy life. To measure this, the FAO has developed an indicator called the *Prevalence of Undernourishment* indicator. The 690 million figure represents the number of people in the world that are undernour-ished by this metric.[2] The vast majority of undernourished live in low-income coun-tries, as you might expect. However, undernourishment of segments of the population is a concern even for the wealthiest countries in the world, including the United States and western Europe.

In addition to the notion of hunger based on undernourishment, the FAO also meas-ures hunger with a concept called *food security*. Food security is similar to undernourish-ment, but it measures hunger in a more sophisticated way. Like undernourishment, it refers to access to food for normal growth and an active, healthy life. However, it measures it at different levels of severity and focuses on both the quantity and quality of food avail-able. The idea is: how certain is access to sufficient food of good quality? At one end of the scale, one might experience mild food insecurity, defined as some uncertainty in the abil-ity to obtain food. At the other end of the scale is severe food insecurity, defined as having no food for at least a day. In between are moderate levels of food insecurity, with food insecurity getting worse the more you have to compromise on getting sufficient food of good quality. The FAO estimates that, in 2018, over one-quarter of the entire population of the world – over two billion people – experienced moderate to severe food insecurity [FAO (2019), pp. 15, 18].

It is also important to keep in mind that the burden of food insecurity is spread une-venly around the world. In North America and Europe, 8% of the population experienced moderate to severe food insecurity and about 1% experienced severe food insecurity in 2018. At the other extreme were Africa and Asia. In Africa, the percentage that experi-enced moderate to severe food insecurity was over 50%, or nearly 700 million people. And though the percentage experiencing moderate to severe food insecurity in Asia was less – slightly under 23% – this still translated to over one billion people. And worryingly, food insecurity appears to be on the rise, especially in Africa and Latin America [FAO (2019), pp. 15, 18, 19].

Food insecurity is a serious problem because it is associated with various health risks and adverse life outcomes. Perhaps most worrying, maternal and child malnutrition are major contributors to infant mortality. Furthermore, malnutrition in the womb directly contributes to low birth weight and early underdevelopment, which have important health consequences later in life. These include stunted physical growth, heart disease, stroke, and diabetes. Perhaps ironically, malnutrition is also associated with increased risk of being overweight, a result of poor eating habits. Overweight and obesity have become serious public health issues in recent years.

## *III. Environmental impacts*

Another set of issues concerning food production are the environmental impacts associ-ated with large-scale production. Among other things, food production:

- Uses a great deal of scarce land and freshwater supplies;
- Contributes to water pollution and threats to human health from fertilizers, herbicides, pesticides, and defoliants;
- Can cause depletion of oxygen (*eutrophication*) in surface waters, resulting in *dead zones*;
- Poses threats to biodiversity; and
- Generates significant amounts of greenhouse gases (GHGs).

Roughly half of all the world's habitable land is devoted to food production, which displaces other uses of the land including urban areas, rainforests, grasslands, and wildlife habitat. As we shall see in chapter sixteen, conflicts with agriculture are an important cause of tropical deforestation. And encroachment on wild areas poses major threats to endangered species, which we will discuss in chapter seventeen. Furthermore, much of world food production relies on irrigation, especially in arid regions, which uses a great deal of water. As a consequence, food production accounts for over two-thirds of total withdrawals of freshwater worldwide.[3]

Another set of environmental threats derives from the use of various chemical methods to enhance large-scale food production, including fertilizers, herbicides, pesticides, and defoliants. The residues from these agents can have negative effects on water quality, local ecosystems, and human health. In one notorious example, the combination of intensive irrigation and liberal use of these chemical methods has resulted in the destruction of one of the largest freshwater lakes in the world, the Aral Sea, in central Asia (see Box 13.1).

---

**Box 13.1: The destruction of the Aral Sea**

For many years, the Aral Sea, located in Central Asia, was the fourth-largest freshwater sea in the world and home to a thriving fishing industry. Then, beginning in the 1960s, the former Soviet Union began to divert the waters from two feeder rivers – the Amu Darya and Syr Darya – to support expanded cotton production. The Aral Sea began to shrink and became increasingly salty. The Soviets also used massive amounts of herbicides, pesticides, and defoliants in order to boost cotton production. These pollutants were then vented back into the rivers, severely degrading water quality. The sea is now a ghost of what it once was. The fishing industry has been destroyed. In addition, the desiccation of the sea caused an increase in local dust storms, severely degrading air quality and leading to increased incidence of respiratory ailments and a rise in infant mortality levels.

---

In less extreme instances, agriculture contributes to soil erosion and runoff into nearby rivers and streams. When fertilizer is used, this runoff can contain nitrates and phosphates, which have negative effects on water quality. In some cases, concentration of nitrates and phosphates downstream can result in *eutrophification*, or the loss of ambient oxygen in the waters. If severe enough, it can result in the creation of dead zones, where there is insufficient oxygen to support life.

Finally, food production results in large emissions of GHGs, contributing to ongoing climate change. Livestock production generates large amounts of methane, a particularly powerful GHG. Other food production processes that generate significant amounts of GHGs include burning of fuels, the release of nitrous oxides from applying fertilizers and manure, and the ploughing and burning of croplands. In all, food production is estimated to account for about 25% of total GHG emissions worldwide.[4]

## Economics of agriculture

A key issue moving forward will be the ability of humankind to produce enough food to feed the world's population, which is projected to reach 9–10 billion by mid-century. In chapter seven, we discussed the importance of technological change in expanding food production. Let us now dig more deeply into this issue.

### I. An important distinction: Extensive vs. intensive food production

For perspective, consider that humankind has gone through two basic phases in its quest to expand food production. In the first phase, new lands are brought under cultivation. This can occur as long as there are new, as-yet untapped, fertile lands to be had by cutting down forests and clearing the land. However, relying on additions to land means that, at some point, one is adding land that is more marginal in productivity or more costly to reclaim. This means there are limits to the amount of increased production farmers can get in this way.

So, at some point, the first phase gives way to a second phase, where expansion in food production occurs by wringing more output out of lands that are already being farmed. This entails adding more other factor inputs – machinery, fertilizer, pesticides, herbicides, and so forth to existing lands. Adding more of these other factor inputs has the effect of increasing plant growth and reducing crop losses to pests and weeds. These two phases are called production growth on the *extensive* and *intensive* margins.

The same idea has been applied in other economic contests. In labor markets, extensive refers to the number of workers, so expanding on the extensive margin means hiring more workers. Intensive means workers on average producing more. When talking about trade, intensive means building on relationships with existing trading partners, while extensive means branching out to new trading partners. But, regardless of context, the basic distinction is always the same.

In agriculture, these two phases did not occur all over the world all at once. The transition from extensive to intensive agriculture happened first in high-income countries like the United States beginning in the early-20th century. In most low-income countries, the transition did not occur until late in the 20th century. Some of the poorest countries today still have not made, or are just beginning to make, the transition [see Ruttan (2002), p. 161].

### II. A key factor: Technological change

In the intensive growth stage, perhaps the most important factor boosting productivity is *technological change*. Historically, we have managed to achieve enormous productivity gains in agriculture through technological advances such as the cotton gin, the mechanical reaper, the thresher, and the harvester. Technological advances also contributed to the development of new and more effective fertilizers, pesticides, and herbicides. And, particularly since the mid-20th century with the advent of the so-called *Green Revolution*, we have managed to enjoy huge increases in crop yields, owing to a variety of scientific and technological advances in genetics and plant breeding.

Many people attribute much of our ability to increase food production in the second half of the 20th century to the efforts of one man – Norman Borlaug. Borlaug was an American agronomist who went to Mexico in the 1940s as part of a research team charged with expanding wheat production. During his 16 years there, he achieved remarkable success in developing new strains of high-yield, disease-resistant wheat. Over the next 50 years, wheat yields in Mexico roughly quintupled. He then moved on to Asia, where he was again able to increase yields of wheat and, when his methods transferred to other crops, to rice. Borlaug is sometimes referred to as the father of the Green Revolution, which dramatically increased food production worldwide beginning in the 1960s. For his work, he received numerous awards, including the Nobel Peace Prize in 1970.

It should be mentioned that, today, Borlaug's legacy is viewed with a bit more ambivalence. There were inevitable limits to the ability of his methods to increase crop yields, for example in sub-Saharan Africa or, more generally, in areas not endowed with sufficient water supplies. And his increased yields have arguably come at a large environmental price in terms of increased use of fertilizers and pesticides, and loss of crop diversity and other local impacts due to promotion of monoculture. There is also an ongoing spirited public debate concerning the entire approach, which promotes the creation of genetically modified organisms (GMOs). For a taste of this debate, see "The GMO controversy" under *Further readings* at the end of the chapter.

### III. Expanding on technological change: Induced innovation

Economists have, however, gone beyond this basic framework to ask the deeper question: Where does this technological advance come from? One important set of answers has to do with the notion of *induced innovation*. Under induced innovation, the development and application of new technologies occurs partly in response to factor prices and resource endowments.[5] To understand this idea, imagine two situations: a country where land is relatively abundant and labor is relatively scarce, and another where the opposite is true. Let us call the first country "United States" and the second country "Japan."

In the United States, the relative scarcity of labor makes labor expensive, which might encourage the application of new farming techniques designed to conserve on labor. These might include things like tractors, threshers, harvesters, and mechanized irrigation systems. In Japan, the relative scarcity of land makes land expensive, which might encourage the application of new techniques designed to conserve on land. These might include chemical fertilizers and herbicides. So the choice of technology depends centrally on resource endowments and on relative factor prices. In essence, the argument is that the demand from farmers for a technology depends on how costly it is to adopt and how much benefit they expect to derive. Just like other factors of production.

The choice of "United States" and "Japan" as representing the labor-scarce and land-scarce economies in this illustration is no accident. The economists Yujiro Hayami and Vernon Ruttan did a detailed comparative analysis of farm technology adoption in the two countries, which are accurately described as relatively labor-scarce and land-scarce. They found sharp differences in technology adoption in ways that conformed to the predictions of the induced innovation model. They concluded that the adoption of farm technology is influenced by the relative scarcity of different factor inputs [Hayami and Ruttan (1985)].

The notion of induced innovation is related to the ideas of Ester Boserup, the Danish economist whom we encountered in the earlier chapter on population. The common theme in both Boserupian growth and induced innovation is the notion that innovation, far from springing out of nowhere, is predictably responsive to market conditions and the circumstances that economic agents find themselves in. This notion has been extremely influential in the field of agricultural economics and has spurred much follow-up research to test and refine its hypotheses. You can find a few representative studies in the *Further readings* section at the end of this chapter.

The induced innovation hypothesis has not been without its critics. Early criticisms centered on the difficulties in knowing whether a given factor-saving decision by farmers constitutes innovation, or merely input substitution toward the more abundant factor. However, multiple subsequent studies devised ways to distinguish the two and, broadly speaking, have confirmed the induced innovation hypothesis.[6]

Others have argued that, in essence, the model is too simple, failing to capture other factors that affect technology adoption. Some have argued that competing interests within the farm sector can negatively affect technology adoption decisions by farmers, resulting in adoptions that are inefficient or inequitable. For example, even in a labor-abundant economy, wealthy farmers might adopt labor-saving technologies.[7] Others have argued that technology adoption decisions are also driven by other factors that affect the relative ease of adoption of different available technologies. This argument is in essence that in addition to demand for technologies, supply-side factors should also be considered [Olmstead and Rhode (1993)]. But, overall, the induced innovation model has stood well the test of time.

## Institutions and property rights

In addition to technology, agricultural productivity is determined by the institutional context in which farmers farm their land, which determine the incentives they face. The question is: How do institutions provide incentives to farmers to work hard and invest in their farming operations?

### I. A famous example: China, 1950s–1980s

For an example of how much institutions can matter, let us consider the case of China. For years, China has had to grapple with the huge challenge of feeding its hundreds of millions of people. In the late 1950s, concerned over its slow progress in boosting food production, it embarked on a forced *collectivization* experiment, requiring all farmers – some 120 million strong – to be members of large agricultural collectives as part of its *Great Leap Forward*. The results were disastrous. Food production fell by 30% within two years (see Figure 13.3), and it is estimated that 30 million people died as a result of this policy [Lin (1990)]. The Chinese government responded by retaining collectivization, but dividing production up among much smaller production teams. This managed to stave off further disaster.

Growth in food production remained sluggish until the 1970s, however, because even within the smaller production teams, rewards to individual farmers were not tied to their effort levels. As a result, in the late 1970s China again tried to reform its agricultural sector. But this time, it moved from its collective farming system to one based on individual household production, in a process known as *decollectivization*.

Under the new system, households were given long-term leases to their lands and permitted to keep most of the fruits of their labor. The result was a dramatic increase in food production (see Figure 13.3). Other changes were occurring at the same time, including pricing reforms, loosened restrictions on trade, and greater reliance on markets. However, the economist Justin Lin has calculated that nearly *half* of the increase in food production from 1978 to 1984 was due to the shift to household production.[8]

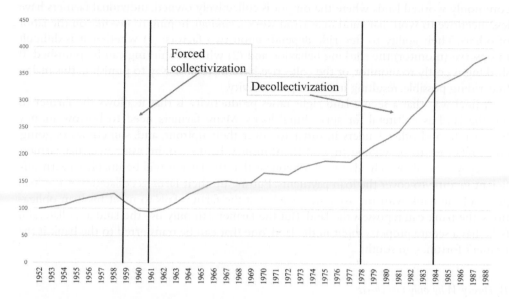

*Figure 13.3* Index of value of agricultural output in China, 1952 to 1988. *Source*: Lin (1990), p. 1233.

The example of China provides some important lessons regarding the central role of institutions in influencing economic performance. Comparing agricultural systems based on collectives to ones based on individual household production suggests that the two systems provide farmers with very different incentives. Under the Chinese system of collectivization, individual farmers had little incentive to work hard to produce food. This apparently changed when the Chinese shifted largely to household production in the late 1970s. The economic interpretation of the difference between the two systems is that they comprised very different ways of defining property rights to farmland and farm output.

## II. What is the effect of good institutions?

### A. Theory: Two cheers for private property rights

Let us now expand on the idea of the importance of private property rights. Economists generally believe that secure private property rights support efficient economic outcomes, by providing the correct incentives for individuals to invest in and use their resources. In the context of agriculture, there are at least three important arguments to keep in mind.

First, farming generally requires investments in the farmed land: building the silos, sheds, outbuildings, and other structures on the land; investing in improvements in soil quality by applying fertilizer; digging irrigation ditches; and so forth. To be willing to undertake these investments, farmers need assurances that they will benefit. A secure property right to the land – one that is safe from being taken away, or *expropriated*, by the government; which cannot be infringed upon by third parties; and which can be sold for the full value of the land if the owner wishes to – encourages farmers to invest in the land, allowing them to maximize its value.

Second, a secure private (individual) land right provides incentive to farmers to work harder than they would work on lands worked in common with other farmers. On commonly worked lands where the output is collectively owned, individual farmers have less incentive to work hard and may avoid work (*shirking*), hoping to *free-ride* on the labor of others. Their ability to free-ride depends upon two factors: (1) whether it is difficult to observe (monitor) the shirking behavior, and (2) whether shirking can be punished. If shirking is costly to monitor, or the collective has no credible way to punish it, this makes free-riding possible, resulting in loss of productivity.

A final way that a secure land right raises productivity is that it allows the farmer to use the land as collateral for agricultural loans. Many farmers need to borrow money from banks and other lenders in order to cover their upfront, and, sometimes ongoing, costs. In order to be willing to lend them money, banks require assurance that farmers will repay these debts. The danger is, of course, that the farmer will be unable to earn sufficient income to cover the loan payments. Putting up their farm as collateral for the loan provides the bank with the assurance needed: if the farmer fails to meet his loan obligations, the bank can repossess the land. But the farmer can only use the land as collateral if (s)he has a secure property right in the land, one that can be transferred to the bank if the farmer's fortunes go south.

### B. Theory: Why only two cheers

Having said all that, there are some instances where there may be costs associated with individual secure rights. In the following discussion, you can think of me trying to make a head-to-head comparison between individual land rights and communal rights of some sort, where land is held in common. Of course, 1950s Chinese-style communes are one possibility, but there are plenty of others. So the question is: Why might communal rights have some advantages relative to individual rights?

First, under certain conditions, there may be economies of scale in farm production, where the best farm size – that is, the one with the lowest average production costs – is larger than small, family-size tracts. If there are transaction costs of coordinating among small private farms, a system of individual small farms may result in higher-than-necessary production costs and a loss of efficiency. A collective farm, or larger farms more generally, may be able to take advantage of economies of scale where smaller farms could not. A related issue is that collectives and larger farms may have access to credit for loans that smaller farms may not.

Second, an important part of the advantages of secure individual rights is that they support the exchange of goods in markets. However, in some settings, particularly in many low-income countries, markets for many things – like outputs, labor, and crop insurance – are incomplete or non-existent. As a result, much of the advantages of individual rights in terms of gains from trade are eliminated or reduced. For example, the thinness of markets may make it difficult for farmers to sell their goods, hire workers, diversify their risk, and obtain crop insurance.

Third, some of the disadvantages of communal arrangements can be mitigated by modifying some of the communal rules of the game. For example, communal systems can, and sometimes do, assign long-term use rights to individual farmers that can even be rented or passed on to heirs. And communal systems may be able to achieve better outcomes by organizing at the right size – a lot smaller than the tens of thousands on the 1950s Chinese communes! – to prevent shirking and promote coordination among members. For example, organizing at a sufficiently small scale to enable face-to-face communication is a key factor in encouraging cooperation.[9] Furthermore, communal arrangements can have certain advantages: for example, they can also diversify risk among its members and provide safety nets for poorer farmers [Binswanger and Deininger (1997), p. 1966].

## C. Some evidence on the importance of secure property rights

Let us now consider some real-world evidence on the importance of secure individual property rights to boost agricultural productivity. Here we will consider low-income contexts, where insecure rights tend to be more of an issue, as we have seen. We will consider two such contexts: low-income countries back in history, and low-income countries in the present day.

### I. HISTORICAL EXAMPLES

Many examples of secure property rights boosting agricultural productivity can be found in past development settings. The economists Terry Anderson and P.J. Hill have concluded that secure property rights in land played an important role in the early development of U.S. agriculture [Anderson and Hill (1976)]. During the colonial era, colonies established a system of private land rights, under which farmers could use their lands as they saw fit, with few government restrictions. After the Revolution, the new government created a system of land disposal that established private land rights, based on a system of rectangular surveys that set clear boundaries for parcels of land. Later on, the invention of barbed wire and its widespread use on the Great Plains in the late 1800s helped enforce property rights to grazing lands, and increased ranching productivity.[10] All of these developments promoted the growth of agriculture in the first century and a half of U.S. history.

A vivid illustration of the incentives provided by secure individual property rights occurred at the very beginning of U.S. history. When Pilgrims on the *Mayflower* landed at Plymouth Rock in 1620, they established a communal property system in which all fruits of the labors of the colonists were to be combined and divided equally. After two disastrous harvests and many deaths, they established a system of private plots of land and everyone was permitted to keep what they grew. They never starved again [see Box 13.2].

---

### Box 13.2: The pilgrims and private land rights

The fall of 1623 marked the end of Plymouth's debilitating food shortages. For the last two planting seasons, the Pilgrims had grown crops communally… But as the disastrous harvest of the previous fall had shown, something drastic needed to be done to increase the annual yield.

In April, [Governor] Bradford had decided that each household should be assigned its own plot to cultivate, with the understanding that each family kept whatever it grew. The change in attitude was stunning. Families were now willing to work much harder than they had ever worked before. In previous years, the men had tended the fields while the women tended the children at home. 'The women now went willingly into the field,' Bradford wrote, 'and took their little ones with them to set corn.' … Although the fortunes of the colony still teetered precariously in the years ahead, the inhabitants never again starved.

[Philbrick (2006) p. 165]

---

Other studies have found evidence that secure property rights in land increased agricultural production in various countries in pre-industrial Europe. In Sweden in the 18th and 19th centuries, crop production increased significantly when performed by freeholders with secure property rights to their land, compared to tenant farmers with insecure rights. Similarly, in early 20th century Russia, a movement from communal to individual farm organization enabled large increases in food production.[11]

Perhaps the most famous, certainly one of the most studied, historical property rights example is the *enclosure movement* that spread throughout Europe beginning in medieval times and well into the 19th century. The enclosure movement involved taking lands that had previously been under common ownership and control and dividing it into individual parcels for private use. Thus, it is a prime example of private land rights creation, with all the associated implications for providing better incentives for farmers to invest in and work the land. And though studies have disagreed about the magnitude of the effect of enclosure on land productivity, most have concluded that there was some, in some cases large, positive effect on crop yields and labor productivity from enclosing farmlands. If you are interested in knowing more about this major institutional change and the scholarly debates over its effects, see *Further readings* at the end of the chapter.

II. PRESENT-DAY DEVELOPMENT CONTEXTS

In the 21st century, insecure property rights is probably mainly an issue in low-income contexts where there is less political stability, more civil conflict, and fewer constraints on governments from simply seizing lands. For example, the U.S. Agency for International Development estimates that, in 2016, over two-thirds of land in low-income countries was unregistered or perceived to be insecure [U.S. Agency for International Development (2017)]. So the issue is not only real but potentially quite important quantitatively, and precisely in those areas of the world most in need of sustained food production.

The question is: What does the evidence tell us about the importance of secure property rights in promoting land productivity in low-income contexts? Summarizing a large literature, it seems fair to say that most studies have concluded that the effect of secure property rights is almost certainly positive. For example, farmers with legal title to their lands have been found to have better access to credit, which encourages greater investment in lands, resulting in higher yields and better land management. And these results hold over a wide range of countries and cultural contexts, including various countries in Africa, Southeast Asia, and throughout Latin America.[12] The bottom line is that secure property rights in land seem to make a positive difference in current low-income contexts.

Regarding the issues of violence and civil conflict, much evidence suggests that they have devastating short-term impacts on food production. Armed conflict and civil war often result in destruction of crops and crop reserves. We interpret this as property rights to both crops and land that are highly uncertain, subject to seizure, and costly to enforce. In 2017, for example, four African countries experienced severe famine, but in only one case was this the result of prolonged drought. In the other three cases, the root cause was violent conflict [Koren and Bagozzi (2017)]. Thus, much depends upon the ability, and inclination, of governments to protect property rights and maintain political stability.

## Agricultural policies and public choice

Let us now turn to a discussion of agricultural policy. A wide variety of policies have been pursued by governments in both low-income and high-income countries. The types and pattern of policies reveals much regarding the economics of farming and the challenges faced by farmers. Furthermore, numerous economic studies tell us much about the political factors that give rise to these policies. These studies speak both to the explanatory power of the public choice approach and its potential limitations in helping us interpret policy outcomes.

## I. Public choice in taxes and subsidies

Let us begin with some evidence that suggests that special interests may play an important role in determining agricultural policies across a broad spectrum of different countries. The evidence can be seen if we break down agricultural policies into two basic types: taxes and subsidies. Let me ask you what you make of the following consistent real-life pattern: *farmers tend to get taxed in low-income countries and subsidized in high-income countries.*[13] Before going on, can you think of a public choice story that would explain this?

Here is a public choice story that many economists believe explains this pattern. Consider the competing demands of two interest groups – farmers and consumers – who are fighting over the distribution of the fiscal resources of the government. The basic idea is that farmers will tend to be less politically effective, and thus less able to secure their favored policies, in low-income countries. Now why would this be?

Well, in low-income countries there tend to be large numbers of geographically dispersed farmers. This means higher transaction costs of political organization, which promotes free-riding on any attempt to organize an effective interest group. Furthermore, farmer stakes in opposing taxes tend to be low because taxes take a tiny bite out of incomes that are themselves relatively low. All of this means that farmers will tend to be less effective in exerting political pressure for fiscal policies that favor them. Furthermore, subsidies to farmers would be costly because of their relatively large numbers. On the other hand, urban consumers have high stakes in low food prices because at low income levels, food is a relatively large part of the household budget. All of this means that in low-income countries, consumers will tend to be successful in political battles against farmers.

On the other hand, in high-income countries the share of food in consumer food budgets is relatively low, meaning that consumers have lower stakes in favorable food policies. At the same time, fewer farmers lowers their transaction costs of political organization, which blunts the free-rider problem. In addition, higher incomes makes them feel the bite of taxes more.[14] These factors swing the political power in favor of farmers, enabling them to gain subsidies. Furthermore, all the factors that tend to make them politically powerful as a group also help them secure tariff protection against foreign imports of food [Gardner (1992), p. 94].

## II. Farm subsidies in the United States

Let us now zoom in on one high-income country, in order to flesh out the political story of how farm subsidies come to be adopted and maintained over time. The agricultural policies of the United States nicely illustrate how economic conditions can give rise to political pressures that generate protective policies like subsidies. In addition, we will see how such policies can come to persist over time, even under changing economic conditions.

### A. Theory

The political demand for farm subsidies in the United States originally stems from what has come to be called the "farm problem." This term refers to a chronic situation of low farm incomes and widely fluctuating incomes and returns on investment over much of the 20th century. The perhaps surprising thing is that U.S. farmers are unsurpassed in the efficiency with which they produce agricultural products, for a variety of reasons including technological change and induced innovation. Farms in the U.S. are some of the most productive in the world. So why should they be: (1) barely squeaking by and (2) suffering under all the uncertainty of farm prices that vary widely from year to year?

On reflection, however, the reasons for the farm problem may not be terribly mysterious. Let us perform some basic economic analysis to understand how it has come about. And let us begin by assuming that the supply and demand of farm products are both quite price-inelastic in the short run. This is a safe assumption, supported by many studies of agriculture in various countries.

Now consider what happens in the short and long term. The short-run situation is shown in Figure 13.4. In the short run, we observe year-to-year shifts in supply: farmers have good years and bad years. In some years we have bumper crops, and the supply curve shifts out to $S_{BC}$. In other years, we have droughts and the supply curve shifts back to $S_D$. Because the demand curve is so inelastic, we observe large swings in prices.

In the long term, we observe steady and rapid technological change, as we have just seen. This means that production costs are declining over time, resulting in outward shifts in the supply curve. And these supply shifts are greater than the outward shifts in the demand curve, since technological change is outstripping population growth. I should mention that there is an implicit assumption here that capital and labor are not exiting the farming sector sufficiently quickly to offset the supply-enhancing effects of technological change.[15] The result is a steady decline in agricultural prices over time, as shown in Figure 13.5.

Data on actual farm prices are consistent with this supply and demand interpretation. Figure 13.6 graphs farm commodity prices in the United States for the better part of the 20th century, from 1910 to 1990. It shows wild swings in farm price levels, with peaks during the two world wars, a dramatic plunge during the 1920s and 1930s, and a steady decline since the WWII peak. It also shows a definite downward trend over the entire time period.

So, historically, the farm problem seems real. I should add, though, that the recent experience – say, over the past 30 years or so – has led some economists to wonder whether the farm problem still exists. Continuing technical innovation, increased labor productivity, labor

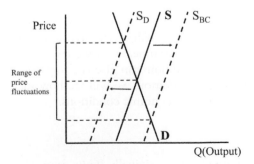

*Figure 13.4* Short-term fluctuations in price of agricultural products

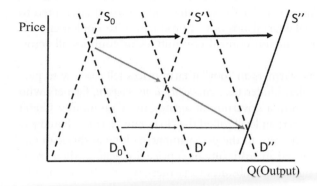

*Figure 13.5* Long-term declines in price of agricultural products

*Figure 13.6* Real prices received by farmers, all commodities. *Source*: U.S. Department of Agriculture. National Agricultural Statistical Service.

movement out of farming, farm incomes at or above non-farm incomes, and U.S. farmers capturing large shares of global markets all suggest that U.S. farming is now doing quite well, thank you [Paarlberg and Paarlberg (2000)]. One question is how we square this view with the evidence we have just seen of steadily declining farm prices to the present day.

### B. The politics of farm subsidies

To begin to answer this question, let us now turn to the question of how this underlying economic situation has translated into an ongoing policy of subsidies for farmers. Direct federal involvement in the operation of U.S. farm markets dates from 1933, when Congress passed the *Agricultural Adjustment Act* (AAA) during the depths of the Great Depression. Examining Figure 13.6, you can see that this was a time of cratering farm prices when farmers were severely distressed. The AAA program had two components, both designed to provide relief to farmers under the circumstances. First, it attempted to prop up prices by paying farmers to take farmlands out of production. To implement this program, it established over 4,000 production control committees in counties all across the country.

Second, it devised a program to give farmers an "out" if farm prices fell too low to permit them to make money on crop sales. Under the *commodity loan program*, farmers who had crops to sell could obtain a loan from a government agency, the Commodity Credit Corporation. If crop prices fell below a certain level called the "loan rate," farmers could pay back their loan simply by forfeiting their crops to the government. As long as this loan rate exceeded the market price, farmers enjoyed an artificially high price for "selling" their crops. The AAA program, which provided significant subsidies to farmers, was the beginning of a decades-long effort by the federal government to improve economic conditions for farmers.

The public choice model might predict that the AAA program was the result of political pressure from farming interest groups, but this was actually not the case. Rather, the program was crafted by a bunch of academics who were part of Roosevelt's inner core of advisors. However, if AAA did not begin life as a special interest baby, it quickly became one in early adolescence. Within a few years, the program came to serve as an organizing focus for lobbying efforts by farmers. The Farm Bureau – the national lobbying organization for farmers – used the local production control committees to triple its membership within seven years.[16]

The commodity loan program subsequently became the central tool of the federal government for providing price support to farmers. Over time, political pressure from farming interests resulted in the ratcheting up of loan rates, and expanded the number of commodities covered under the program. From time to time, periodic slumps in food prices (such as occurred after the end of the Korean War) would cause government costs to increase and bloat government stocks of foodstuffs. Whenever this happened, it generated pressures to reform the programs in order to reduce the size of the subsidies. And there have been periodic other attempts, usually under Republican administrations, to lower the subsidies. For example, both the Nixon and Reagan administrations attempted frontal assaults on the farm price support programs. However, farm subsidies have been remarkably resistant to attempts at downsizing.

How have farming interests managed to stave off attempts at reforming the farm subsidy program? In various ways, including active rent-seeking through things like campaign contributions from agricultural political action committees (PACs) and by building coalitions with other interest groups. One economist found, for example, that PACs representing farmers of specific commodities that make the largest campaign contributions also happen to receive the most program support in Congress. In recent days, farm bills increasingly contain provisions for food stamps and nutrition, habitat protection, and conservation, which gains support from representatives of urban districts, hunters, and environmental groups. And farming representatives in Congress tend to place onto key agricultural committees to promote farm-friendly bills and forestall reform. And they engage in logrolling with non-farm representatives to provide mutual support for each other's programs.

### C. Public choice: Productive vs. predatory policies

For yet another perspective on special interest farm policies, let us consider a different way to distinguish policies. To understand this perspective, it helps to consider the fundamental challenges farmers face in producing food. Most farming occurs in rural areas, often far from markets. Over time, farmers all over the world have faced numerous challenges including poor soil, soil erosion, pests, drought, flooding, and, in many areas, inadequate water supplies for irrigation. Their very livelihoods are based on their ability to grow food and the prices their output can fetch on markets. Production and farm prices can be subject to large fluctuations, as we have seen. At the same time, they have to incur various costs of production for things like seeds, fertilizer, farm equipment, farm laborers, interest on bank loans, and so forth. All of this has meant, for many farmers over the years, a precarious, uncertain existence.

All of which helps explain various policies that governments have set in place. Consider the United States. Over the years, the U.S. government at various levels has undertaken rural mail delivery, rural electrification, and building rural roads and bridges, all of which provide crucial infrastructure servicing farming areas. In addition, it has undertaken soil conservation programs, and established agricultural experiment stations for agricultural research, land grant colleges, and the cooperative extension service. All of these things fill crucial needs of farmers for transportation, communication, and information, all of which have important public good elements and, therefore, are likely to be undersupplied by private companies.

However, the U.S. government has also pursued policies that do not have anything like an obvious public good justification. These include farm price support programs, production quotas, tax breaks, tariffs, export subsidies, and exemptions from antitrust legislation. Rather than serving to enhance the productive capacity of farmers and promote the general welfare, this latter set of programs appear to be designed more to benefit farmers at the expense of others, including consumers, who have to pay higher food prices, and taxpayers, who must bear an increased share of the overall tax burden. Economists have noticed that U.S. agricultural policies seem to fall into these two different categories, what economists have called *productive* and *predatory* policies.[17]

Thinking about this dichotomy may raise the following question in your mind. How does the public choice framework help us to understand the overall pattern of agricultural policy that we observe? With your public choice hat firmly on your head, you may be saying to yourself: I think I understand the predatory policies a lot better than the productive policies. I can see a theory based on special interests helping to explain policies that subsidize farmers at the expense of others. But the productive policies do not have that same flavor. They seem to be more about raising up everybody. So how do we account for *these* policies?

One possibility is that productive policies and predatory policies may be enacted at different times and under different economic and political conditions. In the United States, a number of productive policies were enacted in the 19th and early 20th centuries, for example through acts of Congress that established land grant colleges, agricultural experiment stations and the cooperative extension service [Rausser (1992), p. 134]. Over time, however, a few early policies morphed into policies that became predatory, while other new policies were more clearly predatory from the start. The overall result was a hybrid of the two types of policies.

Some economists have taken the argument a step further and argued that predatory policies may come into existence *as a means of* overcoming political resistance to productive policies. To see the argument, consider Figure 13.7. This shows the market for sugar, which is depicted as having a highly price-inelastic demand curve. The effect of a productive policy is to shift the supply curve outward. Since the demand curve is inelastic,

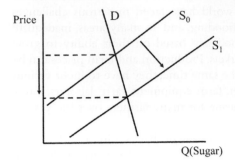

*Figure 13.7* How productive policies may necessitate predatory policies

*Table 13.1* Breakdown of productive vs. predatory assistance by crop

|  | % Productive | % Predatory |
|---|---|---|
| Sugar | 7.9 | 92.1 |
| Milk | 7.8 | 92.2 |
| Rice | 6.4 | 93.6 |
| Wheat | 13.5 | 86.5 |
| Sorghum | 14.5 | 85.5 |
| Barley | 20.9 | 79.1 |
| Corn | 17.7 | 82.3 |
| Oats | 61.6 | 38.4 |
| Soybeans | 74.3 | 25.7 |
| Beef | 55.5 | 44.5 |
| Poultry | 65.0 | 35.0 |
| Pork | 82.5 | 17.6 |

[adapted from Rausser (1992)]

however, there is a sharp drop in the price, inflicting significant costs on producers. If they are politically effective, they will fight the policy. In order to "buy" their acceptance, it may be necessary to institute a predatory policy, such as a price support program [de Gorter et al. (1992); Rausser (1992)].

Some evidence in support of this argument is provided in Table 13.1. If you think about the argument, it implies that farmers who grow crops for which demand is more price-inelastic have more incentive to fight productive policies and push for predatory policies. If they are politically effective, they should tend to succeed in obtaining redistributive relief. All of this implies a negative relationship between the elasticity of demand and the amount of predatory subsidies. Table 13.1 shows the breakdown of public sector assistance into productive assistance and predatory assistance by crop.[18] The crops are arranged in roughly ascending order of elasticity (that is, sugar, milk, and rice are the most price-inelastic). So this table suggests that farmers who grow crops with the most inelastic demand do indeed tend to get the most redistributive subsidies.

The bottom line here is that plenty of evidence supports the view that special interests matter in determining the creation and nature of agricultural policies, both in the United

States and internationally. However, a story based solely on special interest demands for favorable policies is probably too simple. In addition to predatory policies that are largely driven by special interest demands, productive policies have also been pursued – at least in the United States – in order to address demands for public goods.

## Key takeaways

- Food production can be expanded on both the extensive and intensive margins. Historically, expansion on the extensive margin has occurred first, with the transition to intensive agriculture occurring first in high-income countries.
- Historically, technological change has contributed to major increases in agricultural productivity. This occurred especially in the mid- to late 20th century with the Green Revolution.
- Induced innovation is the idea that technological innovation occurs in part as a response to factor prices and resource endowments. There is evidence that this has occurred in food production.
- In the United States, farming has historically been characterized by chronically low average farm incomes and widely fluctuating farm prices. This has been known as the "farm problem." The farm problem may no longer be as serious as it once was, as farmer incomes have risen dramatically since the 1960s. However, widely fluctuating farm prices remains an issue.
- Good institutions, especially secure property rights in land, likely have significant positive effects on food production. However, this depends in part on other factors, such as thin markets for output, labor, and crop insurance.
- The positive effect of good institutions has been empirically supported in historical contexts in the United States and Europe. However, additional research needs to be done to verify the connection in many low-income contexts. In low-income contexts, violence and civil conflict can have devastating short-term effects on food production.
- Internationally, agricultural sectors tend to morph from being taxed to subsidized as economic development occurs.
- In the United States, agricultural policies take two forms: efficiency-enhancing (productive), and redistributive (predatory).

## Key concepts

- Food security
- Extensive vs. intensive food production
- Green Revolution
- Induced innovation
- "Farm problem"
- Productive vs. predatory policies

## Exercises/discussion questions

13.1. Consider a rural region within a low-income country where farming is being carried out in a labor-intensive way for home consumption. What would be the effect of the introduction of new rural infrastructure like roads and bridges on the incentives given farmers to produce and the methods they use to produce?

13.2. As with everything else in life, the expansion of food production in the Green Revolution involved tradeoffs and, therefore, the need to weigh costs and benefits. What do you see as the essential tradeoffs in expanding crop yields using scientific and technological advances in genetics and plant breeding? Do you see any practical ways these tradeoffs might be relaxed?

13.3. Based on the Barrows and Feldman readings (see below), construct: (a) the best evidence-backed argument you can *supporting* continued use of GMOs; and (b) the best evidence-backed argument *opposing* continued use of GMOs.

13.4. Consider the induced innovation hypothesis. Under what economic and political conditions do you think induced innovation would be less likely to occur?

13.5. If a country has abundant workers willing to work for low wages, why might large farming operations adopt labor-saving technologies?

13.6. On communally owned and worked farmlands on a large commune, is shirking by individual farmers inevitable? Aside from dividing up into smaller communes, how might a commune attempt to reduce shirking in order to increase food production?

13.7. In a number of low-income countries in need of humanitarian food aid, political conditions are unstable and the countries are prone to conflict and civil war. Under these conditions, what should be done by international agencies wanting to provide food aid?

13.8. Consider the empirical regularity that farmers tend to get taxed in low-income countries and subsidized in high-income countries. Can you construct a convincing public choice story that explains why this pattern does not seem to hold in certain countries; for example, Australia and New Zealand?

13.9. Reconcile the view of knowledgeable observers like Paarlberg and Paarlberg that U.S. farmers have recently done quite well with the evidence in Figure 13.6 that shows that farm prices continue to show a strong secular decline in recent years.

## Notes

1 The map uses *Landsat* satellite imagery, with 30-meter resolution. See NASA (2017).
2 For more discussion of the Prevalence of Undernourishment indicator, see FAO (2020), pp. 4–7.
3 See, for example, Ritchie (2020).
4 *Ibid.*
5 This idea is generally attributed to the famous British economist John Hicks [Hicks (1932)]. See also Hayami and Ruttan (1981); Ruttan (2002), pp. 163–5.
6 There are many studies of the induced innovation hypothesis, indicating that economists view the issue as being extremely important. For a few studies broadly supporting the hypothesis, see Binswanger (1974); Antle (1984); Chavas et al. (1997); Thirtle et al. (2002); Umetsu et al. (2003); and Paris (2008). For a contrary view, see Olmstead and Rhode (1993). If you pursue this research path, I must give you fair warning: many of the studies are pretty technical.
7 Grabowski (1979). See also the exchange between Hayami and Grabowski in Hayami (1981) and Grabowski (1981).
8 Lin (1992). For a contrary view, see Dong and Dow (1993).
9 This is a very consistent result in a number of studies. See, for example, Ostrom and Walker (1997); Frank et al. (1993); Ostrom (2000).
10 See, for example, the study by Hornbeck (2010).
11 *Sweden*: Olsson and Svensson (2010); *Russia*: Toumanoff (1984).
12 For a few studies that draw these conclusions, see Feder (1987); De Soto (1993); Besley (1995); Trewin (1997); Deininger and Jin (2003); Deininger and Ali (2008); Fenske (2012). For some contrary studies, see Place and Hazell (1993); Kung and Cai (2000).

13 There are a few exceptions to this rule; for example, Australia and New Zealand are two high-income countries that have reduced their subsidies to farmers over time. But the overall pattern is strong and pronounced. See, for example, Anderson et al. (2013), p. 445; Anderson and Hayami (1986), p. 45; Paarlberg (1989), p. 1158; Baliscan and Roumasset (1987), pp. 242–3.

14 See, for example, Rausser et al. (2011), chapter 8; Baliscan and Roumasset (1987), p. 242–3.

15 Economists have puzzled over why capital and labor would not leave farming quickly enough to offset technological change [Gardner (1992), pp. 67–8]. See Johnson and Quance (1972) for an influential theory which they call *asset fixity*, where they argue that there are various adjustment costs that slow the movement of these resources out of the farm sector.

16 See Paarlberg (1989), p. 1160; and Paarlberg and Paarlberg (2000), p. 142 for discussion of these points.

17 For discussion of the public choice interpretation of these policies, see Gardner (1987). For more discussion of productive and predatory policies, see Rausser (1982); Rausser (1992); de Gorter et al. (1992).

18 *Productive*: public good expenditures, information and marketing services, grades and standards inspection, crop insurance, public research, extension services. *Predatory*: Deficiency payments, price supports, trade barriers, storage subsidies, input subsidies, subsidized credit. See Rausser (1992), p. 150.

## Further readings

### I. The GMO controversy

Barrows, Geoffrey; Steven Sexton; and David Zilberman. "Agricultural biotechnology: The promise and prospects of genetically modified crops," *Journal of Economic Perspectives* 28(Winter 2014): 99–119.
*Economists' view of the GMO controversy.*
Feldman, Matthew P.; Michael L. Morris; and David Hoisington. "Genetically modified organisms: Why all the controversy?" *Choices* 15(First Quarter 2000): pp. 8–12.
*Summarizes the GMO controversy, including discussing the benefits and potential risks associated with use of GMOs.*

### II. The enclosure debate

Allen, Robert C. "The efficiency and distributional consequences of eighteenth century enclosures," *Economic Journal* 92(December 1982): 937–53.
*Argues that there were very little gains in efficiency to be had from enclosure. Rather, it simply resulted in a transfer of rents from tenant farmers to landlords.*
Grantham, George. "The persistence of open-field farming in nineteenth century France," *Journal of Economic History* 40(1980): 515–30.
*Argues that enclosure involved a tradeoff between efficiency and equity, benefiting large landowners at the expense of small landowners. In England, the governing class was strong enough to impose the system at the expense of small landowners.*

### III. Public choice in agriculture

Scrimgeour, F.G. and E.C. Pasour, Jr. "A public choice perspective on agricultural policy reform: Implications of the New Zealand experience," *American Journal of Agricultural Economics* 78(May 1996): 257–67.
*Examines major reforms in New Zealand agricultural policy that occurred between 1984 and 1994, in order to derive lessons for other economies' attempts to reform their own agricultural policies. Finds that ideas matter, and that reform is easier if undertaken in the context of broader comprehensive policy reform.*
Van Doren, Terry D.; Dana L. Hoag; and Thomas G. Field. "Political and economic factors affecting agricultural PAC contribution strategies," *American Journal of Agricultural Economics* 81(May 1999): 397–407.

*Investigates political contributions by agricultural political action committees (PACs) to U.S. senators. Finds that contributions went disproportionately to Democratic senators in agricultural states who served on agriculture committees and who happened to be in close election races.*

# References

Alston, Julian M. and Philip G. Pardey. "Agriculture in the global economy," *Journal of Economic Perspectives* 28(Winter 2014): 121–46.

Anderson, Kym and Yujiro Hayami. *The political economy of agricultural protection.* Sydney: Allen & Unwin, 1986.

Anderson, Kym; Gordon Rausser; and Johan Swinnen. "Political economy of public policies: Insights from distortions to agricultural and food markets," *Journal of Economic Literature* 51(June 2013): 423–77.

Anderson, Terry L. and Peter J. Hill. "The role of private property in the history of American agriculture, 1776–1976," *American Journal of Agricultural Economics* 58(December 1976): 937–45.

Antle, John M. "The structure of U.S. agricultural technology, 1910–1978," *Journal of American Agricultural Economics* 33(1984): 414–21.

Baliscan, Arsenio M. and James A. Roumasset. "Public choice of economic policy: The growth of agricultural protection," *Weltwirtschaftliches Archiv* 123(1987): 232–48.

Besley, Timothy. "Property rights and investment incentives: Theory and evidence from Ghana," *Journal of Political Economy* 103(October 1995): 903–37.

Binswanger, Hans P. "The measurement of technical change biases with many factors of production," *American Economic Review* 64(December 1974): 964–76.

Binswanger, Hans P. and Klaus Deininger. "Explaining agricultural and agrarian policies in developing countries," *Journal of Economic Literature* 35(December 1997): 1958–2005.

Chavas, Jean-Paul; Michael Aliber; and Thomas L. Cox. "An analysis of the source and nature of technical change: The case of U.S. agriculture." *Review of Economics and Statistics* 79(August 1997): 482–92.

De Gorter, Harry; David Nielson; and Gordon C. Rausser. "Productive and predatory public policies: Research expenditures and producer subsidies in agriculture," *American Journal of Agricultural Economics* 74(February 1992).

Deininger, Klaus and Daniel Ayalew Ali. "Do overlapping land rights reduce agricultural investment? Evidence from Uganda," *American Journal of Agricultural Economics* 90(November 2008): 869–82.

Deininger, Klaus and Songqing Jin. "The impact of property rights on households' investment, risk coping, and policy preferences: Evidence from China," *Economic Development and Cultural Change* 51(July 2003): 851–82.

De Soto, H. "The missing ingredient: What poor countries will need to make their markets work," in "The Future Surveyed," *Economist* September 8–12, 1993.

Dong, Xiao-yuan and Gregory K. Dow. "Monitoring costs in Chinese agricultural teams," *Journal of Political Economy* 101(June 1993): 539–53.

Feder, Gershon. "Land registration and titling from an economist's perspective: A case study in rural Thailand," *Survey Review* 29(1987): 163–74,

Fenske, James. "Land abundance and economic institutions: Egba land and slavery, 1830–1914," *Economic History Review* 65(May 2012): 527–55.

Frank, Robert H.; Thomas Gilovich; and Dennis T. Regan. "The evolution of one-shot cooperation: An experiment," *Ethology and Sociobiology* 14(July 1993): 247–56.

Gardner, Bruce L. "Causes of U.S. farm commodity programs," *Journal of Political Economy* 95 (April 1987): 290–310.

Gardner, Bruce L. "Changing economic perspectives on the farm problem," *Journal of Economic Literature* 30 (1992): 62–101.

Grabowski, Richard. "The implications of an induced innovation model," *Economic Development and Cultural Change* 27(July 1979): 723–34.

Grabowski, Richard. "Induced innovation, Green Revolution, and income distribution: Reply," *Economic Development and Cultural Change* 30(October 1981): 177–81.

Hayami, Yujiro. "Induced innovation, Green Revolution, and income distribution: Comment," *Economic Development and Cultural Change* 30(October 1981): 169–76.

Hayami, Yujiro and Vernon W. Ruttan. *Agricultural development: An international perspective.* (2nd ed.) Baltimore: Johns Hopkins University Press, 1985.

Hicks, John R. *The theory of wages.* New York: St. Martins, 1932.

Hornbeck, Richard. "Barbed wire: Property rights and agricultural development," *Quarterly Journal of Economics* 125(May 2010): 767–810.

Johnson, Glenn L. and C.L. Quance (eds.). *The overproduction trap in U.S. agriculture.* Baltimore: Johns Hopkins University Press, 1972.

Koren, Ore and Benjamin Bagozzi. "Food access and the logic of violence during civil war," *NewSecurityBeat*, Environmental Change and Security Program, Wilson Center, May 15, 2017, https://www.newsecuritybeat.org/2017/05/food-access-logic-violence-civil-war/.

Kung, James Kai-Sing and Yong-Shun Cai. "Property rights and fertilizing practices in rural China: Evidence from Northern Jiangsu," *Modern China* 26(July 2000): 276–308.

Lin, Justin Yifu. "Collectivization and China's agricultural crisis in 1959–1961," *Journal of Political Economy* 98(December 1990): 1228–52.

Lin, Justin Yifu. "Rural reforms and agricultural growth in China," *American Economic Review* 82(March 1992): 34–51.

NASA. Landsat Science. "New Landsat-based map of worldwide croplands supports food and water security," November 14, 2017. https://landsat.gsfc.nasa.gov/new-landsat-based-map-of-worldwide-croplands-supports-food-and-water-security/

Olmstead, Alan L. and Paul Rhode. "Induced innovation in American agriculture: A reconsideration," *Journal of Political Economy* 101(February 1993): 100–18.

Olsson, Mats and Patrick Svensson. "Agricultural growth and institutions: Sweden, 1700–1860," *European Review of Economic History* 14(August 2010): 275–304.

Ostrom, Elinor. "Collective action and the evolution of social norms," *Journal of Economic Perspectives* 14(Summer 2000): 137–58.

Ostrom, Elinor and James Walker. "Neither markets nor states: Linking transformation processes in collective action arenas," in *Perspectives on Public Choice: A Handbook*, Dennis C. Mueller, ed., Cambridge: Cambridge University Press, 1997.

Paarlberg, Robert. "The political economy of American agricultural policy: Three approaches," *American Journal of Agricultural Economics* 71(December 1989): 1157–64.

Paarlberg, Robert and Don Paarlberg. "Agricultural policy in the twentieth century," *Agricultural History* 74(Spring 2000): 136–61.

Paris, Quirino. "Price-induced technical progress in 80 years of U.S. agriculture," *Journal of Productivity Analysis* 30(August 2008): 29–51.

Philbrick, Nathaniel. *Mayflower: A study of courage, community, and war.* New York: Viking, 2006.

Place, Frank and Peter Hazell. "Productivity effects of indigenous land tenure systems in sub-Saharan Africa," *American Journal of Agricultural Economics* 75(February 1993): 10–19.

Rausser, Gordon C. "Political economic markets: PESTs and PERTs in food and agriculture," *American Journal of Agricultural Economics* 64(December 1982): 821–33.

Rausser, Gordon C. "Predatory versus productive government: The case of U.S. agricultural policies," *Journal of Economic Perspectives* 6(Summer 1992): 133–57.

Rausser, Gordon C.; Johan Swinnen; and Pinhas Zusman. *Political power and economic policy: Theory, analysis, and empirical applications.* Cambridge: Cambridge University Press, 2011.

Ritchie, Hannah. "Environmental impacts of food production," 2020. https://ourworldindata.org/environmental-impacts-of-food

Ruttan, Vernon W. "Productivity growth in world agriculture: Sources and constraints," *Journal of Economic Perspectives* 16(Autumn 2002): 161–84.

Thirtle, Colin G.; David E. Schimmelpfennig; and Robert F. Townsend. "Induced innovation in United States agriculture, 1880–1990: Time series tests and an error correction model," *American Journal of Agricultural Economics* 84(August 2002): 598–614.

Toumanoff, Peter. "Some effects of land tenure reforms on Russian agricultural productivity," *Economic Development and Culture Change* 32(July 1984): 861–72.

Trewin, Ray. "How land titling promotes prosperity in developing countries," *Agenda* 4(1997): 225–30.

Umetsu, Chieko; Thamana Lekprichakul; and Ujjayant Chakravorty. "Efficiency and technical change in the Philippine rice sector: A Malmquist total factor productivity analysis," *American Journal of Agricultural Economics* 85(November 2003): 943–63.

United Nations. Food and Agriculture Organization (FAO). *World food and agriculture – statistical pocketbook, 2018.* Rome, 2018.

United Nations. Food and Agriculture Organization (FAO). *The state of food security and nutrition in the world 2019: Safeguarding against economic slowdowns and downturns.* Rome: FAO, IFAD, UNICEF, WFP, WHO, 2019.

United Nations. Food and Agriculture Organization (FAO). *The state of food security and nutrition around the world in 2020.* Rome: FAO, IFAD, UNICEF, WFP, WHO, 2020.

United States. Agency for International Development. "Securing land tenure and property rights for stability and prosperity," 2017. https://www.usaid.gov/land-tenure

# 14 Water resources

It is difficult to think of a resource that is so ubiquitous and so essential to human well-being as water. We drink it, wash with it, clean with it, swim in it. We use it to grow our food, wash our clothes, water our lawns, cool our power plants, provide habitat for fish and waterfowl, and do a myriad of other activities that are central to supporting our way of life and standard of living. Indeed, we use it for so many things, it is very easy to take it for granted. But it is literally essential to life.

In recent days, water is also one of the resources that has come under the greatest environmental stress. Intensive irrigation taxes surface and groundwater supplies in water-scarce regions of the world. Agricultural runoff deposits silt, nitrates, and phosphates into our rivers and streams, creating algal blooms and hypoxic dead zones. Industrial wastes containing toxic chemicals can leak into neighboring streams or seep into groundwater aquifers. Wastewater from our sinks, toilets, and washing machines enter our septic tanks or municipal wastewater systems, to be deposited back into local rivers and streams.

In low-income countries, many of these problems are heightened because of higher population densities and the relative lack of effective wastewater treatment. In India, the Ganga (Ganges) River is heavily polluted with municipal and industrial wastes discharged from dozens of cities and towns, much of the wastes largely untreated. The challenges are also greater in arid parts of the world where water is scarcer, which includes sub-Saharan Africa, Australia, the Middle East, and the American West.

Addressing these issues effectively is one of the great environmental challenges of our time and will likely continue to be so into the foreseeable future. The solutions we come up with will require the same kind of ingenuity and resourcefulness that we saw in chapter seven on population. They will also require effective institutional responses that permit more judicious and efficient tradeoffs among competing uses, as well as ones that defuse potentially explosive conflicts among competing users.

## Overview of world water resources

Probably the most useful way to think about the world's water resources is fixed supply in the face of steadily growing demand. These is just so much water in the world, and almost all of it – some 96.5% – is tied up in the earth's oceans, which is mostly not usable for human purposes (we will discuss the possibility of desalination shortly). Our best guess is that only 2.5% of the world's total water supply is freshwater, defined as suitable for human purposes such as drinking water and irrigation.

Figure 14.1 illustrates how little of the world's water is actually usable to meet human needs. Compared to the globe that is the earth, the largest bead of water represents the

The World's Water

All water on, in, and above the Earth

Liquid fresh water

Fresh-water lakes and rivers

Howard Perlman, USGS.
Jack Cook, Woods Hole Oceanographic Institution,
Adam Nieman
Data source: Igor Shiklomanov
http://ga.water.usgs.gov/edu/earthhowmuch.html

*Figure 14.1* The world's water. *Source*: US Geological Survey, Igor Shiklomanov, http://ga.water.usgs
.gov/edu/earthhowmuch.html.

amount of water that is actually usable water; that is, contained in non-saline surface water, groundwater, and water vapor in the air (that becomes rain and snow). The smaller bead represents how much of that is actually liquid fresh water. And the barely visible tiniest bead represents how much water is contained in our freshwater lakes and rivers. So, when you think about the world's water supply, it is better not to think about all the water present everywhere on the globe: think instead about the amount that is actually usable to satisfy human needs.

Another important thing to keep in mind about water supply is that it is very uneven, depending on where on the earth you happen to live. Some regions have plenty of naturally occurring water, while others have very little. Figure 14.2 shows worldwide patterns of aridity, using a commonly used measure of aridity: a five-point humid/arid scale. All the darkest areas are called *humid*, which means that, on average, they have plenty of naturally

*Figure 14.2* Patterns of aridity. *Source:* European Commission. Joint Research Centre. *Patterns of aridity*, World Atlas of Desertification. https://wad.jrc.ec.europa.eu/patternsaridity

occurring precipitation to meet most of their needs. The lightest areas are called *hyper-arid*, meaning that they are extremely water-scarce. In between these two extremes are intermediate degrees of aridity, with the lighter colors denoting higher degrees of aridity.

Notice that the most water-scarce regions of the world are in a broad band stretching across northern Africa extending into the Middle East. Elsewhere, we see other water "trouble spots" in central-eastern Asia, Australia, the western United States, and pockets of South America and southern Africa. On the other hand, there is relatively plentiful precipitation across the eastern United States and Canada, most of northern-central South America, almost all of Europe, northern Eurasia, central Africa and the Far East. You will not be surprised to hear that many of the areas of water conflict you read about in the news media today are in the water-scarce regions of the world.

Demand, on the other hand, is determined more by human needs for water, including drinking water and other household uses, irrigation, hydropower, and various industrial uses. In addition, water serves increasingly important environmental demands, such as dilution of pollution in surface waters, providing habitat for fish and waterfowl, and rec-reational demand for boating, fishing, and swimming. All of these demands increase with population and income growth. So the growth in world population and world income that occurred over the latter half of the 20th century both stoked growing worldwide demand for water. For example, the FAO estimates that global acreage equipped for irriga-tion increased by 116% between 1961 and 2006.[1]

As with supply, water demand varies considerably across different regions of the world, depending on population sizes and economic demands. Even though most of Australia is highly water-scarce, its relatively small population allows its relatively sparse water sup-plies to largely meet its economic needs. On the other hand, the relatively plentiful water supplies in western Europe are tapped by a much larger population. As you might guess, the worst combination is low amounts of natural precipitation serving large populations. This describes many countries in central Africa and central Asia and, to a somewhat lesser extent, the southwestern United States.

In assessing the overall water picture, another factor to keep in mind is the quality of available water. Degraded water supplies are unsuitable for many human uses such as drinking water and irrigation, and treatment may be quite costly. Lower water quality widens the gap between demand and supply, which exacerbates the problem of water scar-city. Many areas of the world are confronting serious problems of declining water quality. This is especially true of many low-income countries that lack the resources for effective water treatment. But high-income countries are by no means immune, as evidenced by the recent U.S. experience with degraded drinking water in Flint, MI and Newark, NJ (see chapter ten, Box 10.1).

Projected future trends of growing populations and the demands of economic growth do not bode well for our ability to meet our water demands into the foreseeable future. By mid-century, the same amount of global water is likely to have to serve an additional two to three billion human beings. Notice that Africa, with its relatively scarce water sup-plies along with massive population increases predicted to mid-century, will be especially hard hit.

Furthermore, ongoing climate change is likely to adversely affect water supplies. As the planet heats up, higher temperatures will increase evaporation. At the same time, we will probably experience increasing demand for water for irrigation, cooling, and other uses. Hotter temperatures and increased variability in rainfall will probably significantly reduce water quality by increasing nitrogen, phosphorus, dissolved oxygen content (BOD), and salts in surface waters [World Bank (2019), pp. 59, 68; Dinar (2016), p. 3]. A prominent United States government official in the Obama administration once described the effect of climate change on freshwater supplies as a greater threat to coastal areas than rising sea levels [Olmstead (2014), p. 500].

## The hydrologic cycle

In order to help you understand the discussion to come, I need to explain a very important scientific concept known as the *hydrologic cycle*. The hydrologic cycle describes the circulation of water from land to air and back again (Figure 14.3). In a nutshell, water sitting in bodies of water on land is heated by the sun and evaporates into the air, creating water vapor. This water vapor condenses into clouds that eventually produce precipitation, which falls to earth, replenishing water supplies. And then the cycle repeats again. All of this means that all water on earth is constantly being renewed and replenished in a physically closed system.

Now think about what all this means for humans tapping various sources of water for their needs. When a town diverts water from a river for drinking water, this reduces the amount of water flowing in the river downstream from the town. When a farmer pumps irrigation water from the ground, it reduces the amount of water in the groundwater

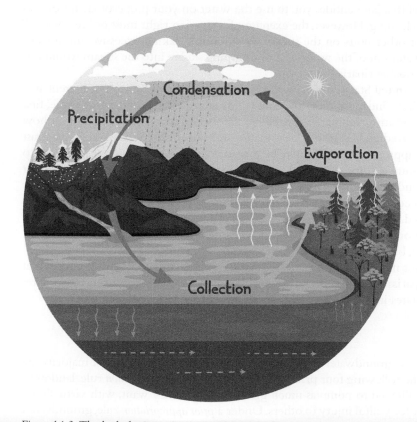

*Figure 14.3* The hydrologic cycle. *Source*: Shutterstock

aquifer, leaving less for other farmers tapping the same aquifer. When farmers apply irrigation water to their fields, some of the water is consumed by the crops and the remainder percolates down into the underlying aquifer. Or it may hit impermeable layers of rock that direct it into nearby rivers and streams. You can think of all of these actions as affecting the hydrologic cycle, which in turn affect water flows and impact other water users.

## Institutional rules governing water use

Because of the value of water as an important natural resource, detailed rules have emerged over time that govern how water may be claimed, developed, and used. These rules are designed to encourage development of water supplies and to provide for peaceful resolution of disputes over water. It turns out that these rules vary quite a bit across different countries and even within countries. To give you a taste of the rules governing water, let me briefly describe the water rights system of the United States, beginning with surface water rights.

### I. Surface water

In the United States, two main systems of rights to surface water have developed over time. In the eastern United States, surface water rights are governed by *riparian law*. Under riparian law, you obtain a right to use the water in rivers and streams by buying adjacent land. Ownership of that land entitles you to use the water on your property, say for drinking, cooking, and cleaning. However, the exercise of a riparian right must be "reasonable" given the needs of other users on the stream. For example, taking excessive amounts of water, exporting water out of the watershed, or dumping sludge into the stream would not be considered reasonable riparian uses.

In the western United States, surface water rights are governed by a very different system: *appropriative law*. Under appropriative law, rights are established on a first come, first served basis. Appropriative rights are to specific quantities of water, depending upon how much water is diverted and used in a "reasonable and beneficial" way.[2] And, in contrast to riparian rights, appropriative rights may be diverted to lands not physically adjacent to the surface water source. "First come, first served" means that if you stake a claim to as-yet unclaimed water and that right is confirmed, you have a right that takes precedence over any claimant who comes along later. Your right is then given a *priority date*, the date your claim was perfected. Later claimants are entitled to claim only what is left over after you have taken the amount you are entitled to. In times of drought, later claimants must cut back their use first, before earlier claimants are required to.

Finally, I will just mention that certain western states have adopted hybrid systems that recognize both riparian and appropriative rights. These include all the Pacific coastal states and some of the states in the central Great Plains region.

### II. Groundwater

In the United States, groundwater rights vary from state to state, but the vast majority are based on one of the following four principles. Under the *absolute dominion* rule, landowners are essentially allowed to pump as much groundwater as they want, with virtually no restrictions except for willful injury to others. Under a *prior appropriation* rule, groundwater is treated like surface water: groundwater may be pumped and applied to a reasonable and beneficial use on a first come, first served basis. Under a *reasonable use* rule, groundwater is subject to reasonable use and may only be used on the overlying tract of land. Finally, under a system of *correlative rights*, landowners enjoy a right to a reasonable share of the underlying groundwater. The shares of different landowners are typically based on the relative size of their landholdings.

You may be able to see the basic property rights system implied by each of these principles. The absolute dominion rule is similar to an *anything goes* rule such as we described in chapter four. In this respect, it is best suited to situations where groundwater is plentiful relative to demand. The prior appropriation and reasonable use rules are analogous to the surface water appropriative and riparian systems, except that they apply to groundwater. Interpreted in this way, they represent different ways to resolve disputes when groundwater is scarce. Finally, you can view the correlative rights rule as a variant of the reasonable use rule, where the size of your landholdings is a good measure of the amount it is reasonable for you to use.

## Some economics of water use

Now that we have some basic facts about water, let us discuss how economists think about water. As we have seen, water has all of the attributes that make it suitable for economic analysis. It has value because it is scarce, and it has multiple uses, which means that there are tradeoffs in how we use water. On the most general level, it is subject to supply and demand pressures. Furthermore, you can think of water supply as being highly price-inelastic at the global level. This means primarily that growing demand yields little increase in water use but, rather, results in steadily increasing water prices over time. Of course, water supply may be price-elastic at the regional or local level as it may be possible to import it from, or export it to, other regions.

Let us now make the picture more realistic by considering the special features of water implied by our discussion of the hydrologic cycle. Recall the idea that water users are interconnected because water flows and migrates from place to place. What does this imply in terms of how water should be used? Let us apply the property rights approach to answering this question.

### I. Surface water

Let me first ask you a question regarding defining rights to surface water. If you wanted to define a right to surface water that promoted efficient water use, how would you do it? To make the argument concrete, refer to Figure 14.4, which depicts a stylized river. Here, there are three water users – A, B, and C – who are located along a river that is flowing down and to the right. So B is located downstream from A, and C is located downstream from B. Each user is pumping an amount $W$ from the river, of which a fraction $R$ is flowing back into the river as recharge. So A is pumping the amount $W_A$ from the river, and $RW_A$ is the amount recharging the river. Similarly for B and C. $R$ is referred to as the *recharge coefficient*. To keep things simple, we assume $R$ to be the same for all users.

The question is: If you were interested in efficient water use along this river, how would you define a water right that could be used and potentially sold to others not on the river? For example, suppose this river is out in the country, and the parched city of Los Angeles wants to buy the right of one of these users and bring this water into the city. If you define A's right as $W_A$, allowing him to sell the water to Los Angeles will inflict

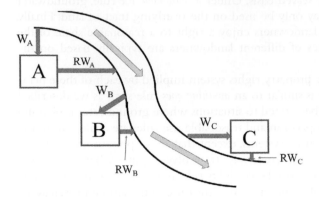

*Figure 14.4* A stylized river

an externality on downstream users. To see this, first notice that selling A's water to Los Angeles reduces the water in the river by $W_A$. In this case, there is no recharge back into the river. However, if A keeps it and uses it, this reduces the water in the river by a smaller amount: $W_A$ (A's diversion) *minus* $RW_A$ (the amount that recharges the river). So selling the water to Los Angeles leaves less water in the river available for downstream users.

This discussion leads naturally to the correct prescription. Instead of defining A's right as $W_A$, it should be defined as $(1 - R)W_A$. This way, if A sells his right to Los Angeles, there will be no impact on downstream users. Defining his right as $(1 - R)W_A$ effectively internalizes the externality. Put differently, the idea is to define his right not to the amount he diverts but, rather, to the amount he consumes. On a river with multiple users, surface water rights should be defined as the quantity of *consumptive use*.[3]

## II. Groundwater

In the case of groundwater, the property rights-based prescription is somewhat different. When a groundwater aquifer is tapped by multiple users, pumping by each user means less water for other users. So, when any user pumps groundwater, this increases the pumping costs of all other users because it forces them to pump from deeper underground. In other words, there is an external cost being inflicted by every user on all other users. Inefficiency may arise when farmers do not take into account these external costs when deciding how much to pump. Here, we can readily see Coase's perspective of mutual costs being inflicted.

One potential solution is to impose a pumping tax on each farmer, so that each bears the full (social) cost of his pumping behavior. There are two practical problems here. The first is knowing where to set the tax, which depends upon the magnitude of the external cost being inflicted. This requires information that may not be available in many cases. The other is that groundwater pumpers may vigorously oppose the imposition of pumping taxes, which may make it politically difficult to enact.

Another possible solution is to place the entire aquifer under the control of one entity, such as a private company or local public district. As we have just seen, the problem with the unregulated aquifer is that no individual pumper has incentive to take into account the costs it inflicts on other pumpers. Having one entity in charge of managing the entire aquifer may address that problem. What is being proposed here is, of course, the unitization solution that we first encountered in chapter eight on fossil fuels.

It turns out that the idea of local entities being empowered to manage local groundwater supplies has been pursued by various states in the western United States. For example, the state of Texas has provided for the creation of groundwater management districts, which may develop plans to manage local groundwater supplies more efficiently. Other states have also provided for the creation of similar entities, including Kansas, Idaho, and California. In 2014, the state of California required, for the first time in its history, mandatory groundwater management by local districts (see Box 14.1).

---

**Box 14.1: The California *Sustainable Groundwater Management Act* of 2014**

For years, the state of California relied upon voluntary efforts by localities to manage their local groundwater supplies, with no statewide system of groundwater regulation in place. Then beginning in late 2011, historic drought conditions rocked the state, which lasted for several years. The three-year period from 2011 to 2014 was the driest on record. In response, in 2014 the state enacted the Sustainable Groundwater Management Act (SGMA), which called for mandatory groundwater management efforts by localities for the first time in state history. Under SGMA, local agencies are required to create a groundwater management plan for their locality, under which groundwater must be sustainably managed to service multiple competing uses over the long term. Farmers criticized SGMA for imposing costs on them and environmentalists criticized it for not requiring full compliance for 20 years (or more). Nevertheless, by making local groundwater management mandatory for the first time ever, SGMA represents a landmark in California water history.

---

## Water policy options

### I. Supply-based policies

When faced with escalating costs due to increasing demand, there are two possible responses. One is to try to alleviate the supply problem by finding new sources of water. Indeed, historically, much water policy in the United States involved finding and developing new sources of water in order to support economic growth. For example, during the entire first half of the 20th century, the federal government constructed massive water development projects throughout the American West, through its main water development arm, the Bureau of Reclamation.

However, over time, the number of suitable water development projects has steadily dwindled, leading policymakers to consider other ways of augmenting water supplies. This has included, for example, miles-long pipelines from Canada and Lake Michigan to the U.S. Southwest, importing water by ocean-going tankers, and even towing icebergs from the Arctic and Antarctic regions. However, long water pipelines have encountered stiff political resistance from regions that would be asked to export their water. And importing water via tankers or icebergs is expensive and unlikely to make much of a dent in the overall water demands of water-scarce regions.

Another supply-based option that you hear about a lot is *desalination*. Desalination involves taking brackish or saline water and subjecting it to a process known as *reverse osmosis*. Under reverse osmosis, salty water is pushed through semipermeable membranes to remove the salt, which is then discarded, leaving potable drinking water. This process has been technologically feasible for some time, but it has suffered from reliability issues that have made it a costly water supply option. Furthermore, it presents various environmental challenges, including the fact that it uses a great deal of energy. And desalination plants located on the coast cause damage to local marine life from the operation of the water intake system and discharges of wastewater with extremely high salt content.

However, technological advances have brought down the cost substantially and, as a result, we are witnessing growing development of desalination as an option to increase water supplies in selected areas, especially arid and coastal areas. Between 1990 and 2008, total global desalination capacity nearly quadrupled. Today, there are desalination plants all over the world, from China to Australia to the Middle East to the southwestern United States. Altogether, more than 300 million people worldwide currently get their water from desalination plants. Many economists and water experts are cautiously optimistic about the prospects for greater global reliance on desalination. However, it seems clear that much research is needed to solve various technological issues before it is cost-competitive at a sufficient scale to address our water needs [Bernat et al. (2010), p. 335; Robbins (2019)].

## II. Demand-based policies

All of these factors make it is unlikely that supply-based solutions alone will be able to solve the world's problem of growing water scarcity. This is why we are seeing a shift in emphasis away from supply-based solutions and towards *water demand management*. Demand management is a catchall phrase that refers to various ways to conserve on water use. It also refers to reallocating water across competing uses in order to increase the overall value derived from water.

A key sector in which water demand management is occurring, and in which there is great potential for future demand reductions, is agriculture. Agriculture is a prime target for concerted demand management efforts because it is a major user of water, exceeding 80% of all consumptive uses in many countries around the world. Thus, modest percentage decreases of water in agriculture could free up a lot of water for competing uses, such as drinking water for cities and habitat protection for wildlife.

To put it concretely, currently California agriculture uses on average 34 million acre-feet of irrigation water per year.[4] This is a lot of water: with this much water, you could cover 31 million football fields with one foot of water. If California farmers could cut water consumption by a measly 10%, the resulting 3.4 million acre-feet of water freed up could provide for the water needs of nearly seven million households. This is well over half of all of the households living in California!

There are a number of ways to manage demand in agriculture. These include using more water-efficient irrigation technologies; pricing reform; water markets; and reducing water losses from leaky canals and pipes. Let us consider a few of these options.

### A. Water-saving irrigation technologies

If you were a farmer who wanted to irrigate your crops, there are three main ways you could do this. The first is *ditch irrigation*, which involves digging a ditch, or canal, from a water source and feeding the water into your fields, either using pumps or gravity. The second is *sprinkler irrigation*, which uses sprinklers housed on long, moving rigs connected to a source of water. Finally, there is *drip irrigation*, which feeds water through hoses to precisely target the root systems of the crops.

Thinking about these irrigation methods like an economist, you can categorize them in the following way. Ditch irrigation is the most water-intensive irrigation method, while drip irrigation is the least water-intensive. That is, ditch irrigation uses the most water and the least other factor inputs, especially capital and energy. Drip irrigation uses the least water and the most other factor inputs. And sprinkler irrigation is somewhere in-between. Thus, where water is more costly or scarce, there is an advantage to adopting the water-saving technologies.

Thinking about these methods in this way, you may not be surprised to hear that ditch irrigation has tended to be used where water is naturally relatively plentiful. Drip irrigation, a more recently developed method, tends to be used where water is scarcer. The extremely water-scarce country of Israel was among the first to adopt drip irrigation on a widespread scale, and is currently a world leader in drip irrigation technology.

But the practical advantages of water-saving technologies go way beyond dealing with the reality of scarce or expensive water. Studies have shown that they can increase yield, reduce soil erosion, and reduce waterlogging of the soil. They can accomplish all this by reducing water applications and permitting irrigation to be more precisely timed and regulated, enabling maintenance of stable soil moisture in the soil. And they can even reduce the need for application of fertilizers and pesticides.[5]

At the same time, there are certain limitations to water-saving technologies. They tend to be more cost-competitive when applied to high-value crops like grapes and fruit trees. They also tend to be more cost-competitive when soils are sandy and when farming is taking place on hillsides. And, ironically enough, under certain conditions they may not even reduce water use. This is because they may increase yield enough to make it profitable to expand production to the extent that farmers actually end up using more water [Dagnino and Ward (2012); Zilberman et al. (2017)]. Generally, however, there is probably great potential for water saving technologies to reduce water use while maintaining crop yields [Jaegermeyr et al. (2015)].

*B. Pricing reform*

As an economic good, the demand for water for irrigation is responsive to changes in its price. When water prices go up, farmers have incentive to use less water. They will tend to use it more carefully, not wasting it and plugging leaks in their irrigation ditches. They will be more likely to adopt water-saving irrigation technologies like drip irrigation. And they will adopt other methods of conserving water, such as leveling their fields using lasers. Conversely, if water is extremely cheap, they will tend to do less or even none of these things. This last point is relevant because the price that farmers pay for water often bears little resemblance to its actual value in productive use. The result can be gross inefficiencies in water consumption.

Consider, for example, irrigation in the western United States as it has developed over time. Currently, farmers throughout the western U.S. pay heavily subsidized prices for water they receive through federal reclamation projects. Our best estimate is that the prices they pay per acre-foot of water to the federal government are only a fraction of the value of the water in production. And the value of water in agriculture is, by every measure, significantly lower than its value in many other uses, especially urban uses such as municipal drinking water. All of this suggests that there might be significant gains from increasing the price of water to farmers, thus freeing up water to flow to the cities where it is much more valuable.

Here, however, the analysis is a bit more complicated than a simple head-to-head comparison of urban vs. agricultural value. To see this, consider Figure 14.5, where D represents agricultural demand for water, and where $P_{Fed}$ represents the (low) price per acre-foot that farmers are currently paying for water. If farmers were receiving as much water as they wanted at the price $P_{Fed}$, they would have incentive to use the amount $Q_{Fed}$. In this case, raising the price to P' would discourage use, as farmers would then have incentive only to use Q' acre-feet of water. This argument is, of course, basic demand analysis.

However, suppose farmers were paying $P_{Fed}$ but only receiving the amount Q' to start with. What would be the effect on water consumption of increasing the price to P' in this case? Here, you should be able to convince yourself that there would be no incentive to reduce water consumption at all. This is because at Q' — the amount of water the farmers are receiving — the value of water on the margin P' is higher than the price the farmers are paying to the government for the water $P_{Fed}$. Raising the price to P' will not discourage consumption because the water remains more valuable on the margin than the price charged by the government.

The important distinction here is that, in the first case, water is being *price-rationed*, whereas in the second case, it is being *quantity-rationed*. When farmers can receive as much water as they want at a fixed price, the *price* determines how much they will use. However, when they are only sold a fixed amount at a price below their marginal value, the *quantity* determines how much they will use. In the latter case, price changes have little effect on their consumptive use. In this case, freeing up water to flow to the cities requires not an increase in the price, but, rather, a reduction in the quantity delivered to farmers.

The question of price-rationed vs. quantity-rationed is, therefore, an important caveat on whether price reform will have the desired impact in terms of discouraging use in order to free up water. In fact, there is some evidence that government water provided to farmers in California is quantity-rationed. If so, this means that pricing reform will have less impact than many might expect in terms of discouraging agricultural use [Kanazawa (1993); Moore and Dinar (1995)].

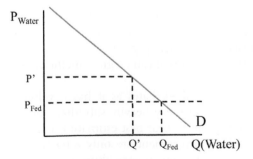

*Figure 14.5* Predicted effect of water pricing reform in agriculture

*C. Water markets*

I. ARGUMENTS FOR WATER MARKETS

These days, if you read much about water scarcity problems in arid areas, it is not long before you encounter the notion of water markets. Water markets are just what they sound like: organized exchanges of water between willing buyers and sellers. The idea is to let supply and demand set prices, and to let voluntary transactions determine where the water is used, for what purpose, and by whom. If all goes well, water should then flow to its highest value uses and be used efficiently: that is, in a way that maximizes societal surplus.

To make all of this more concrete, consider the huge gulf that apparently exists between the value of water in farming in the Central Valley of California and its value to residents of the city of Los Angeles. This implies that there should be all sorts of gains from trade from Los Angeles buying some of the farmers' water. A water market would permit these gains from trade by providing a mechanism for trades to take place. And such a trade would almost surely take place if the water is so much more valuable to Los Angeles than to the farmers.

Let me make my point using some made-up, but realistic, numbers. Suppose an acre-foot of water enables Central Valley farmers to grow an extra $500 worth of almonds, but it is worth $3000 to residents of Los Angeles. A water market would allow LA residents to buy water from the farmers. And LA would be able to make the farmers an offer they couldn't refuse. In this example, LA should be willing to pay up to $3000 for that acre-foot of water, whereas that acre-foot is worth at most $500 to the farmers. Some offer in-between these two amounts should be acceptable to both sets of parties. And the resulting trade would increase the total use-value of that acre-foot from $500 to $3000, making society better off.

Economic studies suggest that there are many instances where there are sizable gains from trade to be had from water transfers.[6] It should thus not be surprising to hear that water markets have been springing up all over the world, in places as varied as India, China, Pakistan, Spain, Australia, Chile, Mexico, Canada, and various parts of the United States, including California, Colorado, Texas, and New Mexico.[7]

In addition to this basic gains-from-trade argument, water markets have other nice features. They encourage users to use water carefully, because there is a greater opportunity cost to wasting the water; namely, the value of the water if they were to trade it on the market. This includes encouraging farmers to adopt water-saving technologies and to grow crops that use less water [Hearn and Easter (1997); Zilberman et al. (2017)]. In addition, they provide a flexible mechanism for allocating water to its highest-value uses under changing conditions, such as shifting demand patterns, or ongoing climate change.

II. CONCERNS OVER WATER MARKETS

Water markets are not an unmixed blessing. One of the biggest concerns is that they may impose costs on other users. We saw this earlier in our discussion of the economics of surface water use. So, a farmer taking his surface water allocation and selling it to a city, for example, deprives his downstream neighbors of some of the water they are accustomed to having. But the story doesn't stop there. There might be ripple effects, as those neighboring farmers cut back on production, affecting suppliers, etc. And there might be environmental impacts as well, as reduced water flows back into local rivers and streams may cause damages to wildlife habitat.

In addition to third-party impacts, water markets may not perform as well if there is only one or a handful of sellers who are able to exercise monopoly power. In this case, the seller could charge high prices and threaten to withhold delivery if buyers do not meet its demands. Possible solutions here are to: expand supply options, help buyers pay their legal expenses, or have a government agency regulate transactions. For example, one economist documents that farmers in Gujarat, India solved the monopoly problem by having pipelines from three or four suppliers connected to their fields. This allowed them to patronize the supplier who offered the lowest prices and best service [Shah (1993)]. More generally, economic studies are mixed on whether monopoly pricing is endemic to water markets, and how much it results in increased prices. See the studies in *Further readings* if you wish to pursue this issue.

Another objection to water markets is that they may disadvantage low-income households and small farmers, who may not be able to pay as much for, and therefore compete for, water with others, such as large farmers or cities. Some economists have argued that this problem is overstated because essential uses like drinking water are high-value uses, meaning that markets will provide for it.[8] However, if you wanted to provide a safety net for low-income users, one possible option would be to use a tiered pricing system, in which all users receive a minimum amount of water at a low price, after which they must pay the market price. This could be done, for example, if there was a fixed set of users and an institutional mechanism in place to administer the market, such as a local water district [Zilberman et al. (2017), p. 2].

III.A KEY ISSUE: TRANSACTION COSTS

Finally, there exist various sources of transaction costs, which may impede the smooth operation of water markets (see Box 14.2). Many of these transaction costs arise because of the way water rights are defined. Riparian law, for example, bases water rights on ownership of adjacent lands, as we have seen. Being legally required to buy lands in order to secure water rights will discourage some parties from trying to buy those rights. In some low-income countries, water-using farms are small and fragmented, making it difficult even to establish individual water rights. In many jurisdictions, groundwater rights are not defined as specific quantities of water, making it difficult to engage in groundwater trades. If nobody knows precisely how much water they are entitled to, it will be difficult to write contracts for exchange that can be enforced.

---

**Box 14.2: Sources of transaction costs in water markets [adapted from Easter et al. (1999)]**

- Inadequate water delivery infrastructure.
- Poorly/incompletely defined water rights.
- Lack of means of enforcing water rights.
- Water rights tied to land holdings.
- Institutional rules that discourage trading.
- Legal challenges by third parties.
- Opposition from farmers and environmental groups.

---

In many cases, potential water users belong to a local organization that contracts for water from a government agency. For example, in California many farmers are members of irrigation districts that purchase water from either the state or the federal government. These districts then turn around and distribute water to their member farmers.

Organizing into districts can help address water rights definition issues by providing an administrative mechanism for defining rights. For example, the district may establish a policy that entitles member farmers to so many acre-feet of water every growing season. The potential downside is that districts may impose various conditions that make transfers more difficult. For example, many districts do not permit farmers to make a profit on sales of their allocation to other farmers. This discourages them from seeking out potential buyers. In other cases, districts simply prohibit farmers from selling water to parties outside their district.

Another key source of transaction costs concerns the impacts of such trades on others. Many jurisdictions have legal rules requiring that any transfers of water not impose impacts on others not directly involved in the transfer. These rules provide legal standing to these others to sue to block these transfers. There is some economic justification for these rules, as they impose costs on potential transferors of water who have no incentive to take into account the impacts of their transfers on others. However, these rules may also block some beneficial trades from taking place.

A final source of transaction costs comes from the opposition of certain groups to water markets. Prominent among these groups are environmental organizations concerned about the environmental impact of water markets, such as the effect of market operations on instream flows that support wildlife habitat. Another group that may oppose water markets are farmers who are concerned about losing access to water, or having to pay higher prices for water. These groups may pose legal challenges to individual transfers, or they may even generate political opposition that keeps water markets from forming in the first place.

A useful way to think about many of these transaction cost challenges is that water markets have some characteristics of anticommons (see chapter five). Recall that anticommons are situations where multiple parties have rights of exclusion to the use of a resource. This is why storefronts in Moscow sit empty and unused, while hundreds of kiosks operate on the city streets in front of them. With water markets, the concept applies because there are many different claimants on an amount of potentially marketed water who can veto that sale. These include the federal government, local districts, other users, and environmental organizations, all of whom have legal standing to challenge transfers and keep them from occurring. This legal standing stems from congressional legislation, irrigation district policy, and legal doctrines such as the public interest and public trust doctrines [see Bretsen and Hill (2009)]. The anticommons idea helps us understand why some proposed, highly valuable transfers of water from rural to urban areas took years to occur (see Box 14.3).

---

### Box 14.3: Anticommons and water markets

Despite evidence of enormous disparities in the value of water, it took years for farmers in the Imperial Irrigation District (IID) and the city dwellers of Los Angeles and San Diego to work out a deal for the IID to sell water to those two cities in southern California. The deal was hampered by internal disagreements in the IID, local opposition to the transfer, and lawsuits from environmentalists and Mexican agricultural interests. A deal for San Diego to receive some IID water only went through when Secretary of the Interior Norton threatened to reduce the state entitlement to Colorado River water if the IID did not transfer water to San Diego. [Bretsen and Hill (2009), pp. 756–60]

In cases where transaction costs pose obstacles to water markets, there are a number of possible policies one could use to try to bring down these transaction costs. To address issues of poorly defined water rights, it might help to legally require that water rights be registered and quantified. This might entail creating an administrative agency with enforcement powers. Also useful would be clear legal rules for resolving conflicts among users when they arise. In addition, creating local water use associations that provide legal information and serve as clearinghouses for water trades would likely also help promote water trading.

Finally, to defuse political opposition from farmers and environmental groups, it might help to undertake a public information campaign to demystify water markets and explain their function and potential benefits. It might also be helpful to establish rules that set aside a portion of the traded water for protection of instream flows. In particularly water-stressed areas such as regions of excessive groundwater overdraft, you could consider imposing caps on water trades.

## Local water governance: An application of SES

The social-ecological systems (SES) framework that we encountered in chapter five is a useful way to gain additional insights into appropriate ways to deal with water problems. In many situations, a source of water is a common-pool resource (CPR), as water use is inherently rivalrous and it is often costly to exclude people from tapping it. The question is whether we should worry about the possibility of a tragedy of the commons: that the local water source is in danger of being severely degraded or destroyed.

### I. A seeming tragedy of the commons: The Ganges River, India

The answer is that it probably varies from situation to situation. In some cases, enormous pressures are put on water sources by economic and population growth, resulting in severe degradation. For years, the Ganges (Ganga) River in India has been one of the most polluted major rivers in the world. The Ganges is some 2,500 kilometers long, with more than two dozen major cities located on its banks. More than 400 million people depend on the Ganges for their livelihoods, including fishing, farming, industrial production, and various small business and domestic uses. Every day, roughly 800 million gallons of sewage are dumped into the river, only a small fraction of it treated. Along one of its major tributaries, the fecal coliform bacteria level is nearly 500,000 times the recommended bathing limit. In recent days, groundwater pumping in the Ganges basin has reduced summer groundwater flows into the river by 50%, causing river levels to dramatically decline. Reduced river flows not only lessen the amount of water available for its various uses, they also exacerbate the river's pollution problem.[9]

Applying the SES framework, the factors that characterized the tragedy of the commons in chapter five seem to apply to the Ganges. Here, you have a resource of major economic value, there are enormous numbers of users compared to its size, and users largely operate independently of each other. In addition, certain religious traditions sustain practices that pollute the Ganges, such as cremating loved ones and throwing their ashes into the river, which is sacred to many Hindus [Rogers (2013)].

In contrast to the pure tragedy of the commons situation, however, there do appear to be governance rules in place. These include a water pollution law – the *Water Act* – passed by India in 1974; a constitutional amendment passed in 1976 that established a national policy of environmental protection; and the creation of a Department of the Environment created in 1980. However, the Water Act includes all sorts of exemptions for discharges into rivers, and state water boards created under the Act have little enforcement power. Overall, the governance structure in place appears to be completely ineffective in maintaining water quality in the Ganges.[10]

## II. Local irrigation projects

In stark contrast to the Ganges, however, there are plenty of examples of local water resources that are managed sustainably. Let us consider one in particular that is important in many agricultural settings: the development of local water resources for irrigation.

Irrigation projects present a number of challenges for farmers wanting to develop a water source and use it sustainably. Local sources of water like rivers and groundwater aquifers are CPRs and, thus, subject to potential degradation if every farmer taps water on his own. This, and the fact that water development is generally subject to economies of scale, provide incentive for farmers to work together. This creates a collective action problem because individual farmers have incentive to hold out, to shirk their responsibilities to contribute to the collaborative effort. If enough individuals do this, the project will not get built and, even if it does, it may not be adequately maintained.

Despite the existence of the collective action problem, there are plenty of examples of successful collaborations to build and manage water projects at the local level. However, success seems to vary depending on a number of factors. First, it helps if farmers have a large stake in the success of the project, and if they view involvement in the project as being for the long term. Second, the ability to communicate face-to-face creates trust among farmers, especially if they have some past history with each other, either direct or by reputation. This makes it easier for them to work out agreements on water allocations, monitoring, and a system of sanctions to punish transgressors. Third, it helps if farmers can enter and exit a project at low cost. If so, they can extricate themselves more easily from situations where they believe they are being taken advantage of. In addition, they are less likely to take advantage of others because they know that others can leave if they believe *they* are being taken advantage of (see Box 14.4 for a summary of these conditions).[11]

---

**Box 14.4: Factors important for cooperation
[adapted from Ostrom (2010)]**

- Project importance
- Long-term time horizon
- Face-to-face communication
- Knowing the other participants
- Ability to exit and/or enter easily
- Mutually acceptable monitoring and sanctioning

---

To place all of this within the SES framework, notice the key differences in the factors that give rise to sustainable management rather than a tragedy of the commons. In both TOC and sustainable management, the value of the resource can be high, and entry and exit are easy. But that is about the end of their similarities. Sustainability appears to be supported by long-term interest in the health of the resource by all parties; seeing other participants as real people as opposed to anonymous scavengers; and being able to communicate with them face-to-face, which establishes trust and makes agreement on management practices – including sanctioning methods – more likely.

## Key takeaways

- Pressures on scarce water resources are growing over time with continuing economic and population growth, and they are likely to be exacerbated by ongoing climate change.
- Policies to augment supplies are becoming increasingly unlikely to yield much additional water, with the possible exception of desalination.
- Current and future policies are likely to rely much more heavily on demand management. Key among these policies will likely be policies to promote more widespread use of water markets. Though water markets have their challenges, in a number of instances they have proved to be an effective means of getting water to its highest-value uses.
- Another likely trend in the future is more heavy reliance on local cooperative arrangements to manage water supplies locally. With continuing research, our understanding of the conditions under which these are likely to succeed is increasing steadily.

## Key concepts

- Hydrologic cycle
- Appropriative water rights
- Riparian water rights
- Absolute dominion
- Correlative rights
- Desalination
- Ditch vs. sprinkler vs. drip irrigation
- Local water governance

## Exercises/discussion questions

14.1. What are the laws governing use of groundwater in your state? Suppose you just bought some farmland out in the country. What legal steps would you have to take to drill a well to start pumping groundwater?

14.2. Comparing riparian rights to appropriative rights, which do you think provides the most secure property rights and why?

14.3. Provide your best intuition for why a surface water right should be defined on the basis of consumption, not diversion, if the goal is to promote efficient use of the waterway.

14.4. Suppose you adopted a policy that subsidized farmers who use drip irrigation. Can you see any way that such a policy might result in increased, not decreased, water use?

14.5. Suppose that the demand for water in the Imperial Irrigation District can be represented as $P = 300 - 0.1Q$, where $P$ is the price per acre-foot and $Q$ is the quantity of water, in thousands of acre-feet. Suppose the price of water delivered by the Bureau of Reclamation is $40 per acre-foot. (a) Calculate the reduction in water use if the price were to raise the price to $80. (b) Now suppose that there is a maximum of 1 million acre-feet on the amount that the Bureau supplies to the district. Reanswer (a).

14.6. What do you think of the argument that because essential uses like drinking water are high-value uses, we can rely on markets to provide it to low-income households and small farmers?

14.7. Assume that, on efficiency grounds, you favor legal rules that permit parties adversely affected by water transfers to be able to sue to block those transfers. If so, this implies you believe that freely operating water markets will not be efficient. Then what is the most compelling argument you can make in favor of water markets?

## Notes

1 FAO (2011), p. 38. "Equipped for irrigation" is not the same as actual irrigated acreage. It is a term used to describe irrigation potential.
2 The phrase "reasonable and beneficial" is standard legal terminology applied by courts. and legislatures. Exactly what this phrase means has varied over time and continues to evolve with changing economic conditions. But you can think of unreasonable uses as ones that excessively damage other users or without good cause. A beneficial use is one that provides substantial benefit to the user. This prohibits, for example, simply wasting water gratuitously, but courts have never been able to provide an exact definition of the word "substantial."
3 See, for example, Johnson et al. (1981).
4 *Water Basics*, California Department of Water Resources. An acre-foot is the amount of water needed to cover an acre of land to the depth of one foot, or about 238,000 gallons.
5 For more information on these effects, see Caswell and Zilberman (1985) and Shani et. al. (2009).
6 See, for example, Easter et al. (1999), p. 99; Grafton et al. (2012).
7 There has been a lot of scholarly interest in this topic. For a few representative studies, see Easter et al. (1999); Grafton et al. (2010), p. 3; Carey and Sunding (2001).
8 For this argument, see Mendelsohn (2016), p. 3.
9 For more on the details of this paragraph, see Mallet (2015); Ghosh (2018).
10 See, for example, Robinson (1987); Sharma (2014).
11 These are a series of consistent conclusions of Ostrom and her co-authors. See, for example, Ostrom and Gardner (1993); Lam (1998); Ostrom (2010); Lam and Ostrom (2010); Ostrom (2011).

## Further readings

### I. Institutional analysis of water markets and water management

Boadu, Fred O.; Bruce M. McCarl; and Dhazn Gillig. "An empirical investigation of institutional change in groundwater management in Texas: The Edwards Aquifer case." *Natural Resources Journal* 47 (Winter 2007): 117–63.

*Examines an important institutional rule change governing groundwater in Texas, which moved Texas groundwater law from a rule of capture to a permit system. This rule change had important efficiency and equity consequences and also provided important environmental benefits.*

Carey, Janis M. and David L. Sunding. "Emerging markets in water: A comparative institutional analysis of the Central Valley and Colorado-Big Thompson Projects," *Natural Resources Journal* 41 (Spring 2001): 283–328.

*Studies the differential pace of adoption of water markets over time in California and Colorado. Concludes that different institutions have imposed greater transaction costs on trading in California, which has discouraged, in a path-dependent process, the development of water markets.*

## II. The SES framework and local water governance

Naiga, Resty; Marianne Penker; and Karl Hogl. "Challenging pathways to safe water access in rural Uganda: From supply to demand-driven water governance," *International Journal of the Commons* 9(March 2015): 237–60.
*This study uses the SES framework and Ostrom's design principles to analyze the impact of a move in 1990 by Uganda from centralized government water supply to reliance on local water governance.*


## III. Public choice and water policy

Bishop, Bradford H. "Drought and environmental opinion: A study of attitudes toward water policy," *Public Opinion Quarterly* 77(Fall 2013): 798–810.
*Examines the effect of drought conditions on public attitudes toward regulation of water use and concerns over water supply.*


# References

Bernat, Xavier; Oriol Gibert; Roger Guiu; Joana Tobella; and Carlos Campos. "The economics of desalination for various uses," in *Re-thinking water and food security*, Luis Martinez-Cortina, Alberto Garrido, and Elena Lopez-Gunn (eds.), Leiden: CRC Press, 2010.

Bretsen, Stephen N. and Peter J. Hill. "Water markets as a tragedy of the anticommons," *William & Mary Environmental Law and Policy Review* 33(2009): 723.

California. Department of Water Resources. *Water Basics: Agriculture*. https://water.ca.gov/Water-Basics/Agriculture.

Carey, Janis M. and David L. Sunding. "Emerging markets in water: A comparative institutional analysis of the Central Valley and Colorado-Big-Thompson projects," *Natural Resources Journal* 41(Spring 2001): 283–328.

Caswell, Margriet and David Zilberman. "The choices of irrigation technologies in California," *American Journal of Agricultural Economics* 67(1985): 224–34.

Dagnino, M. and F.A. Ward. "Economics of agricultural water conservation: Empirical analysis and policy implications," *International Journal of Water Resources Development* 28(2012): 577–600.

Dinar, Ariel. "Dealing with water scarcity: Need for economy-wide considerations and institutions," *Choices* 31(3rd Quarter 2016): 1–7.

Easter, K. William, Mark W. Rosegrant, and Ariel Dinar. "Formal and informal markets for water: Institutions, performance, and constraints." *World Bank Research Observer* 14(February 1999): 99–116.

Ghosh, Sahana. "As India's Ganges runs out of water, a potential food shortage looms," *Mongabay*, September 17, 2018.

Grafton, R. Quentin; Clay Landry; Gary D. Libecap; and Robert J. O'Brien. "Water markets: Australia's Murray-Darling Basin and the U.S. Southwest," Working paper no. 15797, National Bureau of Economic Research (March 2010).

Grafton, R. Quentin; Gary D. Libecap; Eric C. Edwards; Robert J. O'Brien; and Clay Landry. "Comparative assessment of water markets: Insights from the Murray-Darling Basin of Australia and western U.S.A," *Water Policy* 14(2012): 175–93.

Hearn, Robert R. and K. William Easter. "The economic and financial gains from water markets in Chile," *Agricultural Economics* 15(1997): 187–99.

Jaegermeyr, J.; D. Gerten; J. Heinke; S. Schaphoff; M. Kummu; and W. Lucht. "Water savings potential of irrigation systems: Global simulation of processes and linkages," *Hydrology and Earth System Sciences* 19(2015): 3073–91.

Johnson, Ronald N; Micha Gisser; and Michael Werner. "The definition of a surface water right and transferability," *Journal of Law and Economics* 24(1981): 273–88.

Kanazawa, Mark. "Pricing subsidies and economic efficiency: The U.S. Bureau of Reclamation," *Journal of Law and Economics* 36(April 1993): 205–34.

Lam, Wai Fung. *Governing irrigation systems in Nepal: Institutions, infrastructure, and collective action.* Oakland: ICS Press, 1998.

Lam, Wai Fung and Elinor Ostrom. "Analyzing the dynamic complexity of development interventions: lessons from an irrigation experiment in Nepal," *Policy Sciences* 43(March 2010): 1–25.

Mallet, Victor. "The Ganges: holy, deadly river," *Financial Times*, February 13, 2015.

Mendelsohn, Robert. "Adaptation, climate change, agriculture, and water," *Choices* 31(3rd quarter 2016): 1–7.

Moore, Michael R. and Ariel Dinar. "Water and land as quantity-rationed inputs in California agriculture: Empirical tests and water policy implications," *Land Economics* 71 (November 1995): 445–61.

Olmstead, Sheila M. "Climate change adaptation and water resource management: A review of the literature," *Energy Economics* 46(2014): 500–9.

Ostrom, Elinor. "Beyond markets and states: Polycentric governance of complex economic systems," *American Economic Review* 100(June 2010): 641–72.

Ostrom, Elinor. "Reflections on 'Some unsettled problems of irrigation'," *American Economic Review* 101(February 2011): 49–63.

Ostrom, Elinor and Roy Gardner. "Coping with asymmetries in the commons: Self-governing irrigation systems can work," *Journal of Economic Perspectives* 7(Autumn 1993): 93–112.

Robbins, Jim. "As water scarcity increases, desalination plants are on the rise," *YaleEnvironment360*, June 11, 2019. https://e360.yale.edu/features/as-water-scarcity-increases-desalination-plants-are-on-the-rise.

Robinson, Nicolas A. "Marshalling environmental law to resolve the Himalaya-Ganges Problem," *Mountain Research and Development* 7((1987): 305–15.

Rogers, Brett. "Life and death on the Ganges," *National Geographic*, May 29, 2013.

Shah, Tushaar. *Groundwater markets and irrigation development: Political economy and practical policy.* Bombay: Oxford University Press, 1993.

Shani, U., Y. Tsur; A. Zemel; and David Zilberman. "Irrigation production functions with water-capital substitution," *Agricultural Economics* 40(2009): 55–66.

Sharma, Dinesh. "New Ganges plan raises concerns," *Frontiers in Ecology and the Environment* 12(August 2014): 318.

United Nations. Food and Agriculture Organization (FAO). *The state of the world's land and water resources for food and agriculture: Managing systems at risk.* Oxford: FAO, 2011.

World Bank. "Quality unknown: the invisible water crisis," Washington: World Bank, 2019.

Zilberman, David; Rebecca Taylor; Myung Eun Shim; and Ben Gordon. "How politics and economics affect irrigation and conservation," *Choices* 32(4th Quarter 2017): 1–6.

# 15 Fisheries

In November 2006, a study in the respected journal *Science* predicted the collapse of all seafood ocean fisheries by the year 2050. This study, authored by an international team of ecologists and economists, argued that the loss of marine biodiversity was causing the world's oceans to lose its capacity to produce seafood, filter pollutants, and respond to ongoing stresses like overfishing and climate change. "Unless we fundamentally change the way we manage all the ocean species together as working ecosystems, then this century is the last century of wild seafood," said Stephen Palumbi, a professor of marine sciences at Stanford University and one of the co-authors of the study [Dean (2006)].

Studies predicting the collapse of the world's fisheries are heard with disturbing, and some would say alarming, frequency. In 2008, a report from the United Nations warned of fishery collapse as a result of climate change, ocean pollution, and overfishing. In 2016, a report by 80 scientists in a dozen countries warned that ocean fisheries were "on the verge of collapse" because of climate change and overfishing. In 2019, an international study assessed the global situation with regard to freshwater fisheries and came up with findings equally grim, with large freshwater species such as sturgeons, salmon, and catfishes experiencing a massive population decline in recent years.[1]

## Overview of world fisheries

Fisheries are an extremely important sector of the world food economy. The United Nations Food and Agricultural Organization (FAO) estimates that, in 2017, fish consumption accounted for over one-sixth of all of the animal protein consumed by the global population. This reliance on fish to meet basic food requirements is even higher in low-income countries, especially in Asia, where it is a fundamental food staple. In addition, the fishing industry is a source of employment for nearly 60 million people worldwide, including over eight million women [FAO (2020), p. 5, 36].

Worldwide fish production has been steadily increasing over time. Between 1961 and 2017, total world fish production increased at an average rate of 5.6% per year, well outpacing population growth. Over the same period, worldwide per capita fish consumption more than doubled, an annual growth rate of 1.5% per year. China is the largest fish-producing country in the world and since 2002 has been the largest world exporter of fish and fish products. Other major exporters include Norway and Vietnam. Over half of all imports of fish and fish products are by the European Union, the United States, and Japan [FAO (2020), p. 80].

At the same time, there are warning signs that current levels of ocean fishing are probably not sustainable. This means that fish stocks are being harvested faster than the ability of the fish stocks to replenish themselves through natural reproduction. According to the FAO, the proportion of fish stocks worldwide that are fished at biologically unsustainable levels has increased from one-tenth in 1974 up to over one-third in 2017. Overfishing has caused many fish populations to decline dramatically in size. In some cases, freshwater fish populations have been rendered extinct. The FAO has called the persistence of overfished stocks "an area of great concern" [FAO (2018), p. 6].

In addition to overfishing, marine fish populations are being stressed by increasing pollution levels including: toxic discharges, ocean disposal of plastics, increased sedimentation, and reduced oxygen availability (*eutrophication*) in lakes, rivers, and various localized regions of the ocean. Climate change has further exacerbated the situation, by heating up the oceans, thus stressing coral reefs and various heat-sensitive fish species. Finally, illegal, unregulated. and unreported (IUU) fishing has become a serious problem in recent days. By some estimates, the total annual IUU catch amounts to roughly one-fifth of the global reported catch [Flothmann et al. (2010), p. 1235].

Partly in response to the steady decline of wild fish populations, we have observed an important development over the past 30 years or so: the rise of *aquaculture*, or fish farming. Aquaculture involves cultivating fish populations under controlled conditions as opposed to harvesting fish in the wild, either in coastal or inland waters. Figure 15.1 shows the time trend of global fishing since 1950, both wild (capture) harvesting and aquaculture production. Until the early 1980s, aquaculture was virtually nonexistent. Since then, however, it has increased dramatically, accounting for virtually all growth in total fish production over the past 30 years. This figure shows that aquaculture has played a key role in enabling continued growth in total fish output worldwide. The steady rise in per-capita fish consumption in recent years has only been enabled by the growth of aquaculture in the face of stagnant wild harvesting levels.

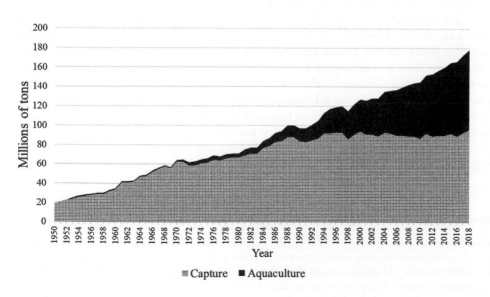

*Figure 15.1* World fish harvesting, 1950 to 2018. *Source*: Food and Agriculture Organization. Fisheries Division. *Yearbook of Fishery and Aquaculture Statistics*. Global Capture Production, 1950–2018.

## Institutions governing fisheries

Currently, international law consists of various principles that attempt to coordinate the use of the world's oceans by different countries. Under the *United Nations Convention of the Law of the Sea* (UNCLOS), most countries have the exclusive right to develop all natural resources in a zone that extends 200 nautical miles from their shores out to sea. These zones are called *exclusive economic zones* and include the rights to all fish stocks within the zone. These exclusive economic zones stretch much further out than a country's territorial waters, which typically end at 12 nautical miles. Beyond these zones, the oceans are open to all countries, and no country can exercise sovereignty over any part of open ocean waters.

Current international law emphasizes the importance of managing fish stocks sustainably. The *Sustainable Development Goals* of the United Nations include Article 14.4, the objective of "regulating harvesting, ending overfishing and restoring stocks to levels that can produce maximum sustainable yield in the shortest time feasible." In 2017, the United Nations hosted the Ocean Conference, the first ever global conference dedicated to the health of the world's oceans. In this conference, delegates representing nations, UN entities, academics, nongovernmental organizations, and private sector firms met to, among other things, discuss and set in place ways to implement Article 14.4. Since 2016, the oceans have been included in general UN discussions on climate change [FAO (2018), p. 78].

Within the European Union (EU), fisheries are regulated under the *Common Fisheries Policy* (CFP), which was officially enacted in 1983 to manage fish stocks on a coordinated basis for all member countries of the EU. The CFP sets fishing quotas for all member countries based upon two things: (1) an EU-wide permitted total harvest (*total allowable catch*, or TAC) set by its Council of Ministers, and (2) each country's historical share of the total. However, the CFP permits member countries to use different management approaches to attain their quotas, including: issuing fishing licenses, restricting entry into fisheries, regulating fishing gear, mandating minimum fish catch sizes, and closing down fishing areas. Monitoring and enforcement of these policies in order to ensure compliance is the responsibility of individual member countries.

Evidence on the effectiveness of the CFP is decidedly mixed. Fish stocks in the North Atlantic have declined dramatically since its enactment, though this recent trend may be merely a continuation of a longer-term trend of fisheries depletion over time. And some countries like Norway and Iceland have elected to stay out of the EU in part to avoid being subject to the provisions of the CFP. At the same time, it appears that some of its provisions have resulted in wasteful fishing practices. In addition, enforcement is a major issue, as much illegal fishing continues to occur in North Atlantic waters. Finally, heavy lobbying by elements of the fishing industry has caused certain of the provisions of the CFP to be watered down, such as a ban on deep-sea trawling proposed by the European Commission in 2011 (see Box 15.1).

---

### Box 15.1: Rent-seeking and the Common Fisheries Policy

The controversial fishing practice of deep-sea, or bottom, trawling involves dragging weighted nets along the ocean floor, harvesting everything in their path. It is a widely used method, annually accounting for over 30 million tons of fish catches worldwide, more than any other single fishing method [Watson (2018)]. However, bottom trawling is highly environmentally destructive: it damages seabed ecosystems and poses risks to deep-sea coral beds, sea turtles, and various marine mammals. Furthermore, it is especially threatening to deep-sea fish populations, which are slow to recover from overharvesting because these species tend to reproduce slowly.

In 2011, in response to intense public pressure, the European Commission proposed putting an end to the practice. In the European Parliament, the ban gained the support of the Environment Committee but encountered strong opposition in the Fisheries Committee. When it came time for the Parliament to vote on the measure, it narrowly went down to defeat in the face of strong opposition from countries where bottom-trawling was widely practiced. This included France and Spain, where the practice occurs most commonly. A highly placed source was quoted as saying: "It is no secret that France, and to a lesser extent Spain, have blocked the discussion." [Rabesandratana (2013); see also "European Parliament rejects ban."]

In the United States, management of deep-sea fishing is primarily governed by the *Magnuson-Stevens Fishery Conservation and Management Act* [Magnuson-Stevens Act]. The objectives of the Magnuson-Stevens Act are to address overfishing in U.S. waters, replenish fish stocks, increase societal benefit from fisheries, and ensure a sustainable seafood supply. To achieve these objectives, the Act created a series of eight fishery management councils, each responsible for fishery management in its own region. These councils were required to design and submit fishery management plans to the Secretary of Commerce, and to update them periodically as the need arose. Later amendments called for measures for habitat preservation, especially spawning grounds.

The Magnuson-Stevens Act has had limited success in achieving its fishery conservation objectives [U.S. NMFS (1995)]. This lack of success has occurred despite experimentation with a wide range of policy options. For example, one of the regional councils – the New England Fishery Management Council – has applied a wide range of regulatory policies including catch quotas, limitations on fish size and mesh size, regulation of spawning grounds, limiting catch season lengths, a moratorium on the issuance of new licenses, and a voluntary fishing vessel buy-back program. None of these approaches has managed to significantly improve fishing conditions in the region [Sutinen and Upton (2000); Wang and Rosenberg (1997)].

In the United States, another important governance issue has arisen from attempts by states to regulate fisheries in their coastal waters. Under U.S. maritime law, states have jurisdiction over their coastal waters up to three miles out to sea. Granted this authority, many states have attempted to enact legislation that would impose entry restrictions on fishers from other states. When these laws have been subject to legal challenges, the courts have consistently struck them down, on the basis that they promote the monopolization of local fisheries [U.S. NOAA Fisheries (nd); Johnson and Libecap (1982), pp. 1006–7]. As you can perhaps guess, these rulings exacerbate the common-pool resource problem, by making it more difficult for fisheries to exclude other fishers. We will return to this point later when we discuss local arrangements to manage fisheries.

## Economics of the fishery[2]

### I. A bionomic model of a fishery

In economic terms, a particular fish population is an example of a renewable resource. Left to itself, it will tend to reproduce and grow, at least under favorable natural conditions; say, clean, food-filled waters containing few predators. Figure 15.2 shows how a fish

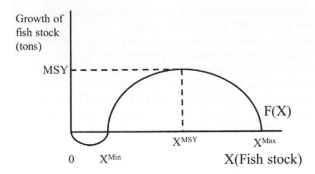

*Figure 15.2* Growth dynamics in a simple fishery

population will tend to grow as a function of its size. This is captured in the S-shaped growth function F(X). Here, F is the growth of the fish population, measured in tons of fish, and X is the size of the fish stock, also measured in tons. F(X) looks kind of complicated. What is going on here?

To interpret F(X), let us first focus on the point where it is the highest. At this point, the fish stock is a certain size, which I am calling $X^{MSY}$ (more about *why* I am calling it $X^{MSY}$ shortly). Here, the fish stock is growing as quickly as possible, measured in physical tons of fish. At all other fish stock sizes, it is growing more slowly. Notice, however, that to the left of $X^{MSY}$, fish stock growth is *increasing* as X increases, while to the right of $X^{MSY}$, growth is *decreasing* as X increases. This indicates that growing conditions are becoming increasingly favorable to the left of $X^{MSY}$ and increasingly unfavorable to the right of $X^{MSY}$.

Intuitively, why is this? Well, to the left of $X^{MSY}$, relatively small fish stocks mean there is plenty of food while, simultaneously, fish stocks are sufficiently large that they afford fish plenty of opportunities to find mates and reproduce. This is the case all the way to $X^{MSY}$. However, past $X^{MSY}$ conditions for the fish are becoming less conducive to growth. More and more fish are competing for food, which makes it more difficult for fish to get enough nutrients to reproduce. Growth rates thus start to fall. At $X^{Max}$, growth stops altogether. Thus, we will give it a special designation: $X^{Max}$ is called the *carrying capacity* of the fishery.

So far, so good. But what about the region of X where the growth function dips below the horizontal axis, where the fish stock size is less than $X^{Min}$? This is the region of negative growth. When the fish stock falls below $X^{Min}$, it starts shrinking and will eventually end up at zero. Why is there negative growth in this region? Intuitively, this is because the fish stock is so small that encounters between fish for mating purposes have become so rare that not enough fish are being spawned to make up for the numbers that die. Without any human intervention, this fish population is doomed to extinction.

Notice that there are three stable fish stock sizes: 0, $X^{Min}$, and $X^{Max}$. This is because these are the only fish stock sizes where the growth rate equals zero. At all other fish stock sizes, growth is either positive or negative. If growth is positive, the fish stock is growing, and if growth is negative, the fish stock is shrinking. So suppose a fish stock happens to be at $X^{MSY}$. You would not expect it to stay there for very long, because the positive growth rate at that point means the fish stock is still growing. It will continue to grow all the way to $X^{Max}$, where growth stops. Thus, 0, $X^{Min}$, and $X^{Max}$ are the only fish stock sizes you would expect to observe for any length of time. We will call these fish stock sizes the *bionomic equilibria*.

Notice, however, an important difference between 0 and $X^{Max}$ on the one hand, and $X^{Min}$ on the other hand. If a fish stock is at either 0 or $X^{Max}$, it tends to stay there. Once a population is extinct, there are no natural processes that will launch it toward positive growth. And once a fish stock is at $X^{Max}$, it will never fall below that level because, if it does, positive growth will push it right back to $X^{Max}$. Thus, 0 and $X^{Max}$ are called *stable bionomic equilibria*.

The same thing is not true for $X^{Min}$. If a fish stock is at $X^{Min}$ and starts to fall a little below that level, negative growth kicks in and it continues to decline in size, again, all the way to zero. On the other hand, if a fish stock is at $X^{Min}$ and starts to rise above it, it keeps on going. Thus, small disturbances in either direction get magnified by the natural growth process. For this reason, we call $X^{Min}$ an *unstable bionomic equilibrium*.

The entire analysis so far has left harvesting completely out of the picture. Let us fix this by adding a harvest function. In Figure 15.3, this function is denoted $H(X, E_0)$, where H is the harvest, in tons, and X is again the size of the fish stock, in tons. In addition, the harvest is a function of E, the amount of effort expended by fishers. You can think of effort as representing some measure of the total factor inputs used by the fisher. The harvest function shown in Figure 15.3 assumes that a particular amount of effort, $E_0$, is being expended.

You can think of this harvest function as similar to a regular production function, meaning it represents harvest *potential*, under the natural conditions in which the fishery exists. In line with intuition, we assume that larger harvests are possible when the fish population is larger. This is why we draw H as upward-sloping. To keep things simple, however, we will assume H is a simple linear function of X.

The question you should ask yourself is: Under these conditions, what fish stock size would you expect to observe in equilibrium? Keeping in mind our earlier discussion, the question is: At what stock size would growth be zero, that is, neither positive nor negative? Here, we have added harvesting to the picture, which represents, in addition to natural predation and fish dying off for other reasons, a subtraction of fish from the fishery. In this case, equilibrium occurs when net growth is zero, where net growth (NG) is defined as:

$$NG(X) = F(X) - H(X, E) \tag{15.1}$$

That is, the natural growth of the fishery minus the amount harvested by fishers. When harvesting equals natural growth, we would expect the fish stock size to be stable.

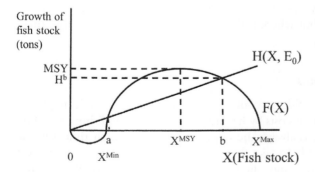

*Figure 15.3* Fishery model, growth and harvest functions

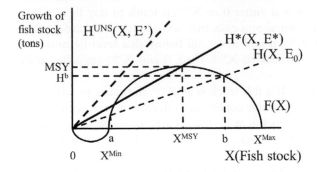

*Figure 15.4* Maximum sustainable harvesting

With this interpretation in mind, we can see that there are again three equilibria: 0, a, and b. And keeping in mind our earlier discussion, only two of these are stable: 0, and b. This means that if you see a fishery actually operating in the real world, you can interpret it as being at point b. And the associated harvest is interpreted as $H^b$.

I should point out that the fishery depicted in Figure 15.3 is not producing to its sustainable maximum potential. Its maximum potential would be realized if the harvest function H cut the growth function F right at its peak, the point of maximum growth, as in Figure 15.4. Here, the harvest function is denoted $H^\star$, which assumes a larger amount of effort $E^\star$. If the harvest function was $H^\star$ rather than H, the fishers could be harvesting the amount MSY, obviously a greater amount than $H^b$. Furthermore, it could do this period after period. Intuitively, this is because it is harvesting right at the point where the growth potential of the fish stock is at its greatest. This maximum harvest amount that can be sustainably harvested is called the *maximum sustainable yield*, hence the label MSY.

To make one more point, notice the harvest function denoted $H^{UNS}(X, E')$ shown in Figure 15.4. Here, so much effort is being expended that the harvest function is always above the growth function F. When harvesting exceeds natural growth, what happens? The fish stock declines in size. Here, it will decline all the way to zero: this amount of harvesting effort will drive this fish population to extinction.

Wow. So the harvesting function seems crucial in determining how good an outcome you can expect to observe for a given fishery: everything from maximum sustainable yield to extinction and everything in between. The question you might be asking yourself is: What determines the position of the harvest function? The other question you might be asking is: This is all well and good, but where is the economics? Let us now turn to these important questions.

### II. Adding economics to fishery dynamics

Like every other economic activity, fishing confers benefits on fishers, while requiring them to incur various costs. They incur costs by having to buy a boat and fishing nets, hire workers, and so forth. And there is, as always, the opportunity cost of their time. On the other hand, they benefit by being able to take their catch, sell it in a market, and collect revenues. To bring these considerations into the picture, let us consider the question not of how much fish to harvest but, rather, how much *effort* to expend in order to harvest those fish. This will allow us to connect the profit-maximizing choices of the fishers to the dynamics of fish growth. It will also allow us to draw some important conclusions regarding the likely fishery outcomes under different assumptions about access to the fishery.

Consider again an ocean fishery, and assume for the moment that there is one fishing company with an entire fleet of boats at its command. Given the natural productivity of the fishery, the amount of fish that can be harvested is a function of the amount of effort expended. This is shown on the left-hand side of Figure 15.5 as the *effort-harvest function*.

Notice that what is being measured on the horizontal axis is *effort*, not the size of the fish stock, as in the previous graphs. This accounts for the slightly different shape of the curve. To be concrete, think of effort as the number of boats used by the company. So the effort-harvest function starts at the origin because if the company sends no boats out into the fishery, it is not likely to catch any fish. As it uses more boats, harvest increases, which makes sense. Notice, however, that it increases at a decreasing rate. This is because of diminishing productivity: the second boat sent out into the fishery is able to catch less fish than the first, the third less than the second, and so forth.

At some point, the company has sent out so many boats that an additional boat adds nothing to the total harvest (the effort level $E_0$). Past that point, boats start interfering with each other, actually reducing the size of the total harvest. Finally, one can perhaps imagine a fishery *so* crammed with boats that any harvesting is impossible. This is where the effort-harvest function crops back down to zero.

Let us now figure out how much effort a profit-maximizing fishing company would expend. The answer depends upon conditions of access to the fishery, so let us distinguish two important cases: *closed access*, and *open access*.

## A. Closed access

Let us begin with closed access conditions. This simply means the fishing company has exclusive rights to fish and in particular, it does not have to worry about other companies coming along and horning in on its territory. If it helps, for intuition you can imagine the company has been granted exclusive rights to fish and the granting authority is able to enforce its rights at zero cost. From our earlier discussion of institutions, this probably sounds unrealistic, so think of it as a base case, to which we will compare the more realistic open access case.

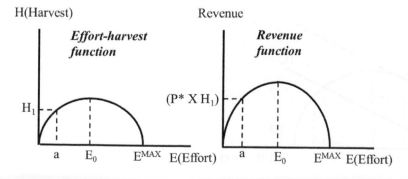

*Figure 15.5* Effort-harvest function and revenue function

To see what this company will do under these conditions, we need to represent fishing revenues and costs as a function of effort. To keep things simple, let us assume the price of fish is fixed, at $P^\star$ per ton. For intuition, you can assume that the company is selling its catch on a competitive market at the market price $P^\star$. Then fishing revenues are simply the total harvest, in tons, say, times the market price per ton:

$$\text{Total revenue} = P^\star \times H \tag{15.2}$$

The revenue function is shown on the right-hand side in Figure 15.5. Notice that it is the same basic hill shape as the effort-harvest function, which makes sense because, at every effort level, revenue is simply harvest times the price $P^\star$. The revenue "hill" is simply the harvest "hill" multiplied by a constant, the price of fish.

Let us now turn to the cost side of things. To keep things simple, let us assume that every additional unit of effort costs the same. That is, the fifth boat costs the same as the first boat, or the company can hire as many workers as its wants at a going wage rate. We are assuming, of course, constant marginal cost of effort. This assumption is not critical, but it makes the graphical analysis easier and, at the same time, it is fairly intuitive. If it holds, then we can graph total costs as a function of effort as a straight line from the origin. Figure 15.6 adds the total cost function to the revenue function of the previous graph.

Assuming that the company is profit-maximizing, how much effort will it expend to catch these fish? Recalling that profit equals total revenue minus total costs, the answer is the effort level where the vertical distance between TR and TC is the greatest. In Figure 15.6, this is the effort level $E^\star$. At $E^\star$, the TR curve is exactly parallel to the TC line (the dotted line tangent to TR at $E^\star$ is drawn as parallel to TC). This is, of course, the point at which the *marginal* cost and *marginal* revenue of effort are equal.

Before moving on, I should point out that this analysis suggests that this company will never expend effort greater than $E_0$. The reason is that with an upward-sloping TC curve, the only candidates for profit-maximization are effort levels where TR is also upward-sloping. This is, of course, a nice result because all of the effort levels greater than $E^{Max}$ are obviously inefficient. For example, the effort level b in Figure 15.6 wastes a lot of resources: it permits the fisher to realize only $R_1$ in revenues, when he could realize that same amount of revenues by only expending an effort level of a. That entire extra effort difference $(b - a)$ does the fisher absolutely no good in terms of adding revenues but

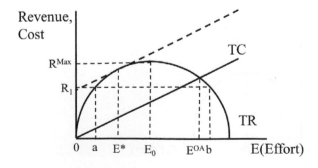

*Figure 15.6* Total cost, total revenue

obviously requires him to bear all sorts of costs. The same is true of every effort level greater than $E_0$.

*B. Open access*

With all that in mind, consider an open access situation, where any fishing company with a boat can enter the fishery and start to fish. Obviously, if the existing company was expending $E^\star$, other companies would have incentive to enter because there are all sorts of profits to be made. So entry would begin to occur and total effort would increase past $E^\star$. How long would entry continue to occur? Until all excess profits from the fishery are exhausted, or where total revenues equal total costs. On Figure 15.6, this amount of effort is labeled $E^{OA}$, which wastes resources, as we have just seen. This is the basic result: overfishing occurs under open access conditions. And the process whereby it occurs is said to result in *rent dissipation* – the loss of rents due to the competition for resources. This is sometimes more colloquially put as "money wasted chasing after money."

## Traditional policy prescriptions

So, what can be done? In this section, we will examine traditional policies designed to curb overfishing: limits on catches, and regulation of various kinds (including taxes). We assume that open access conditions are present, in line with our theoretical model and our earlier discussion of real-world institutions governing fishing. In the next section, we will relax this assumption and consider property rights ways to restrict access.

### I. Catch limits

One possibility, perhaps the most obvious, is to simply restrict the total amount of fish that can be harvested from a given fishery. This is sometimes called the *total allowable catch*, or TAC. This policy requires having a regulating authority monitor the fishery, keep track of how much is caught by all companies, and, then when the TAC is achieved, close down the fishery for the season. This policy is shown in Figure 15.7. Here, the optimal effort from the previous analysis, $E^\star$, is shown along with the associated harvest. This $E^\star$ being the optimal amount to be caught from this fishery, the authority simply sets the TAC at the associated harvest level.

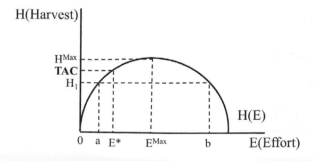

*Figure 15.7* Total allowable catch

There are two important challenges to successfully implementing this policy. Perhaps the most obvious is that it will generally be extremely difficult to fully monitor all fishing activity that goes on, including the sizes of the catches of individual companies. This will especially be the case in large, remote, open-sea fisheries where the roving bandits problem is likely to be particularly acute. It will also be the case in fisheries containing multiple species, where there might be different TACs for each species. In these cases, we say that there are transaction costs associated with enforcing these policies.

Another challenge is the fact that a TAC policy may well elicit excessive levels of effort in order to catch the targeted amount. Recall that under open access conditions, there are profits to be made at the effort level E★. Indeed, this is precisely why the open access regime creates problems. Because of this, companies have incentive to enter and expend effort on fishing before the TAC is reached. Importantly, this may be done in an inefficient manner, with oversized boats, too many crew members, and the like. So, even if the TAC can be achieved, it may be achieved in a wasteful manner, with potentially far more effort than is necessary.

## II. Regulation

In addition to catch limits, fisheries can be, and are, regulated in many other ways. Regulating authorities can shorten the length of the fishing season, or limit the number of days during the week when fishing is permitted. They can also impose limits on the size and horsepower of fishing boats. In order to allow smaller fish to grow and reproduce, they can impose permissible minimum fish lengths or prohibit fine-meshed fishnets. They can regulate the type of gear that is permissible for fishing, including prohibiting the use of bottom-trawling nets, cyanide, or dynamite to kill or capture fish. Authorities have many weapons in their regulatory arsenal. And often they can and do use different types of regulation in tandem.

All of these various ways to regulate fisheries may be interpreted as increasing the cost of fishing. Figure 15.8 shows the effect of cost-increasing regulation, as a pivoting upward of the TC curve. Under open access conditions without regulation, the open access level of effort, $E^{OA}$, is expended. Under these conditions, the best regulation would pivot the TC curve all the way out to TC★, where the optimal effort level E★ is being expended.

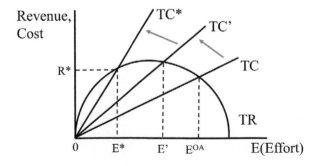

*Figure 15.8* Effect of cost-increasing fishery regulation

Just as with catch limits, the success of these regulations depends upon the ability of the authorities to successfully monitor fishing activity. It is not simply a matter of being able to observe fishing behavior and sanctioning it when it violates the regulations. Whenever you impose a regulation on one dimension of fishing behavior, you are providing incentive to fishing companies to exploit other possible, unregulated, margins. For example, if you impose a limit on permissible boat horsepower, you may simply encourage companies to make more trips to harvest more fish. This is a general problem with regulations of various kinds: there are typically ways around them that companies are given every incentive to exploit.

## A property rights perspective on fisheries

For additional insights into understanding fisheries and what sorts of policies might be called for, let us now view fisheries through the lens of property rights. If you think about the policies of catch limits and regulation, they are basically command-and-control systems. They are therefore subject to all the economic concerns about command-and-control that we encountered in earlier chapters. This includes not providing incentive for individual fishers to do two important things: to engage in cost-minimizing behavior, and to seek out opportunities to engage in mutually beneficial exchanges with other fishers.

A property rights–based policy that promotes both of these behaviors is to create individual fisher quotas for a certain amount of the catch, and then permit fishers to buy and sell their quotas. This is a system based on so-called *individual transferable quotas*, or ITQs. The way this would work is as follows. First, a regulating authority sets a catch limit (TAC), which is divided up into individual quotas. These quotas are allocated to individual fishers somehow. The authority then permits fishers to buy and sell quotas, while keeping track of all transactions. This is, of course, so that it can tell whether any fisher is in violation of their permitted catch limit. It thus has to monitor the ongoing performance of the system, to make sure everyone is abiding by the rules.

If you think about this system, it is basically a cap-and-trade system like the ones we encountered in the chapters on climate change and air and water quality. It thus has essentially the same rationale and potential issues. The basic rationale is to place the right to fish in the hands of the lowest-cost, most efficient fishers, who will tend to outbid others for the quotas because they are most valuable to them. This promotes fish harvesting in the most efficient manner.

The potential issues will also sound familiar. First, the entire system may be costly to monitor, particularly if the number of participating fishers is large. Second, it may be a challenge to determine the appropriate overall TAC; that is, the one that has the best chance of maintaining a profitable fishery in a sustainable way. The general problem is that there are all sorts of sources of uncertainty: factors such as predator populations, weather patterns, and climate change, all of which may affect the growth dynamics of the fish population.

Third, the performance of a system of ITQs will depend upon the magnitude of transaction costs in the fishery. In many situations, transaction costs may be quite high. One key factor is that there may be legal prohibitions on restricting access into the fishery. As we have seen, this turns out to be true in the United States, where the courts have consistently ruled that agreements to restrict access to fisheries are in violation of U.S. antitrust laws.[3] This one factor alone is likely to increase transaction costs significantly because many fishers will be unwilling to buy an ITQ if there is no guarantee that the fish will be there in the future.

However, there are also other sources of transaction costs. In many ocean fisheries, many characteristics of the local fishery are unknown. These characteristics include the size of the fish stock, the effect of fishing activity on the fish population, and the effect of fishery regulation on the fish population. In addition, in some cases the fish species is migratory over long distances. All of these factors may make it difficult for fishers to assess the benefits of negotiating among themselves over quota shares in a fishery, which is likely to make negotiations challenging.

Finally, it may be difficult to devise a politically acceptable mechanism to get the individual quotas into the hands of fishers in the first place. There are various allocation mechanisms that could be used; for example, auctions, randomized distribution, or distribution based upon historical fishing activity. Each of these mechanisms advantages some fishers over others, making conflict likely. Since a quota is valuable, the regulating authority will have to consider which set of fishers will receive a potentially large windfall profit. Depending upon the mechanism, the policy may generate sufficient political opposition to make it difficult or impossible to enact. Conversely, certain mechanisms may be more politically acceptable than others which might, all else equal, argue in their favor on the grounds of political feasibility. We will return to this issue shortly.

## Managing local fisheries

In order to systematize the discussion of transaction costs, let us turn again to the social-ecological systems (SES) framework, which has been applied to local fisheries management. As we saw in chapter five, tragedies of the commons are neither universal nor inevitable. Similarly, fisheries depletion may not be either. What it will depend upon is the combination of factors in the SES framework that are present in specific fisheries.

### I. The roving bandits problem again

Applying that framework, we would expect tragedy of the commons outcomes to occur under the specific conditions that underlie the roving bandits problem. In the fisheries context, this would be where fishers take an extreme short-term view, have no historical relationship with each other, it is extremely difficult to keep them out, and it is difficult to monitor and sanction predatory behavior. This unfortunately characterizes many local fisheries in our currently globalized world [Berkes et al. (2006)].

However, scholars have also been successful in using the SES framework to identify and characterize situations where fisheries may be managed sustainably. For a good example of how the roving bandits problem can be avoided, let us consider the lobster fishing industry in Maine.

### II. Overcoming the roving bandits problem: Lobster fishing in Maine

The Maine lobster story is a striking example of successful local management over the long term, based upon cooperative arrangements designed by local lobster fishers [Acheson (1997)]. From its low point in the 1920s and 1930s, when the local lobster population was nearly destroyed, the industry expanded to thrive and become extremely profitable, totaling over $186 million in sales by the year 2000. This happy success story is important because it carries some important lessons for possibly helping us to avoid roving bandit situations in some cases. Let us see this using the second-tier variables of the SES framework, which you can find in Table 5.2. Let's go through an exercise similar to the one we did in chapter five.

The first thing to mention here is that many Maine lobster fishers, in stark contrast to our stereotypical roving bandits, have long-time historical attachments to their industry and to the communities in which they live. In many cases, lobster fishers have lived in their communities for generations and thus, have deep roots there (A3, A4). This has allowed them to build trust among themselves and to develop norms of reciprocity (A6). In addition, certain respected lobster fishers have assumed leadership positions in the communities (A5). Finally, over time they have accumulated much useful knowledge regarding local lobster conditions (A7). All of these factors encourage the lobster fishers to take a long-term perspective in management of the fishery.

However, even all of these factors may not have been sufficient if it were not for certain laws passed by the state of Maine that took advantage of some unique characteristics of the lobsters themselves. Specifically, the state established a fund to buy females of egg-bearing age from fishers. It then had lobster wardens punch a hole, and later a V-notch, in lobster tails and made it illegal to sell lobsters with these markings (RU6). Interestingly enough, over time the fishers themselves began to voluntarily notch lobsters that they caught, and to refrain from buying notched lobsters from other fishers. This practice evolved into an informal norm that spread among the fishers (A6) [Acheson and Gardner (2011)]

Finally, it is important to note that this norm may not have been effective if not for other characteristics of the lobsters. If, for example, it was not possible to return lobsters to the ocean to continue to grow and reproduce (RU2), then fishers would have lost a lot of incentive to notch them and return them to the ocean. It probably also helped quite a bit that lobsters did not migrate, again allowing fishers to realize returns from the notching norm (RU1).

There are several lessons of the Maine lobster fisheries. Various factors led the fishers to take a long-term perspective on management of the fisheries, including longstanding residence in the communities, intimate accumulated knowledge of the lobster fishery practices, and effective leadership that channeled fisher efforts toward greater sustainability. This provided them greater incentive to monitor fishing and to sanction individual fishers who violated local fishing norms. Cooperation was aided by the fact that lobsters could be notched, which probably significantly reduced enforcement costs. Finally, the fact that lobsters largely stayed put and grew over time made it more likely that fishers would benefit from notching lobsters and returning them to the sea. If any of these characteristics had not been present, it is less clear whether the reciprocal norms would have been respected or, even, whether they would have emerged in the first place [Ostrom (2007)].

## Public choice in fishery management

Fishery management also provides a nice example to study because it illustrates a number of principles of public choice and how fishery outcomes can be affected. We saw a hint of this earlier in the chapter in the example of the Common Fisheries Policy of the European Union, where political pressures resulted in a watering down of some of its regulatory provisions. At this point, let us take an extended look at fisheries policy in the United States, which is governed by the Magnuson–Stevens Act.

The purpose of fishery management is, of course, to address overfishing. Unfortunately, the same conditions that tend to lead to overfishing may also tend to result in political conditions that defeat attempts to regulate overfishing. To see why, consider the situation in many ocean fisheries. In many such fisheries, there are typically a relatively small number of commercial fishing operations, each with a large stake in being able to fish without interference. Our earlier discussion of public choice suggests that these two features would tend to make fishers a powerful interest group.

Adding to their political influence is the fact that fishers tend to be concentrated geographically and, thus, have particularly powerful sway with their congressional representatives. As a result, those representatives have strong incentive to be selected onto congressional committees where they can exercise extra influence in supporting their fisher constituencies. And they have historically done precisely this. For many years, the House Merchant Marine and Fisheries Committee, which exercises jurisdiction over fisheries legislation, was dominated by representatives from fishing districts [Dana (1997), p. 837].

On the other hand, ask yourself who would have a stake in regulating the fishery in order to *prevent* overfishing. My guess would be consumers (who would oppose depletion of fish stocks) and environmentalists (who are concerned with supporting natural wild populations). These two groups, however, would tend to be large and group members would tend to have relatively small individual stakes. Public choice theory tells us that free-riding within each group would tend to reduce the political influence of these groups on this issue. All in all, the political odds seem to be stacked against fishery regulation.

And yet, you may be asking yourself, the Magnuson-Stevens Act did indeed get enacted, which would seem not to be predicted by public choice theory. So what gives? The answer seems to lie in a few factors. First, in many cases, fishers are not a uniform, united constituency. For one thing, in many fisheries, in addition to commercial fishers there are also sport fishers. Sport fishers tend to have different incentives and are in general more supportive of measures to manage fisheries sustainably. For another thing, commercial fishers in a particular fishery are sometimes at odds with each other, for example, if they are based in different ports or harvest different types of fish. So here is a possible example of the public choice principle that interest group heterogeneity may diminish the political effectiveness of an interest group.

Second, it needs to be recognized that in many ways, the Magnuson-Stevens Act turned out to be relatively toothless. This is because of certain provisions of the Act, which may well have been inserted there at the behest of commercial fishers. One of these provisions placed regulatory power in the hands of regional fishery management councils, which were given much discretion to set fishery management policy within their regions. It then created a process whereby these councils would wind up being composed almost exclusively of commercial fishers. And, not surprisingly, the councils have often not enacted effective regulatory management measures.

The Act did call for the National Marine Fisheries Service to exercise oversight over the regional councils. However, it did not contain language directing the service to take specific actions in cases where regional council plans failed to adequately address overfishing. This made it more difficult for the service to act independently of congressional oversight committees. Remember them? The ones that were stacked with representatives of fishing constituencies? Again, it is not surprising that the councils often did not effectively regulate fisheries in their regions.

## Key takeaways

- The open access conditions of many ocean fisheries and the failure of effective regulatory oversight has resulted in severe overfishing in many locations. At the same time, it has sometimes been possible to come close to sustainably managing certain fisheries.

- The SES framework provides us some insight into factors that make is possible to do so, as we saw with the Maine lobster industry.
- A key question moving forward will be: How unique is the lobster industry, and how much do its lessons apply to other fisheries?
- Public choice theory provides a political story for the creation and implementation of the Magnuson–Stevens Act.

## Key concepts

- Aquaculture
- Exclusive economic zones
- Total allowable catch
- Carrying capacity
- Bionomic equilibrium
- Individual transferable quotas

## Exercises/discussion questions

15.1. As we have seen, the courts have repeatedly struck down state laws designed to restrict fishers from other states from entering state fishing grounds, on the grounds that those laws permit monopolization of local fisheries. Isn't it efficient to restrict monopolies from operating? Why are these rulings not efficient? Or are they? Why?

15.2. Explain intuitively why the only stable bionomic equilibria occur where the harvest function intersects the growth function from below, and not from above.

15.3. Consider a single halibut fishery, which is one among many fisheries supplying halibut to the market. Suppose that conditions are such that fish growth over time is the following function of the size of the fish stock: $F(X) = -75X + 20X^2 - X^3$ where $F(X)$ = growth rate, in tons of fish; $X$ = size of fish stock, in *thousands* of tons. To keep things simple, assume that the harvest function is independent of the level of effort by local fishers, and can be expressed as follows: $H(X) = 2X^2 - 3X$ where $H(X)$ = harvest rate, in tons of fish. (a) Calculate all bionomic equilibria and their associated fish stock sizes. In addition, determine which ones are stable and which ones are unstable. Explain your answer.

NOW: Let's bring economics into the picture. Assume that this fishery is mined under *closed access* by one fisher who can harvest fish under the following harvest function: $H(E) = 10E - E^2$ where $E$ = amount of effort, in person-hours. Furthermore, suppose that the price of halibut is $4 per ton but that effort is bought under increasing marginal costs, at: $MC = 2E$. (b) Calculate the profit-maximizing harvest and effort levels, and any stable equilibrium fish stock size. Explain your answer.

Now suppose instead that this fishery is mined under *open access*, under the same harvest conditions. (c) Recalculate the harvest and effort levels, and any stable equilibrium fish stock size. Explain your answer. (d) Calculate the optimal unit taxes on harvest and effort. Explain your answer.

15.4. Explain intuitively why imposing minimum fish lengths or prohibiting fine-meshed fishnets have the effect of shifting up the cost curve for fishing.

15.5. What do you think are the lessons of the Maine lobster fishing example that could apply to other fisheries containing different fish species in terms of promoting long-term fishery sustainability? Or do you think the Maine lobster example is a unique case? Why or why not?

15.6. The discussion suggests that the public choice can help explain the enactment of the Magnuson–Stevens Act. However, it also describes an Act that has been relatively ineffective. It would seem to be a waste of time to enact an Act to regulate fisheries that does nothing to really regulate fisheries. In public choice terms, why would Congress do this?

## Notes

1  For more on the 2008 UN report, see "UN Report" (2008); for the 2016 report, see Ferrie (2016); for the 2019 report, see Lovgren (2019).
2  Economic models of the fishery owe much to two classic contributions: Gordon (1954), and Scott (1955).
3  See, for example, Johnson and Libecap (1982), p. 1006; Libecap (1989), p. 18.

## Further readings

### I. Institutional analysis

Sundstrom, Aksel. "Corruption in the commons: Why bribery hampers enforcement of environmental regulations in South African fisheries," *International Journal of the Commons* 7(August 2013): 454–72.
*Investigates the effect of bribery on the extent to which fisheries comply with management regulations of fisheries in South Africa.*

### II. Fisheries and SES

Frawley, Timothy H.; Elena M. Finkbeiner; and Larry B. Crowder. "Environmental and institutional degradation in the globalized economy: Lessons from small-scale fisheries in the Gulf of California," *Ecology and Society* 24(May 2019).
*Examines the impacts of increased globalization on local fisheries using social-ecological system concepts, with a detailed look at small-scale fisheries in the Gulf of California.*
Partelow, Stefan; Paula Senff; Nurliah Buhari; and Achim Schluter. "Operationalizing the social-ecological systems framework in pond aquaculture," *International Journal of the Commons* 12(2018): 485–518.
*Operationalizes the social-ecological framework to model and understand community-based pond aquaculture in Indonesia.*

### III. Public choice in fisheries management

Brzezinski, Danielle T.; James Wilson; and Yong Chen. "Voluntary participation in regional fisheries management council meetings," *Ecology and Society* 15(September 2010).
*Examines participation in public hearings before the New England Fishery Management Council by various stakeholders. Concludes that there may be a correlation between the costs of participation and likelihood of attendance at these hearings, possibly biasing Council policies in favor of stakeholders with lower participation costs.*
Dana, David A. "Overcoming the political tragedy of the commons: Lessons learned from the reauthorization of the Magnuson Act," *Ecology Law Quarterly* 24(1997): 833–46.
*Examines the political tragedy of the commons in the context of fisheries management, which states that government agencies designed to combat overfishing may be subject to political capture by those with an interest in continuing overexploitation. Suggests some possible ways to address this, based upon the battle over reauthorization of the Magnuson Act in 1996.*

# References

Acheson, James M. "The politics of managing the Maine lobster industry: 1860 to the present," *Human Ecology* 25(March 1997): 3–27.

Acheson, James M. and Roy Gardner. "The evolution of the Maine lobster V-notch practice: Cooperation in a prisoner's dilemma game," *Ecology and Society* 16(March 2011).

Berkes, F. et al. "Globalization, roving bandits, and marine resources," *Science*, New Series 311(March 17, 2006): 1557–8.

Dana, David A. "Overcoming the political tragedy of the commons: Lessons learned from the reauthorization of the Magnuson Act," *Ecology Law Quarterly* 24(1997): 833–46.

Dean, Cornelia. "Study sees 'global collapse' of fish species," *New York Times*, November 3, 2006.

"European Parliament rejects ban on deep-sea bottom trawling," *Pew*, January 14, 2014.

Ferrie, Jared. "A perfect storm: climate change and overfishing," *The New Humanitarian*, September 19, 2016. http://www.thenewhumanitarian.org/feature/2016/09/19/perfect-storm-climate-change-and-overfishing

Flothmann, S. et al. "Closing loopholes: Getting illegal fishing under control," *Science* 328(2010): 1235–6.

Gordon, H. Scott. "The economic theory of a common-property resource: The fishery," *Journal of Political Economy* 62(April 1954): 124–42.

Johnson, Ronald N. and Gary D. Libecap. "Contracting problems and regulation: The case of the fishery," *American Economic Review* 72(December 1982): 1005–22.

Libecap, Gary D. "Distributional issues in contracting for property rights," *Journal of Institutional and Theoretical Economics* 145(March 1989): 6–24.

Lovgren, Stefan. "Earth's largest freshwater creatures at risk of extinction," *National Geographic*, August 8, 2019. https://www.nationalgeographic.com/animals/2019/08/freshwater-animals-risk-extinction/

Ostrom, Elinor. "A diagnostic approach for going beyond panaceas," *PNAS* 104(September 25, 2007): 15181–7.

Rabesandratana, Tania. "Industry lobbying derails trawling ban in Europe," *Science, New Series* 342(November 1, 2013): 544–5.

Scott, Anthony. "The fishery: The objectives of sole ownership," *Journal of Political Economy* 63(April 1955): 116–24.

Sutinen, Jon G. and Harold F. Upton. "Economic perspectives on New England fisheries management," *Northeastern Naturalist* 7(2000): 361–72.

United Nations. Food and Agriculture Organization (FAO). Fisheries and Aquaculture Department. *The state of world fisheries and aquaculture 2018: Meeting the sustainable development goals*. Rome: FAO, 2018.

United Nations. Food and Agriculture Organization (FAO). Fisheries and Aquaculture Department. *The state of world fisheries and aquaculture 2020: Sustainability in action*. Rome: FAO, 2020.

"U.N. report warns of global fisheries collapse," *SeafoodSource*, February 24, 2008. https://www.seafoodsource.com/news/environment-sustainability/u-n-report-warns-of-global-fisheries-collapse

United States. Department of Commerce. National Marine Fishery Service (NMFS). *Our living oceans. Report on the status of U.S. living marine resources*, NOAA Technical Memo. NMFS-F/SPO-19: 1995.

United States. Department of Commerce. National Oceanic and Atmospheric Administration (NOAA) Fisheries. "Understanding fisheries management in the United States," (nd) https://www.fisheries.noaa.gov/insight/understanding-fisheries-management-united-states

Wang, Stanley D.H. and Andrew A. Rosenberg. "U.S. New England groundfish management under the Magnuson-Stevens Fishery Conservation and Management Act," *Marine Resource Economics* 12(Winter 1997): 361–6.

Watson, Reg A. "Mapping nearly a century and a half of global marine fishing: 1869 – 2015," *Marine Policy* 93(July 2018): 171–7.

# 16 Forests and deforestation

In November 2019, fires burned in the Brazilian Amazon as deforestation surged to its highest rate in more than a decade. Data from Brazil's National Institute for Space Research showed that the loss of rainforest increased nearly 30% from August 2018 through July 2019, compared to the 12 months prior. It marked the highest rate of deforestation since 2008 and amounted to a cleared area larger than Yellowstone National Park. The fires, deliberately set in order to clear land for farming and cattle ranching, sent smoke throughout the region and caused the skies to darken all the way to Sao Paolo [Irfan (2019)].

For a number of years, the forests of the world have been shrinking, a process called *deforestation*. The loss of forests has occurred all over the world, with steady economic growth and dramatic increases in the world's population over the past century. In recent days, however, the decline has been particularly pronounced in tropical areas, where much of the world's rainforests are found. In 2019, for example, the total loss of rainforests worldwide equaled an area the size of the country of Belgium [Magnusson (2019)]. And some recent evidence suggests this process may be accelerating.

The loss of the world's forests presents important challenges for humankind, as they comprise a natural resource that provides a great many economic and environmental benefits. Among other things, forests:

- Supply food, fuel, and wood for housing and other products.
- Are sources of naturally occurring ingredients for a variety of pharmaceuticals.
- Provide habitat for a wide variety of bird, animal, and plant species, many of them endangered.
- Are home to, and form the basis for the livelihoods of, many of the world's rural poor, including many Indigenous populations.
- Absorb and store significant amounts of carbon dioxide, which is an important factor in combatting climate change. Deforestation and degradation of forests currently account for roughly 11% of global greenhouse gas emissions.

The issue of deforestation is intimately tied up in the issue of biodiversity, because the world's forests are home to the vast majority of terrestrial plants and animals. Indeed, scientists are increasingly blurring the line between forests as trees and forests as the ecosystems of which trees are a part. In this textbook, I will be preserving the blurry distinction between forests and biodiversity by discussing forests in this chapter and biodiversity in chapter seventeen. Both are important topics, and both deserve separate treatment in their own right.

# Forests and deforestation: A world overview

The forests of the world are distributed widely across all continents except for Antarctica (see Figure 16.1). However, over half of the world's forests are located in five countries: Brazil, Canada, China, Russia, and the United States. Currently, the total forested area of the world is estimated to be about 3.9 billion hectares, which is roughly 9.6 billion acres, or an area about equal to four Chinas, give or take a few *mǔ*. There are two main types of forests: *temperate*, and *tropical*. The temperate forests lie primarily at higher latitudes across North America, Europe, and northern Asia. Tropical rainforests lie at much lower latitudes in South America, Africa, Southeast Asia, and Australia and New Zealand.

With continuing economic and population growth, the world's forests have been shrinking over time, though the time profile looks very different for temperate vs. tropical forests. For the long view, Figure 16.2 shows the estimated extent of world deforestation going all the way back in time to the period prior to 1700 AD. Prior to 1700, almost all deforestation occurred in temperate forests. Since then, there has been a steady decline in deforestation of temperate forests, with very little occurring, on net, since 1950.

The time pattern for tropical deforestation looks quite different: it apparently began to emerge as a serious issue in the 1920s, rose to a peak after the mid-20th century, and then began to experience a modest decline toward the end of the 20th century.[1] In absolute terms, however, tropical deforestation is still an important issue. And, currently, the vast majority of deforestation is concentrated in two areas: Africa and South America. From 2010 to 2020, Africa and South America lost nearly 6.6 million hectares of forest – an area nearly the size of Ireland – *every year*. Every other major region of the world – Asia, Europe, North and Central America, and Oceania – experienced modest increases in forest area, or little to no loss [FAO (2020), p. 12].

The destruction of tropical forests over the past hundred years or so has largely resulted from the felling of trees in order to clear land for farming and ranching. Much of this has been accomplished through the common practice of *clear-cutting*. Clear-cutting simply means that all of the trees in an area are harvested. This practice is commonly used in harvesting tropical forests. The other main alternative – *selective cutting* – involves harvesting only trees that satisfy certain criteria, like size, age, or species.

Because of the pattern of deforestation shown in Figure 16.2, economists have hypothesized that deforestation may exhibit a pattern known as an *environmental Kuznets curve*. The environmental Kuznets curve describes a characteristic pattern of environmental quality as a function of economic development. Environmental quality tends to be good at low levels of development, gets worse as countries move to intermediate stages of development, and then improves again at more advanced stages of development. The U-shaped environmental Kuznets curve pattern for deforestation is shown in Figure 16.3, where environmental quality is graphed against gross domestic product (GDP).

In the context of deforestation, the environmental Kuznets curve may be interpreted as follows. In the early stages of economic development with simple small-scale farming, little deforestation occurs. The onset of large-scale industrial farming leads to clear-cutting of forests to make way for large industrial farms. Then further development reduces reliance on large-scale farming, easing the pressure to cut down forests. This basic pattern seems to be consistent with the deforestation patterns in Figure 16.2, where temperate forests – which tend to be located in higher-income countries – are subject to less

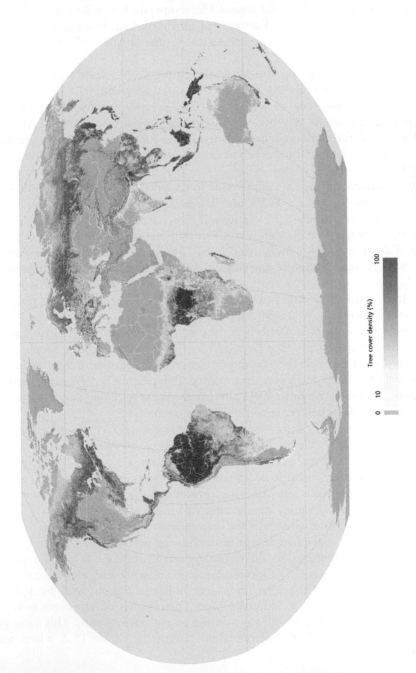

*Figure 16.1* Forests of the world. *Source:* Roser, Max. "Forests," *Our World in Data.* https://Ourworldindata.org/forests

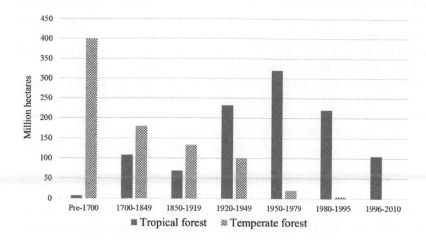

*Figure 16.2* Deforestation trends over time. *Source*: Roser, Max. "Forests," *Our World in Data*. https:// Ourworldindata.org/forests

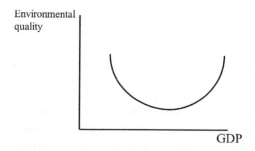

*Figure 16.3* Environmental Kuznets curve

deforestation than tropical forests, which tend to be located in lower-income countries. The same argument also applies to individual countries or regions of the world as they progress through various stages of development over time.

Evidence from economic studies suggests that the environmental Kuznets curve may be present in tropical countries, but this may vary across different regions of the world. Several studies have focused on tropical forests in Latin America, Africa, and Asia.[2] These studies tend to document the existence of an environmental Kuznets curve in at least some of these regions. However, these studies also often find that the level of development where deforestation starts to decline can be significantly higher than current levels of GDP. If so, this suggests that for these regions of the world, deforestation is only going to get worse before it gets better.

## International institutions for forest management

A number of related initiatives form the basis for global action on deforestation. In coordinating international actions to fight deforestation, the United Nations plays a central role. The United Nations has been working on fighting deforestation ever since the Earth Summit in Rio de Janeiro in 1992, when more than 178 countries adopted *Agenda 21*, a comprehensive plan for sustainable development, including a detailed plan to fight deforestation contained in Chapter 11. Since then, plans and specific targets for policies to deal with deforestation have been set forth in a series of multinational agreements.

Most recently, two agreements are particularly noteworthy. In 2015, the United Nations enacted the *2030 Agenda for Sustainable Development*, a broad framework that contains 17 goals for sustainable development. These include socioeconomic goals such as addressing global poverty and hunger; promoting education, health, and wellbeing; achieving higher standards of living; and lowering income inequality. It also includes environmental goals such as climate action, building sustainable cities, and combatting deforestation.

Plans targeted more specifically at deforestation were adopted in 2017, when the UN Forum on Forests drew up the *Strategic Plan for Forests*, which contains a series of goals and targets for forest management and recovery, also to be achieved by 2030. Among these is Goal 1, which aims to reverse the loss of forest cover worldwide by taking various steps to manage forests sustainably and prevent forest degradation. The target of this goal is to increase the area of global forests by 3% by 2030. This is an area more than twice the size of France.

Since 2008, global action on deforestation has been folded in with efforts to address climate change, as the importance of forests in combatting climate change has come to be increasingly recognized. This effort is primarily governed by the United Nations *Programme on Reducing Emissions from Deforestation and Forest Degradation*, more commonly known as UN-REDD. UN-REDD is a collaborative initiative of several United Nations organizations including the FAO and the UN Development Program. It was created in 2008 in response to the United Nations Climate Conference that took place in Bali in 2007.

The objective of UN-REDD is to combat global climate change while contributing to sustainable economic development. It does this by partnering with low-income countries, providing them with advisory and technical support tailored to their individual country needs and circumstances. It also operates through the associated program REDD+, which provides payments to countries for meeting emissions reductions objectives [United Nations (2016)]. Since its creation, participation in the program has expanded and, currently, the program has over 60 official partner countries.

## Modeling forest resources

In this section, we develop a simple economic model of forest management, in order to help understand and interpret forest harvesting behavior. The model essentially comes down to weighing the costs and benefits of harvesting trees for production of things of value to society. A key element of the analysis is to account for the fact that forests naturally replenish themselves. Therefore, the tradeoff centers on the fact that there are benefits associated with waiting to harvest: if we harvest now, we forego harvesting in the future. And because of natural growth, future trees may be larger, and there may be more of them. By harvesting now, we lose the potentially larger value from waiting to harvest until later.

If this argument sounds familiar, it should. We just encountered similar issues in chapter fifteen on fisheries. This déjà vu stems from the fact that both fisheries and forests fall into the category of renewable resources, and so they raise conceptually similar issues. However, we will also see some subtle differences, which we will highlight because of their importance in helping us prescribe effective policy solutions.

## I. Forest renewability

To begin with, let us think a bit about how forests renew themselves. Like human beings, trees tend to grow fast when they are young and then, as they get older, their growth rate declines. The result is something like the profile of wood volume vs. tree age shown in Figure 16.4. In Figure 16.4, *Volume* represents the total volume of wood in a forest, and *Age* represents the age of the trees in the forest. The trees in young forests tend to grow rapidly and, in doing so, add quickly to total forest volume. However, as you move to older and older forests, the trees grow less and less rapidly, which causes forest volume to level out.[3] This basic biological pattern is assumed in all economic models of forest growth.

## II. Economic model of a forest

One important implication of this volume-age profile is that, as a forest ages, at some point it makes less sense to postpone harvesting. When a forest is young and growing rapidly, there is greater benefit to holding off, in expectation of much more harvested wood later. To put it concretely, it makes no economic sense to plant a bunch of saplings and then immediately harvest them. You plant them and let nature take its course, expecting to harvest much larger trees some years down the line.

However, as volume growth slows down, postponing harvesting adds less additional benefit. So, if you think about this argument, it seems to be saying that the marginal benefit of postponing harvesting (eventually) decreases as a forest ages. At some point, it will fall below the marginal cost of postponing. At this point, it will make economic sense to harvest. All of this suggests that there may be an optimal forest age at which to harvest, the point where the marginal costs and marginal benefits are equal.

Let us make all of this more rigorous with a little algebra. Consider the decision that a forest manager would have to make every year: either to harvest the trees in a forest or leave them standing for another year. To cut or not to cut: that is the question. Suppose that if he harvests now, in time period $t = 0$, he realizes total revenue of $R_0$. If he waits a year, he realizes revenue of $R_1$. Assume that there is a constant cost of harvesting in either period, which we shall call C. In addition, since we are talking about comparing harvesting value in different time periods, we need to define the discount rate, which we will call r.

Now let us consider whether it makes sense to harvest now or wait until next year. If the trees are harvested now, the forest manager receives a net benefit of $(R_0 - C)$. If he waits and harvests them next year, he receives $(R_1 - C)$ next year. For a fair comparison,

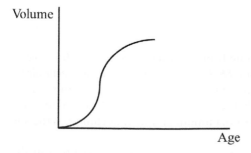

*Figure 16.4* Volume-age profile

we need to discount the future net benefit, as always. So the question of harvesting now or later comes down to whether $(R_0 - C)$ is greater or less than $(R_1 - C)/(1 + r)$.

Now: to understand the argument, it is important to keep in mind the volume-age profile of the forest. This profile suggests that $R_0$ starts out small (that is, when forests are young and growing rapidly) relative to $R_1$, but that the gap between $R_0$ and $R_1$ grows smaller as the forest matures (grows less and less rapidly). So, when the forest is young, it makes sense to wait. However, as forest growth slows down as the forest ages, at some point $R_0$ becomes large enough relative to $R_1$ that it makes sense to harvest. This implies the following equilibrium condition:

$$(R_0 - C) = (R_1 - C)/(1 + r) \tag{16.1}$$

When this condition is satisfied, the forest manager should be indifferent to harvesting now or waiting until next year.

Notice what is happening in equilibrium: *the net benefit (R − C) from harvesting the forest is growing at the rate r.* It is perhaps easier to see this if we rearrange equation (16.1):

$$NB_1 = NB_0(1 + r) \tag{16.2}$$

Here, of course, we have simply substituted in the equivalent expressions for $NB_0$ and $NB_1$ and multiplied both sides of equation (16.1) by $(1 + r)$. What does equation (16.2) tell us? That the forest manager should treat the forest as an investment, where the required return on the forest should equal the rate of return on alternative investments.

This is probably the key insight from the forest model, and, if it sounds familiar, it should. We saw a very similar result before, in our earlier discussion of exhaustible resources. In that discussion, depletion of an exhaustible resource caused the price of the resource to increase over time at the rate r in what was called a *Hotelling price path*. In that case, we interpreted the price increase over time as the owners of the resource taking their returns entirely in added resource value, what economists call *capital gains*. Here, we are tweaking the analysis to allow for the self-replenishment of the forest, but the same economic forces are driving the outcome.

### III. Applying the forest model

It turns out that this model is also useful for helping determine the effects of various factors on the optimal harvest time. To see this, let us return to equation (16.1) and rearrange it as follows:

$$(R_0 - C)r = \Delta R \tag{16.3}$$

Here $\Delta R = (R_1 - R_0)$, or the increase in revenue from period 0 to period 1. (*Before going on, it may help to make sure you see why equations 16.1 and 16.3 are algebraically equivalent*). The right-hand side represents the marginal cost of harvesting now, because this is how much extra revenue you forego by waiting to harvest. Given the shape of the volume-age profile, $\Delta R$ should be hill-shaped, increasing to a maximum and then declining, as the age of the forest increases.

It follows that the left-hand side represents the marginal benefit of harvesting in the current period. You can interpret this expression as representing the return on taking

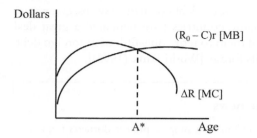

*Figure 16.5* Marginal benefit, marginal cost of harvesting

first-year net benefits and investing it at the market interest rate. Again, keeping in mind the shape of the volume-age profile, we would expect this MB expression to increase as the forest gets older, but at a decreasing rate. These MB and MC curves are plotted in Figure 16.5 as a function of the age of the forest. The point where they intersect provides the optimal age at which the forest should be harvested, or A*.

We can use this model to predict how the optimal harvest age would be affected by changes in various factors. See the exercises at the end of the chapter for practice in working with this model.

## Forest management policy tools

The fact that timber harvesting may inflict negative externalities, and in some cases may destroy native lands and livelihoods for Indigenous populations, provides justification for policies to counteract overharvesting on both efficiency and equity grounds. In this section, let us examine several policies that have been proposed as means of addressing overharvesting and deforestation. These policies are: *debt-for-nature swaps, payments for ecosystem services, land trusts,* and *conservation easements.* All of these policies have received a good deal of public attention and deserve a closer look.

## I. Debt-for-nature swaps

The idea behind debt-for-nature swaps is that it may be possible to encourage actions in low-income countries to protect the environment, when those countries happen to owe large amounts of debt to banks or external funding agencies. To see how this might work, consider a situation where the debt of a low-income country is held by a private commercial bank. In a typical debt-for-nature swap, a non-governmental organization (NGO) buys the debt at a discount from the bank, and then turns around and sells the debt back to the low-income country for local currency. This currency is then given to a local environmental group in the country to use to fund an environmental initiative of some kind. To combat deforestation, for example, the group might use the funds to protect certain forestlands from being cut down.

The idea of debt-for-nature swaps originated in the 1980s, when a number of low-income countries were using debt to deal with severe balance-of-payment deficits, and there was an imminent danger of numerous defaults. Since the first ever debt-for-nature swap took place in Bolivia in 1987, many low-income countries have participated in debt-for-nature swaps, mostly in Africa and Latin America. From 1987 to 1992, 12 countries – mostly in Africa and Latin America – undertook debt-for-nature swaps, generating more than $61 million in local currencies. Some countries have benefited a great deal from these swaps. Costa Rica, for example, used them to retire one-fifth of its foreign debt, while spending over $100 million to protect its forests [World Bank (1993)].

---

**Box 16.1: Debt for nature swaps in the news**

*News item*: "US and Indonesia announce $28.5 million debt swap to protect Borneo's tropical forests," *WWF*, September 29, 2011.

"The Nature Conservancy and WWF are joining with the Indonesian and US Governments today to sign a debt-for-nature swap agreement that will result in a new US$28.5 million investment to help protect tropical forests in three districts of Kalimantan, Indonesian Borneo.

The deal will create models for forest conservation and sustainable economic development in Borneo, the third-largest island in the world, home to unique species such as orangutans, gibbons, clouded leopards, 'pygmy' elephants, hornbills, and up to 15,000 flowering plants.

The districts of Berau and Kutai Barat in East Kalimantan Province and Kapuas Hulu in West Kalimantan Province each contain carbon-rich tropical forest and vast biodiversity under threat from unsustainable natural resource extraction. These forests can serve as examples of sustainable development to the rest of Indonesia and the world."

---

You may be asking yourself why a bank would be willing to sell its debt at a discount to the international organization. The answer is that the bank may believe that it will be difficult for the government of the low-income country to repay its debt, so that defaulting on its loans may be a real possibility. The bank may see participation in a debt-for-nature swap as a way to get something back, which is a lot better than getting nothing. Given this financial reality, debt-for-nature swaps may provide a win-win situation for the banks, low-income countries' governments, and environmentalists.

I should mention some concerns that have been raised about debt-for-nature swaps. First, there has been a lack of clearly articulated, systematic criteria for deciding which countries should receive the assistance. As a result, it is not clear that the funds available from debt-for-nature swaps are going where they are most needed. Some have argued, for example, that Costa Rica has been a major beneficiary of debt-for-nature swaps, whereas other countries with equally pressing environmental problems have received nothing. Another concern is that they may risk causing inflation in the low-income country, because they often inject significant amounts of currency into the country's economy. Finally, some have warned that we should not view debt-for-nature swaps as a panacea for combatting deforestation. Rather, they need to be supplemented with additional reforms such as promotion of democracy and strengthening of property rights. We will return to these arguments in the next section.

## II. Payments for ecosystem services

In recent days, the notion of *ecosystem services* has gained increasing currency. The idea of ecosystem services is that ecosystems provide a variety of unpriced, unpaid-for services of significant societal value. These include: habitat for species protection, oxygen production, soil retention, air purification, water filtration, nutrient and waste cycling, and pollination.

The fact that ecosystems provide all of these benefits provides the rationale for payment for ecosystem services (PES) programs that call for payments to landowners who maintain lands in their natural state, including forest cover. In many cases, payments are accompanied by technical assistance and resources for management planning. In principle, these programs provide subsidies to landowners that bring the private value and social value of protected lands into closer alignment. Doing so should result in a more efficient level of forestland protection.

The practice of subsidizing the protection of land cover has been around in various guises since the 19th century. One of the earliest programs actually called a PES program was undertaken in the 1990s by the city of New York, which provided payments for land management in the Catskill Mountains' watershed in order to protect its drinking water [Thiel (2019)]. In recent years, PES programs have become quite popular. A study published in 2018 identified over 550 active PES programs worldwide, with total estimated transactions in the range of $36–42 billion annually [Salzman et al. (2018)]. Activities covered by these funds include carbon capture, water purification, and conservation of biodiversity. When economists have carefully examined these programs, they have found that they can be very effective in reducing the loss of forest cover (see Box 16.2).

---

**Box 16.2: The Payments for Hydrological Services program in Mexico**

For a sense for how PES programs work, consider a recent example: a program instituted by Mexico in 2003 called the Payments for Hydrological Services program. The objective was to maintain hydrological services provided by the forests, including higher water quality, reduced soil erosion, and reduced hazards from flooding, as well as carbon sequestration. The areas where the program was implemented had been experiencing severe deforestation. The program also had explicit goals of maintaining rural incomes and reducing poverty.

Under this program, landowners were given annual payments to maintain forest cover on their lands over a period of five years. Forest plots were monitored and payments would be suspended upon evidence of logging or conversion of forests to agriculture. The payments given to participating landowners were substantial, comprising anywhere from 8–12% of household income.

One economic study found that the program was highly effective in promoting the maintenance of forest cover. Though it did not completely stop forest cover loss, the study found that the program likely reduced the loss by roughly 40–50%. On the other hand, it found that the program alleviated poverty, but that the effects were small [Alix-Garcia et al. (2015)].

---

At the same time, we should recognize that there are also costs associated with these programs. Perhaps the most obvious cost being the opportunity costs, in terms of the value of the land devoted to alternative uses. For example, for Latin America, one study estimates an opportunity cost of anywhere from $1.50 to $8.00 per ton of carbon dioxide associated with reducing deforestation by 10% [Kindermann et al. (2008)]. These estimates, which assume the alternative is large-scale commercial agriculture, drop significantly if the alternative is assumed to be small-scale subsistence agriculture.

On the other hand, implementation of these programs is accompanied by transaction costs of various kinds, including the cost of setting up, implementing, and monitoring the programs. One recent study of six projects in Peru designed to stem deforestation in the Peruvian Amazon found that transaction costs averaged nearly $300,000 per year, which translated into roughly $0.71 per ton of carbon dioxide [Thompson et al. (2013)].

If accurate, these modest numbers suggest that the transaction costs of implementing PES programs should not pose insuperable obstacles.

### III. Land trusts and conservation easements

A third type of policy for protecting forests is the use of land trusts and conservation easements. A conservation easement is an agreement that legally binds a private landowner to abide by certain restrictions regarding how the land is used. This typically means that uses are restricted to ones that preserve or enhance the environmental amenities existing on the land. These amenities include things like: wetlands, river corridors, watersheds, endangered species habitat, scenic views, and recreational trails. Often the restrictions are permanently binding on the landowner.

A land trust is an organization, typically non-profit, that administers the easement, to ensure that the landowner abides by the restrictions. Land trusts are common in the United States, and they are present in all 50 states. As of 2015, there were nearly 1,400 land trusts in the United States, which protected roughly 56.4 million acres of land [Land Trust Alliance (2015)]. This is a total area roughly the size of the state of Idaho. Land trusts can also own land outright, but this is a less common arrangement and one that has decreased over time.

The growth of conservation easements in recent years is probably due to two factors. First, there are significant tax incentives at both the federal and state levels that encourage parties to enter into conservation easement agreements. Second, a number of states have enacted laws that make it easier to enforce conservation easement agreements [Parker (2004), pp. 491–6].

The large number of land trusts using conservation easements in the U.S. suggests that this can be an effective means of protecting forests and other environmental amenities. And they are definitely attractive in certain ways. For example, entering into the agreements is voluntary, and it is probably cheaper for a land trust to purchase an easement than entire tracts of land.

On the other hand, conservation easements come with certain disadvantages. The fact that there are various restrictions on land use means that monitoring will be necessary, to make sure landowners are abiding by the restrictions. Furthermore, the fact that conservation easements are permanent creates some risk that the restrictions will become socially undesirable at some point as economic and environmental conditions change. With easements, this is because of the difficulty, once property rights to land have been divided up, of piecing them back together again [Mahoney (2002), p.744].

Finally, the United States experience with conservation easements may not transfer well to the context of low-income countries where tropical forests dominate. In many low-income country contexts, the rule of law may not be as well developed, making it more difficult to monitor and enforce the terms of easements. Furthermore, some systems of law may embody certain legal principles that create obstacles to the successful operation of conservation easements. On the other hand, in recent years we have observed some progress in the growth of conservation easements internationally, suggesting the obstacles may not be insuperable [Korngold (2011)].

## A property rights perspective on forests and timber

Now let us dig a little deeper. Another way to think about forest management is through the lens of property rights. In the forest model we developed earlier, there was no explicit mention of property rights. However, we were making some implicit assumptions. First, we were assuming that harvesters enjoyed what we earlier called secure property rights. In this context, this means that the rights to derive benefits from harvesting the timber could be enforced at zero cost. There was no concern, for example, that outsiders could enter and start harvesting timber with impunity, roving bandit-like, at the expense of the harvesting party. And there was no concern that the forestlands could be taken away by, say, an unscrupulous government, or that there were disruptive influences such as war, invasion, or revolution that might render property rights uncertain. In short, there was nothing to keep the harvesting party from fully exercising her right to do with the timber whatever she pleased, including selling it off if she saw fit.

In many settings, however, property rights to forest stands are insecure because of difficulties in enforcing those rights. When this occurs, it tends to reduce incentives to harvest forests sustainably. Over the course of history, for example, effective forest management has tended to occur only in stable societies with a well-established rule of law. We can find evidence for this as far back as the ancient Greek and Roman civilizations, which were times of relative peace and stability. Both the Greeks and the Romans established effective forest conservation programs that forbade the unauthorized cutting down of forests. Their policies included enclosure of woodlands, tree planting, and enforcement of forest management programs, including patrolling the forests and levying fines for illegal harvesting.

On the other hand, there are numerous examples from history where major deforestation has occurred during times of war, revolution, anarchy, and enemy occupation. During these times, there is generally greater uncertainty about the security of claims to forestlands, which reduces the incentive to manage forests. To make matters worse, wars and revolutions have often been destructive of forests in various ways. During early medieval times, for example, large swathes of forests in Lithuania were cut down and used for defensive purposes, to try to hinder invasion by the Crusaders. Entire forests were set on fire by invading armies as a weapon against the Indigenous population in 13th century England and 14th century Russia. During the late 18th century French Revolution, political turmoil and the abolition of existing institutions, including revoking feudal forest privileges and ecclesiastical claims to forest tracts, led to major reductions in forestland areas [Deacon (1999)].

Currently, insecure property rights to forestlands are common in many settings, which may exacerbate the problem of deforestation. This may especially be a problem in many low-income countries, where systems of property rights are often incomplete and war and political instability tend to be more prevalent. It is, of course, precisely in these countries that tropical rainforests predominate and have been steadily shrinking in recent years. All of this may make it difficult to keep out intruders, giving harvesters incentive both to harvest quickly and to not replant after cutting has been done.

In some cases, government policies in awarding property rights to harvesters can create perverse incentives that can cause greater deforestation. Specifically, governments trying to encourage frontier settlement sometimes require settlers to clear forestlands in order to receive a property right to the land. The settlement policies of the government of Brazil exemplify this policy, which is designed to discourage land speculation and to encourage smallholders to improve the wilderness [Mendelsohn (1994), p.751].

The importance of secure property rights in providing appropriate incentives for timber harvesting has an important practical implication. Namely: policies that work in certain contexts may not work in others. Forest conservation methods that promote effective husbanding of forests in countries with strong rules of law and effective rights enforcement may fail miserably in countries that lack these things. One lesson is that we should not view forest management policies in institutional isolation. We may do far more good if we encourage stable government and general respect for the rule of law rather than narrowly focusing on policies like land trusts, payment for ecosystem services, and debt-for-nature swaps.[4]

## Social–ecological systems in forestry management

Now let us dig even deeper. The discussion in the last section suggests that simply providing more secure property rights will promote better management of forestlands. However, the discussion in the previous section had suggested that, even then, timber harvesters may have insufficient incentive to take negative externalities into account. Furthermore, in many contexts forest management is a collective effort that requires cooperation among landowners to establish local governance arrangements for forests held in common. With all this in mind, let us return to the social–ecological systems (SES) framework, which provides us with a way to think about the question of the factors that will contribute to successful local governance.

A number of studies have examined the effect of local governance arrangements on forest management. These studies typically examine situations where there is a local community of users who are able to band together in order to manage a local forest. They then examine various characteristics of the local governance structure and correlate them with some index of forest management success, such as measures of forest density. The question is: What factors tend to make the governance arrangements more and less effective? For reference, you can consult the second-tier variables of the social–ecological systems framework found in Table 5.2.

A key factor, replicated in repeated studies, is the adoption of an effective system of monitoring to ensure compliance with the governance rules. When local users are able to set in place a reliable ongoing system for monitoring user behavior, local forests tend to do better. These same studies indicate that users are more likely to create such a monitoring system when they enjoy rights to harvest, which gives them incentive to consider the long-term health of the forests.

However, merely having a long-term stake in the forest is not sufficient in itself to ensure positive management outcomes. Other studies have found that the following factors matter also: forest size, commercial value, user dependence on the forests for subsistence, the degree of rule-making autonomy, and user beliefs regarding the legitimacy and fairness of the governance system. One general consistent finding is that forests generally do better if users are given greater latitude to make their own rules, are actively involved in designing the governance system, and believe that the system they come up with is legitimate and fair.

For certain factors, however, it matters whether they are present in combination with other factors. Small- to medium-sized forest stands tend to be better managed, especially forests of lesser commercial value and ones that users do not depend heavily upon for subsistence (for example, reliance on the forest for firewood). Conversely, large forests of high commercial value that users rely upon heavily for subsistence tend to be subject to greater overharvesting and degradation. However, the probability that overharvesting and degradation occurs depends upon the degree of enforcement of governance rules. The stricter enforcement is, the better are the forest outcomes for both small and large forests alike [Chhatre and Agrawal (2008)].

An important general conclusion of the SES scholarship on forests is that it is difficult to predict forest management outcomes simply on the basis of the property regime. Regardless of whether forests are government owned, privately owned, or communally owned, they may be well managed or poorly managed. In order to predict whether overharvesting and degradation will occur, we need to dig deeper into the details of the local system. As with the other common-property resources we have studied – water, grazing commons, and fisheries – forest outcomes may depend upon a complex combination of factors. And, as with these other resources, nailing down this combination remains a work in progress.

## Public choice and deforestation

Let us conclude our discussion of forest management and deforestation by considering some issues of public choice. When we consider forest policies set by elected officials, we must be aware of the practical realities of interest group politics, and how these realities might influence the policies we observe. Doing so allows us to understand how these policies came to pass in the first place, and how to interpret these policies in terms of how well they serve societal objectives.

Let us begin by identifying the likely interest groups regarding tropical deforestation policy. As we have seen, the cutting down of tropical forests generally occurs as forests vie with alternative uses of the land; namely, farming, logging, and grazing. Farmers, loggers, and ranchers have a large financial stake in forest policy and, thus, they constitute likely influential interest groups who we would expect to support liberal policies on logging. Consumer groups for products like meat and timber would also have a stake in liberal policies. However, because meat and timber from low-income countries tends to be a small fraction of total world trade, their interest in these types of policies would tend to be less urgent [Katzman and Cale (1990)].

On the other hand, tropical deforestation harms certain citizens of the countries where it occurs, who are often Indigenous peoples with little political power. In addition, it also may harm citizens of high-income, industrialized countries by exacerbating climate change, potentially destroying the ingredients for valuable pharmaceuticals, causing the loss of existence value (see chapter three) associated with the existence of tropical rain-forests, and insulting ethical, moral, and environmental sensibilities. Some of these citizens may be able to affect deforestation outcomes, for example, by engaging in debt-for-nature swaps. And the world community may be able to apply political and economic pressure through trading blocs and international organizations. However, there are limits to how much external pressure can be effectively exerted, for two reasons.

One is that tropical countries have sovereignty, and there are strong international norms to respect the sovereignty of all countries, even if there are seemingly overwhelming reasons to intervene. Countries in the international community have various tools at their disposal to exert political and economic pressure, such as economic sanctions, expulsion of diplomats, and freezing of financial assets. But these all stop short of physical intervention. The experiences of the world community with Brazilian President Bolsonaro in 2019 vividly illustrate the challenges of bringing international pressure to bear to combat deforestation in an individual country (see Box 16.3).

---

### Box 16.3: Bolsonaro and the Amazon rainforest

*News item*: "Bolsonaro shrugs off German aid cuts, as deforestation surges," *Climate Home News*, December 8, 2019.

"Brazilian President Jair Bolsonaro on Sunday said his country has 'no need' for German aid aimed at helping protect the Amazonian forest, after Berlin said it would suspend some payments because of surging deforestation …

'They can use this money as they see fit. Brazil doesn't need it,' Bolsonaro, a far-right populist, told journalists in Brasilia.

His comments came after Germany on Saturday said it would block payment of €35 million ($40 million) to Brazil for forest conservation and biodiversity programs until the Amazon's rate of decline attained encouraging levels once again."

---

The second reason has to do with the difficulties of international coordination to exert pressure, given various geopolitical realities. A few countries like the United States are economically powerful enough that they can impose sanctions on their own that are large enough to hurt. However, their ability to do so depends upon internal political debates, which may be sharply divided on the severity of measures that should be taken or, even, whether they should be taken at all. Most countries must try to exercise influence through international organizations such as the European Union or the United Nations. Here, the challenges of coming to common agreement may be even greater. You may perceive that the argument here is the by-now familiar one that transaction costs may pose significant obstacles to meaningful concerted action.

Thus, given the nature of the interest groups and their likely stakes in deforestation policy, we would not predict that the politics would favor meaningful, effective policy to curb deforestation. In general, public choice theory predicts more intense rent-seeking from higher-stakes groups, all else being equal. And some evidence suggests that these groups have indeed been able to extract favorable policies from the governments of low-income countries, including tax concessions and other subsidies of various kinds [Mendelsohn (1994), p. 750; Katzman and Cale (1990), p. 828]. These sorts of policies tend to encourage forest harvesting and accelerate the rate of deforestation.

At the same time, other evidence is more difficult to interpret in terms of the pure public choice model. One study found, for example, that in the awarding of title to property rights in the Brazilian frontier, those titles tended to go to tracts of land that were low-value and distant from population centers. At the same time, more valuable, less remote tracts tended to go untitled [Alston et al. (1996), p. 58]. This pattern does not suggest that larger, more wealthy claimants are capturing the process of land titling, as would be predicted by the basic public choice model. The overall bottom line is that the public choice approach helps explain a good deal about forest management policies. However, more research needs to be done to refine its predictions and generate hypotheses, in order to better understand and interpret the policies that we observe in the real world.

### Key takeaways

- The steady loss and degradation of valuable forest resources, especially in tropical climates, poses serious threats to the homes and livelihoods of many of the world's rural poor, including many Indigenous peoples. It also threatens the world's biodiversity through the destruction of vital habitat for many animals, birds, and plants. And it exacerbates the serious and possibly existential problem of ongoing climate change.

- Economic analysis suggests two key reasons why forests tend not to be managed sustainably. One is the negative externalities associated with cutting down forests, including the loss of a carbon sink to absorb and store carbon dioxide. The second is insecure property rights that results in overly rapid harvesting of trees and insufficient incentives to replant trees to take their place.
- Significant transaction costs impede bargaining and, thus, internalization of the negative externalities. And the political and social instability of many settings where deforestation is occurring poses major challenges to policies to firm up property rights.
- The SES framework provides guidance regarding governance provisions that tend to promote sustainable forest usage by local users. However, we have a way to go before we can feel confident that we understand the complexities of effective governance.
- Political realities suggest that effective policies to address deforestation may be a long time coming.

## Key concepts

- Clear-cutting
- Environmental Kuznets curve
- Debt-for-nature swaps
- Payments for ecosystem services
- Land trusts
- Conservation easements

## Exercises/discussion questions

16.1. Explain in your own words why the data in Figure 16.2 suggest that deforestation may be subject to an environmental Kuznets curve.
16.2. Show that equation (16.1) and (16.3) are algebraically equivalent.
16.3. Consider the forestry model graphically depicted in Figure 16.5. Show, using this model, the effect on the optimal age of harvest under these conditions:
   - Market interest rates increase.
   - Market prices of timber increase.
   - The forests become habitat for endangered red-cockaded woodpeckers.
   - Forests are found to significantly reduce GHG emissions.
16.4. The SES framework specifies that the resource sector (RS1, see Table 5.2) matters. Suppose you are investigating a local forest CPR. In what ways would your choice of other SES factors to examine be different than if you were investigating a fisheries CPR?
16.5. Consider the criticism of debt-for-nature swaps that Costa Rica has benefited disproportionately. Is this a fairness or efficiency argument? What evidence would you collect to help you answer this question?
16.6. We have seen that conservation easements bind private landowners from developing lands as they see fit, in order to preserve environmental amenities on the lands. We have also seen the general argument that permitting freedom of choice tends to promote efficient resource allocation. Does that apply here? Why or why not? How

would you make an *evidence-based* argument that conservation easements promote efficient use of the lands?

16.7. Can you think of a good public choice argument for why low-value tracts of land distant from population centers would be titled while higher-value tracts nearer to population centers would not?

## Notes

1   Notice that the fifth set of bars for the years 1950 to 1979 summarizes a *30- year* period, while the sixth and seventh set of bars for 1980 to 1995 and 1996 to 2010 summarize *15-year* periods. This means that compared to 1950 to 1979, there was actually more deforestation *per year* in the 1980–1995 period and a modest decline in the 1996–2010 period.

2   See, for example, Cropper and Griffiths (1994); Barbier (2001); Bhattarai and Hammig (2004).

3   To be clear, in constructing this profile, we are making an important simplifying assumption. In real-world forests, of course, there is not one tree age but, rather, a mixture of trees of different ages. It is not even fair to think of this *Age* variable as being the average age of trees in a forest, because different age mixtures could all have the same average age. In drawing this volume-age profile, economists are essentially asking you to think of forests as being a collection of identical trees.

4   See, for example, Deacon (1999), p. 357.

## Further readings

### *I. Forests and the SES framework*

Delgado-Serrano, Maria del Mar and Pablo Andres Ramos. "Making Ostrom's framework applicable to characterize social ecological systems at the local level," *International Journal of the Commons* 9(September 2015): 808–30.
*Attempt to more fully operationalize the SES framework to make it more useful for diagnosing local environmental systems. Applies the framework across three different types of resource systems: biodiversity and water management, forest management and land use, and marine and coastal management.*

### *II. Institutions and forest management*

Coleman, Eric A. and Scott S. Liebertz. "Property rights and forest commons," *Journal of Policy Analysis and Management* 33(Summer 2014): 649–68.
*Examines the effect of secure property rights on the success of forest management. Documents their influence on household decision-making in several low-income countries.*

Szulecka, Julia and Laura Secco. "Local institutions, social capital and their role in forest plantation governance: Lessons from two case studies of smallholder plantations in Paraguay," *International Forestry Review* 16(April 1, 2014): 180–90.
*Examines the effect of institutional differences in the management of forests in Paraguay. Highlights the importance of local institutions and social capital as important determinants of successful forest management.*

## References

Alix-Garcia, Jennifer M.; Katharine R.E. Sims; and Patricia Yanez-Pagans. "Only one tree from each seed? Environmental effectiveness and poverty alleviation in Mexico's payments for ecosystems services program," *American Economic Journal: Economic Policy* 7 (November 2015): 1–40.

Alston, Lee J., Gary D. Libecap and Robert Schneider. "The determinants and impact of property rights: Land titles on the Brazilian frontier," *Journal of Law, Economics, and Organization* 12(April 1996: 25–61.

Barbier, Edward B. "The economics of tropical deforestation and land use: An introduction to the special issue," *Land Economics* 77(May 2001): 155–71.

Bhattarai, Madhusudan and Michael Hammig. "Governance, economic policy, and the environmental Kuznets curve for natural tropical forests," *Environment and Development Economics* 9(June 2004): 367–82.

Chhatre, Ashwini and Arun Agrawal. "Forest commons and local enforcement," *Proceedings of the National Academy of Sciences* 105 (September 9, 2008): 13286–91.

Cropper, Maureen and Charles Griffiths. "The interaction of population growth and environ-mental quality," *American Economic Review* 84 (May 1994): 250–54.

Deacon, Robert T. "Deforestation and ownership: Evidence from historical accounts and contemporary data," *Land Economics* 75(August 1999): 341–59.

Irfan, Umair. "Brazil's Amazon rainforest destruction is at its highest rate in more than a decade," *Vox.com* November 18, 2019. https://www.vox.com/science-and-health/2019/11/18/20970604/amazon -rainforest-2019-brazil-burning-deforestation-bolsonaro

Katzman, Martin T. and William G. Cale. "Tropical forest preservation using economic incentives," *BioScience* 40(December 1990): 827–32.

Kindermann, G. et al. "Global cost estimates of reducing carbon emissions through avoided deforestation," *Proceedings of the National Academy of Sciences* 105(2008): 10302–7.

Korngold, Gerald. "Globalizing conservation easements: Private law approaches for international environmental protection," *Wisconsin International Law Journal* 28(2011): 585–638.

Land Trust Alliance. National land trust census. (2015) https://www.landtrustalliance.org/census-map/

Magnusson, Niklas. "Deforestation wipes out an area the size of Belgium," *Bloomberg.com* April 24, 2019. https://www.bloomberg.com/news/articles/2019-04-25/how-bad-is-deforestation-two-conne cticuts-were-lost-last-year

Mahoney, Julia. "Perpetual restrictions on land and the problem of the future," *Virginia Law Review* 88(June 2002): 739–87.

Mendelsohn, Robert. "Property Rights and Tropical Deforestation," *Oxford Economic Papers* 46(October 1994): 750–6.

Parker, Dominic P. "Land trusts and the choice to conserve land with full ownership or conservation easements," *Natural Resources Journal* 44(Spring 2004): 483–518.

Salzman, James; Genevieve Bennett; Nathaniel Carroll; Allie Goldstein; and Michael Jenkins. "The global status and trends of payment for ecosystem services," *Nature* (June 15, 2018).

Thiel, Anne. "The market for payment for ecosystems services is growing up," *GreenBiz* (December 17, 2019) https://www.greenbiz.com/article/market-payment-ecosystems-services-growing

Thompson, Olivia R. Rendon; Jouni Paavola; John R. Healey; Julia P.G. Jones; Timothy R. Baker; and Jorge Torres. "Reducing emissions from deforestation and forest degradation (REDD+) transaction costs of six Peruvian projects," *Ecology and Society* 18(March 2013).

United Nations. "About REDD+," *UN-REDD Programme Fact Sheet*, February 2016.

United Nations. Food and Agriculture Organization (FAO). *The state of the world's forests 2020: Forests, biodiversity, and people*. Rome: FAO, 2020.

World Bank. "Debt-for-nature swaps," *World Bank Information Briefs*. C.05.4-93., 1993.

World Wildlife Federation (WWF). "US and Indonesia announce $28.5 million debt swap to protect Borneo's tropical forests," *WWF*, September 29, 2011.

# 17 Wildlife, endangered species, and biodiversity

In spring of 2019, an agency of the United Nations issued a comprehensive report warning of "accelerating" rates of species extinctions worldwide and "unprecedented" declines in natural environments. This report, issued by the *Intergovernmental Science-Policy Platform on Biodiversity and Ecosystem Services* (IPBES), was the product of three years of work by hundreds of experts from 50 countries, and it represented the most comprehensive assessment of world biodiversity levels ever produced. The picture it paints is sobering.

Among other things, it finds that about three-quarters of the entire terrestrial environment has been "severely altered" by human activity, with a loss of nearly half of all global ecosystems. Worldwide, over 40% of all amphibian species, nearly 33% of all ocean corals, and over 33% of all marine mammals are threatened with extinction. Roughly one million species of plants, animals, birds, and insects are in danger of going extinct in the next few decades. Many of these trends are driven by human activities, especially agriculture, deforestation, mining, urbanization, and climate change.

In this chapter, we examine the issue of biodiversity and endangered species through the economist's lens. The key questions we will focus on are:

- How do we understand the issues of biodiversity and endangered species in economic terms?
- What sorts of policies have been set in place, how well have these worked, and why?
- What sorts of policies make economic sense?
- And, finally, what policies can we expect to observe given various political realities?

## Wildlife and endangered species: A general overview

The 2019 IPBES report is only the latest in a series of assessments of trends in world biodiversity. Since 1964, the *International Union for Conservation of Nature* has published a list of endangered species worldwide, known as the *Red List of Threatened Species*. The *Red List* is a comprehensive listing of *taxa*, or groups of organisms that scientists have decided have enough similarities to be considered one related biological unit. The notion of a taxon is broader than that of a species and also includes higher classifications such as genus, family, and so forth – all the way up to kingdoms, if you remember your high school biology class, the class mammals and the genus African elephants, for example, are both considered taxa. For the remainder of the discussion in this chapter, you may assume that we are treating species as the level of the taxa when describing different populations of plants and animals.

In order to measure threats to individual species, scientists have come up with a system with classifications reflecting how endangered those species are. At one extreme, species are considered *extinct*, meaning that there are no remaining individuals left in the world. Two of the more well-known examples of species whose extinction has occurred in the recent past are the dodo and passenger pigeon. In the 19th century, the passenger pigeon was perhaps the most abundant bird in North America and, possibly, the world. In 1871, for example, there were some 136 million breeding adults in Wisconsin alone. However, its numbers subsequently plummeted because of overhunting and habitat destruction. The last known passenger pigeon died in 1914 in the Cincinnati Zoo [Yeoman (2014)].

The remaining categories found in the Red List, in descending order of threat to survival of a species, are: critically endangered, endangered, vulnerable, near-threatened, and least concern. A species is considered *endangered* when its numbers have fallen to dangerously low levels, are in rapid decline, and/or inhabit habitat that is in rapid decline or is severely fragmented. Some examples of *critically endangered* species are the black rhino, red wolf, and the Yangtze river dolphin. The difference between endangered and critically endangered has to do with how imminent scientists believe extinction is.

A species is considered *vulnerable* if scientists consider there to be a high risk of it becoming endangered in the foreseeable future. *Near-threatened* means there is some non-negligible risk of extinction for a species, but its numbers and habitat extent are such that the risk is not considered serious right now. Finally, *least concern* means that there is little risk of extinction under current conditions.

The Red List lists literally tens of thousands of taxa found around the world, along with a scientific assessment of the level of risk of extinction for each taxon. Figure 17.1 shows the current assessment of the Red List for species worldwide, of which over 32,000 species are in danger of extinction. Figure 17.1 also shows that the number of endangered and vulnerable species has been steadily increasing over the past 20 years, and it is now at roughly triple what it was at the turn of the century.

The rate at which species are being threatened with extinction varies around the world. Species living in regions with rapid population growth, such as central Africa and Southeast Asia, are especially vulnerable. And certain parts of the world are home to especially rich

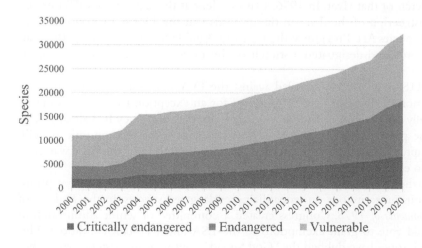

*Figure 17.1* Endangered and threatened species, Red List, 2000 to 2020. *Source*: IUCN Red List, version 2020-2: Table 2. Last updated, July 9, 2020.

and diverse animal and plant populations, such as the tropical rainforests of South America, Central America, and Southeast Asia. Continuing encroachment on habitat due to farming, ranching, and logging in those areas likely pose particularly dangerous threats to local populations.

## Institutions governing species protection: The United States and Europe

### I. United States

Since the 1970s, governments in the United States and Europe have taken important steps to protect wild plants and animals and their habitat. In 1973, the United States Congress passed the *Endangered Species Act* (ESA), which established national policy for conservation of endangered and threatened species and the ecosystems in which they exist. The ESA recognized that species have "ecological, educational, historical, recreational, and scientific value." It also established a set of administrative procedures for the U.S. Fish and Wildlife Service and the U.S. National Marine Fisheries Service to follow in order to protect endangered species. Current procedures call for these agencies to list species when they become endangered or threatened, to designate critical habitat for their survival, to prohibit activities that threaten their survival, and, if all goes well, to delist species when they are no longer in danger. Under the ESA, private parties may not "take" a listed species, or they risk fines or imprisonment.

Since its enactment in 1973, the ESA has since been the subject of court challenges, which have clarified how to interpret some of its provisions. Two U.S. Supreme Court rulings have been extremely important: *Tennessee Valley Authority v. Hill*, decided in 1978, and *Babbitt v. Sweet Home Chapter of Communities for a Great Oregon*, decided in 1995.

### A. Tennessee Valley Authority v. Hill

In 1973, the tiny snail darter fish was discovered in the Little Tennessee River. At the time, the Tennessee Valley Authority (TVA) was in the process of constructing the Tellico Dam along a stretch of that river. In 1976, a law student at the University of Tennessee sued to halt construction of the dam on the grounds that the TVA was in violation of the Endangered Species Act. Previously, the Fish and Wildlife Service had listed the snail darter as endangered and designated a stretch of the Little Tennessee River as critical habitat.

In 1978, the U.S. Supreme Court ruled against the TVA and issued an injunction on further construction of the dam. The TVA argued for an exception to the provisions of the ESA essentially on economic grounds: that large amounts of money had already been expended on the project and that the dam would provide a lot of economic benefit. However, the Supreme Court ruled six to three that the language of the ESA was clear and permitted no exceptions.

In response to the TVA ruling, Congress quickly amended the ESA to permit exceptions to its unambiguous proscriptions on jeopardizing endangered species. In the *Endangered Species Act Amendments of 1978*, Congress created a committee of senior administration officials that could vote to grant exemptions to the ESA under certain conditions. This committee, which some have dubbed the "God Squad," continues in existence today, but it has rarely been invoked to decide cases of particular endangered species.

*B. Babbitt v. Sweet Home Chapter of Communities for a Great Oregon*

In *Babbitt v. Sweet Home*, the issue was how to interpret that portion of the Endangered Species Act that defined what it meant to "take" a species. The original ESA had defined "take" using a variety of terms including, importantly, causing "harm" to a species. In 1975, the Secretary of the Interior expanded the meaning of "take" to include degradation of the habitat in which species live. Sweet Home Chapter was a land developer that wanted to log lands that were habitat to northern spotted owls and red-cockaded woodpeckers. These two species of birds were listed as threatened under the ESA. The Interior Department filed suit to halt development on the basis that it would harm the habitat of these birds.

In 1995, the U.S. Supreme Court ruled six to three that harm could include "significant habitat modification or degradation" that affect endangered species. So, even though the activities of Sweet Home were not harming the species directly, the Court ruled that habitat modification could inflict harm indirectly. The significance of this ruling is that it expanded the legal grounds on which the Interior Department could protect endangered and threatened species under the ESA.

## II. Europe

In Europe, the general framework for wildlife protection is provided by the *Bern Convention on the Conservation of European Wildlife and Natural Habitats* (Bern Convention), which was enacted in 1979 and went into effect in 1982. The Bern Convention was the first international treaty that coordinated the efforts of different countries to protect endangered species and their habitats. Like the ESA, the main objective of the Bern Convention is the conservation of wild plants and animals and their habitats, especially ones designated as endangered or threatened. Countries that officially agree to its principles – so-called *signatory nations* – are obligated to take "all appropriate measures" to ensure conservation of wild plant and animal species. These include national conservation policies and coordination of national policies with those of other countries to protect migratory wildlife. As of 2018, almost all European countries had signed the Bern Convention.

Actual implementation of wildlife and habitat protection policy in most European countries is provided for by a series of directives enacted over time by the European Union. In 1979, the European Union enacted the so-called *Birds Directive*, which targeted conservation of wild birds. This was followed in 1992 by enactment of the *Habitat Directive*, which broadened the focus of conservation efforts to include natural habitats and wild plants and animals. These two directives form the backbone of EU biodiversity conservation policy. Under these two directives, more than 26,000 protected sites covering over one-sixth of the land area of the EU have been set aside [Evans et al. (2013), pp. 97–8].

## Institutions governing international trade in endangered species

One issue with important implications for endangered and threatened species is the existence of a booming international trade in goods derived from wild plants and animals. It turns out that international trade in these goods is a multi-billion dollar industry. This trade includes food products, timber and wood products, specialty tourist goods, leather goods, and medicines. In response to the potential threat this trade poses to the survival of many species, the *Convention on International Trade in Endangered Species of Wild Fauna and Flora* (CITES) was enacted in the early 1970s and entered into force in 1975. CITES is an international agreement among countries whose objective is to ensure that international trade in plant and animal goods does not threaten the survival of endangered plants and animals. Currently, some 5,800 animal species and 30,000 plant species are protected under CITES.[1]

## The economic approach to endangered species protection

As you might expect, the economic approach to protection of endangered species is fundamentally based on an assessment of costs and benefits. To the extent possible, information available on the costs and benefits of protecting species from extinction should at least enter into consideration when proposing policy. In addition, economic principles can provide insight into the likely private responses to policies, which can sometimes alter our assessment regarding likely outcomes and the desirability of those outcomes. We will expand on this latter point later on in the chapter.

In general, it is recognized that there are important costs and benefits associated with protection of endangered species of plants and animals and their habitat. The costs are typically expressed as foregone benefits from development activities such as logging, farming, grazing, or expanding cities, which are commonly the alternative uses of the lands inhabited by endangered species. The magnitude of these costs are relatively easy (though by no means trivial!) to get a handle on, because foregone timber, cattle, or crops are commonly sold in markets, where their market price provides a reasonable measure of their societal value.

The benefits of protection, on the other hand, are often less clear or, at least, less easily quantified. Recalling the earlier discussion in chapter three of valuing the environment, we can think of there as being *use value*, *non-use value*, and *existence value* benefits of protection. Use value benefits might derive, for example, from the pharmaceutical value of wild plants, or from ongoing sustainable management of commercial fisheries. Non-use value benefits might derive from ecotourism, wildlife viewing, or preservation of cultural heritage. Finally, existence value benefits might derive from the psychic satisfaction of knowing that certain species exist (see chapter three for a further discussion of the points made in the last two paragraphs).

## A useful metaphor: The Noah's Ark problem

A useful way to model the economics of species preservation was proposed by the late economist Martin Weitzman (yes, him again!). For Weitzman, the preservation of endangered species was analytically similar to the problem faced by the biblical Noah when bringing animals into the Ark prior to the Great Flood. Bringing in the animals two by two essentially ensured that each species would survive. The problem, of course, was that the Ark was only big enough to house so many animals. If Noah wanted to maximize the total value of species brought into the Ark, how would he go about doing it? Weitzman called this the *Noah's Ark problem* [Weitzman (1998)].

To understand the Noah's Ark problem, Weitzman asks us to consider a library full of books. Now ask yourself: What is it that we value about a library? Well, we value the books themselves for a variety of reasons including the pleasure of reading them and the information they contain. Most people would agree that better libraries contain a wider variety of types of books. No matter how much you like your favorite fiction writer, you would probably not consider a library very good if it contained only that author's works. And not just other fiction writers but also non-fiction: books on current affairs, histories, biographies, travel books, and so forth. The best libraries give readers plenty to choose from.

Similarly, ecosystems and wilderness areas are "better" if they contain a diversity of wild plants and animals. The wild world would be much less interesting if it contained only pigeons, pine trees, and ground squirrels! That is to say: humans generally value species diversity for its own sake, aside from how we feel about individual species like snow leopards and cuddly pandas. And, the more a species adds to diversity, the more we tend to like it. In a drab world of pigeons, pine trees, and ground squirrels, most of us would probably greatly prefer to add brightly colored tropical macaws rather than a few chipmunks.

None of this is to argue that biodiversity is only about having variety. Biologists would rightly argue that it is also about paying attention to ecosystem dynamics, keystone species, and other important biological variables. Weitzman is being the economist that he is, by following the traditional economic approach of drastically simplifying, and, hopefully, not oversimplifying, reality in order to gain insights into actionable policy.[2]

The particular value of this notion of diversity is that we can add it to economic analyses as a factor to consider that is important to most people. We have already alluded to another factor: the direct value humans derive from individual species, which is akin to the joy we experience when we read a rollicking novel. On the other hand, species protection comes at a cost, for example in terms of lost opportunities for economic development. Finally, evaluating any economic policy for species protection must consider the effectiveness of that policy in actually increasing the chances of species survival. Equation (17.1) brings all of these factors together.

$$\text{Value of species protection} = \left[ P \times (B + U) \right] / C \tag{17.1}$$

In equation (17.1), the benefits of protecting a species is given by B and U, where U is the value we derive from gazing at a macaw or a panda and B is the value of a species in increasing overall biodiversity. P is the increased probability that a policy will enable a species to survive and, thus, represents the effectiveness of that policy. So the entire numerator represents the expected benefits to be derived from a policy to protect an endangered species. Finally, C is the dollar cost of that policy. Dividing expected benefits by the cost transforms the expected benefits into expected benefits per dollar spent on the policy.

Equation (17.1) thus provides a framework to think economically about how best to provide protection for a particular endangered species. It implies that the desirability of a policy to protect an endangered species will be greater:

- The more (consumption) value we derive from that species;
- The more that species adds to overall biodiversity;
- The more effective the policy is in improving that species' chances of survival; and
- The less the policy costs.

We can use this basic framework to help us evaluate specific species protection policies, based on these four factors.

## Economics of the ESA

Now that we have an analytical framework to work with, let us now turn to an assessment of the Endangered Species Act. The ESA has been called one of the most important environmental laws ever enacted in the United States, and there appears little doubt that it has had an important impact in protecting endangered species. However, economists have raised various concerns regarding whether it has done so in an economically sensible manner. And, indeed, the original ESA did not recognize costs and benefits as being factors germane to whether or not species should enjoy protection, as we have seen. In this section, we will focus on two related issues. The first is whether the correct decisions are being made with regard to which species get protected under the ESA. The second concerns the impact of protection on the species themselves.

### I. Which species get protected?

Weitzman and his co-author Andrew Metrick have performed an analysis of the listing decisions by the U.S. Fish and Wildlife Service (FWS) to examine the extent to which they conform to good practices, as suggested by the Noah's Ark framework [Metrick and Weitzman (1998)]. The basic question is whether FWS listing policies target species according to expected benefits and costs. They find mixed evidence on whether species that are more endangered receive more protection. On the one hand, they find that species tend to get listed when they are more endangered and also, interestingly enough, when they happen to be the only species in their genus. They interpret the latter finding as suggesting that species that contribute more to species biodiversity (variable B) tend to receive more protection, at least when it comes to making the list of species that are considered endangered.

On the other hand, they also find that after they are listed, species that are more endangered do not actually receive greater spending of public moneys for protection and recovery efforts. In a related study, they argue that non-economic characteristics of the species probably play an important role in affecting decisions to protect different species [Metrick and Weitzman (1996)]. For example, certain species such as elephants, pandas, and lions and tigers – examples of so-called *charismatic megafauna* – simply have more emotional appeal to humans. The result is that we tend to overspend on protecting them, relative to the extent to which they are actually endangered or contribute to overall biodiversity.

Further suggestive evidence that emotional appeal matters is shown in Table 17.1. This table lists the ten animal species on which the most spending for protection had been lavished in the United States by the mid-1990s, when their study was published. These ten

*Table 17.1* Top ten species by total spending, mid-1990s

| Common Name | Spending ($millions) | Cumulative spending (Percent) |
| --- | --- | --- |
| (1) Bald eagle | 31.3 | 9.9 |
| (2) Northern spotted owl | 26.4 | 18.3 |
| (3) Florida scrub jay | 19.9 | 24.5 |
| (4) West Indian manatee | 17.3 | 30.0 |
| (5) Red-cockaded woodpecker | 15.1 | 34.8 |
| (6) Florida panther | 13.6 | 39.1 |
| (7) Grizzly (or brown) bear | 12.6 | 43.1 |
| (8) Least Bell's vireo | 12.5 | 47.1 |
| (9) American peregrine falcon | 11.6 | 50.7 |
| (10) Whooping crane | 10.8 | 54.2 |

*Source*: Metrick and Weitzman (1996), p. 2.

species received over half of total spending on species protection, and the top species – the bald eagle – enjoyed nearly one tenth of total spending all by itself.

You can see that the list consists entirely of relatively large mammals and birds, the type of animals that evoke the greatest emotional response in humans (as opposed to, say, fish and small reptiles). At the time, several of these species such as the bald eagle, northern spotted owl, and grizzly bear were not in danger of actually becoming extinct. Furthermore, several of the species – the northern spotted owl, Florida scrub jay, and grizzly bear – had closely related subspecies that were not in danger. They thus added relatively little to overall biodiversity.

## II. The impact of protection

Let us now turn to an assessment of what may be the most common policy employed by the FWS for endangered species protection: habitat set-asides. As we have seen, under the ESA, endangered species protection can occur through setting aside land for habitat. On the surface, it may seem like a simple matter for the ESA: just designate lands to be protected as habitat and, thus, spare endangered species from damages from economic development.

One problem is that the operation of the set-aside program provides perverse incentives for landowners to preemptively destroy habitat once they realize lands will be listed. There are numerous examples of this, some of which have been rigorously documented by economic studies. In one case, mere days before the golden-cheeked warbler was set to be listed by the FWS, a firm hired workers to saw down hundreds of acres of forests comprising warbler habitat. In other cases, timber was harvested prematurely to avoid ESA regulations over protection of northern spotted owls and black-capped vireos. If this behavior occurs with regularity, implementing the provisions of the ESA could actually lead to a long-term decline in habitat and populations of endangered species.[3]

In a rigorous economic study, economists Dean Lueck and Jeffrey Michael investigated the issue of preemptive timber harvesting for the case of red-cockaded woodpeckers (RCWs) in North Carolina [Lueck and Michael (2003)]. Using data on timber harvesting and the presence of local populations of RCWs, they found that timber tended to be harvested more quickly and trees were harvested at a younger age when RCWs were present. These results are consistent with preemptive harvesting of timber in order to avoid costly ESA regulations on timber harvesting. This study provides more compelling evidence of preemptive harvesting than accounts in the news because of its careful experimental design. In North Carolina, RCWs are found only in mature stands of southern pine, so that there was a clear connection between the listing decision and harvesting behavior. Furthermore, other endangered species were not present in these pine stands to muddy the interpretation of the findings.

---

**Box 17.1: Preemptive harvesting in the news**

*News item:* "Town, woodpeckers fight over nest eggs," *NBCNews.com*, September 26, 2006.

   In 2006, NBC news reported that landowners in Boiling Spring Lakes, NC were clear-cutting thousands of trees to prevent them from becoming homes for red-cockaded woodpeckers. This occurred when the Fish and Wildlife Service announced that it was working on a plan to set aside land to protect 15 active clusters of woodpeckers. Landowners reportedly "swarmed" city hall to apply for land-clearing permits [Breed (2006)].

One important open question of the Lueck and Michael study concerns the magnitude of the preemptive harvesting effect. They estimate that the premature harvesting by itself resulted in the loss of thousands of acres of mature pine stands and the loss of a few dozen RCW colonies over a six-year period. Furthermore, landowners may have harmed RCWs in other ways, including directly killing them. On the other hand, listing of RCW habitat by the ESA probably protected some pine stands from being harvested. It is unclear whether the positive effects of these habitat set-asides outweighed the negative effects of preemptive harvesting and private takings of RCWs.

## Managing wild elephant populations in Africa

One animal species that has received a lot of attention from economists and policymakers is the wild elephant, which is probably the classic example of charismatic megafauna. Both the Asian elephant and the African elephant have seen major declines in wild populations over time. Once numbering in the tens of millions, the number of African elephants in the wild likely fell below one million for the first time in the 1980s.[4] Since then, its numbers have continued to fall, to the point that there are currently fewer than half a million on the entire continent.

A number of factors have been responsible for the decline in wild elephant populations, including conflicts with growing human populations, and loss of habitat through conversion of land to farms and ranches. However, illegal poaching is believed to be the main factor in reducing elephant numbers. Elephant tusks are valued for their ivory, which is used to make a variety of consumer products that are marketed all over the world but are especially prized in much of Asia. Fluctuations in the international price of ivory have large effects on poaching activity. When prices increase, poaching tends to increase.

There have been three main policy responses to these dramatic declines in wild elephant populations. One is to set aside reserves such as national parks where elephants enjoy protected status, sometimes called the *fortress approach*. The second is to encourage community-based management on the local level, sometimes called the *participatory approach*. Finally, the international community has stepped in to ban international trade. Let us examine these policies in turn.

### I. The fortress approach

A number of countries in Africa have set aside large protected areas for elephants, including Botswana, Zimbabwe, Tanzania, South Africa, Kenya, and Zambia. In order to be effective in protecting the elephants, these policies require governments to expend many resources to enforce protections against illegal poaching. This has turned out to be a tall order for many countries, whose parks extend over a large geographic expanse, which makes it difficult to guard again roving bandit-like attacks on the elephant populations. One response has been to increase the penalties for illegal poaching. The countries of Botswana, Kenya, Namibia, and Tanzania have all called for draconian measures to combat poaching in recent years. For example, in 2013 the Tanzanian minister of natural resources and tourism called for poachers to be shot on sight [Smith (2013)].

Despite these measures, illegal poaching continues to create significant problems for countries trying to protect their wild elephant populations. Beginning around 2010, illegal poaching in Africa spiked and remained at high levels for several years. Figure 17.2 shows CITES data for elephant killings for all of Africa. This figure shows the percentage of

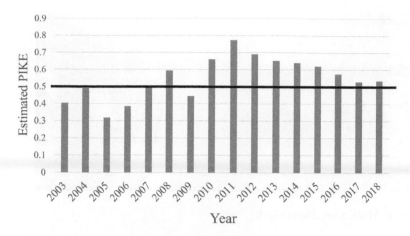

*Figure 17.2* Illegal elephant poaching in Africa. *Source*: CITES, Data for "Report on monitoring the illegal killing of elephants (MIKE)," CoP18 Doc. 69.2, CITES Secretariat. https://cites.org/sites/default/files/eng/cop/18/doc/E-CoP18-069-02.pdf

illegally killed elephants (PIKE) by year [CITES (2019)]. To help interpret these numbers, a PIKE number above 0.5 (shown as the horizontal line) is considered unsustainable for maintaining elephant populations. Figure 17.2 thus suggests that African elephant herds have been managed unsustainably for most of the 2010s. This increase in illegal poaching is particularly pronounced in Central and West Africa and is probably attributable to spikes in the local black market price of ivory [Wittmeyer et al. (2014)].

### II. The participatory approach: Community-based management

The fortress approach is an example of what is sometimes referred to as a *top-down* approach. That is, it is a policy created by a government and implemented in a centralized fashion. And as such, the fortress approach brings with it certain issues endemic to top-down approaches in general: less investment in and ownership of the policy by local communities, less familiarity with local conditions, and potentially high costs of monitoring.

An alternative would be a *bottom-up* approach, in which decision-making power is at the local level. We have already seen this approach in action in managing different resources, including forests, fisheries, and local water supplies. The basic idea here is the same: to empower local communities to manage their own natural resources sustainably: in this case, elephant herds. That is, members of the community actively participate in herd management, presumably for the betterment of the community.

Much of the potential for local communities to successfully manage elephant herds comes from a booming worldwide ecotourism industry (see Figure 17.3). Because of tourist fascination with charismatic megafauna in general, local communities have incentive to maintain and protect local herds in order to attract tourists. That is, as long as they can realize returns on their investments in management and providing protection against poachers. This essentially means two things.

First, communities must be given lands they can manage to service ecotourist demands. This might entail governments transferring reserve lands to communities equipped to manage elephant herds effectively. Second, communities must be allowed to retain the

*Figure 17.3* Ecotourists in Africa. *Source*: Shutterstock

proceeds from providing ecotourism activities, which in practice has meant mostly hunting and photographic safari trips. To put it in terms that we are by now familiar with, they should be given broadly unconditional property rights to the elephant herds, including the right to enjoy the stream of returns on the investments they make.

The *Communal Areas Management Programme for Indigenous Resources* (CAMPFIRE) in the country of Zimbabwe provides a good example of a participatory program that works reasonably well.[5] CAMPFIRE was created in 1989 by the government of Zimbabwe to support community-based wildlife management. The program allows local communities to grant access to local wildlife to safari operators, whereby communities earn revenue mostly through the operation of safaris and sale of animal products. It is true that there have been allegations of corruption and concerns about uneven distribution of revenues in different areas. However, in the first 12 years of its operation, the program generated more than 20 million dollars for the participating communities [see Frost and Bond (2008)]. In addition, wildlife numbers and habitat have stabilized and, in some areas, even increased.

The participatory approach is a relatively new approach to wildlife management, which raises the important question of how to move from the traditional fortress approach to the participatory approach. If you think about it, a move from the fortress approach to the participatory approach constitutes a change in the rules governing wildlife management. It can thus be viewed as an institutional change aimed at reducing transaction costs. These transaction costs especially consist of costs of enforcing exclusion of poachers from wildlife reserves. Switching to the participatory approach reduces enforcement costs if local communities are able to enforce anti-poaching efforts more easily than a central agency can. They may be able to accomplish this by taking ownership of local efforts and getting buy-in to management efforts by members of the local community. Participatory wildlife management is thus similar to management of other natural resources like forests and water.

Nevertheless, movement to the participatory approach has been uneven, succeeding pretty well in certain countries like Zimbabwe and not as well in others. The question is why. The answer seems to be that other sources of transaction costs can impede the move to the participatory approach.

The differing experiences of Peru and Tanzania illustrate how political factors can be decisive in determining the success of participatory efforts. In the early 2000s, both countries attempted to establish wildlife reserves that were to be managed by local communities. In Peru, local communities mostly supported the establishment and operation of these reserve areas. The result was the creation of a successful wildlife reserve, the Amarakaeri Communal Reserve, within Manu National Park. In Tanzania, on the other hand, community-based reserves largely failed to take hold in the face of indifference and even outright hostility from local groups [Haller et al. (2008)].

Several factors explain the difference between these two countries in local support for establishment of these reserves. In Peru, Indigenous groups viewed the establishment of the reserves as offering the prospect of gaining more secure land rights in the face of intrusions by outside groups such as miners, settlers, and oil and gas companies. In Tanzania, however, local groups did not view the management program as providing them greater control over the lands, which they viewed as continuing to be controlled by the government and NGOs. In addition, local communities commonly had a more antagonistic relationship to local wildlife, which they viewed as pests and potentially dangerous. Finally, government agencies imposed numerous regulations on local communities in how they managed wildlife herds, lowering their stakes in managing wildlife and taking away incentives for them to participate [Nelson et al. (2007)].

### III. International bans on ivory trade

As we saw earlier, the international community has responded to threats to the survival of various species by regulating trade in certain animal and plant products. This includes outright bans on trade in certain species, including Asian elephants and some populations of African elephants. CITES imposed a ban on international trade in ivory in 1989, in response to serious declines in elephant populations in a number of counties, plus evidence that international trade in ivory was an important contributing factor. This decision was confirmed in 1997 when signatories to CITES agreed to continue to list African and Asian elephants as critically endangered.

Economists differ in their opinions regarding the effect of an ivory trade ban in protecting wild elephant populations. On the surface, banning trade in ivory would seem to have the effect of protecting elephants, by reducing the benefits from harvesting ivory. However, the issue is considerably more complicated.

For one thing, banning trade in ivory reduces the supply in ivory markets, which exerts upward pressure on ivory prices. If prices increase significantly, this can create incentives for illegal poaching, which puts pressure on wild elephant herds if elephant protection is costly to enforce. We have seen much evidence that enforcement is costly under the fortress approach. Furthermore, allowing legal trade in ivory might increase incentives to manage elephant herds in a sustainable manner. And any revenues gained from a legal ivory trade could be used to combat illegal poaching. Overall, perhaps surprisingly, it is somewhat unclear whether trade bans will tend to increase or decrease elephant herds: it depends on the relative magnitude of these different effects.

As an alternative to a trade ban, the economists Michael Kremer and Charles Morcom have proposed a novel policy that they argue might be effective in protecting wild elephants. For governments that face relatively low costs of enforcement, they recommend focusing on anti-poaching efforts. However, when enforcement costs are high, they recommend stockpiling ivory and threatening to release it onto ivory markets if elephant populations fall. This would have the effect of crashing ivory prices, which would discourage illegal poaching. The stockpiled ivory, then, is wielded like a cudgel against prospective poachers, who have to think twice about poaching ivory [Kremer and Morcom (2000)].

## Public choice issues surrounding endangered species protection

So far, we have seen some evidence that political factors can determine how policies get implemented, or even if they get enacted in the first place. Let us now take a more systematic approach to understanding the influence of politics in endangered species protection by applying the insights of public choice theory. In this subsection, we shall return to the context of the United States and the Endangered Species Act. The basic issue is the extent to which political factors have influenced the creation and implementation of the ESA.

As we saw in chapter six, under the public choice approach we think of government policies as being fundamentally driven by interest group pressures. Interest groups "demand" policies, which are then "supplied" by legislators. Supply occurs through the enactment of legislation designed to accomplish specific goals, such as endangered species protection. Under the U.S. congressional system, in order to enact legislation, bills have to be crafted by legislators, make it out of committee, endure floor debate, and then, eventually, be voted on by legislators. Only bills that muster majority support among congresspersons and senators can make it to the president's desk for her/his signature. And, once passed, in many cases, government agencies such as the EPA or the Fish and Wildlife Service are then charged with implementing the policy, under the oversight of an appropriate committee.

This description suggests two key points where we can focus an analysis in order to investigate how politics affects policy outcomes: (a) on the floor of Congress, and (b) in the behavior of agencies such as the Fish and Wildlife Service that are charged with implementing policy. I should add that administrative agencies are themselves also often subject to political pressures from interest groups. Let us take these one at a time, in the order in which they appear in the legislative process.

### I. Congressional lawmaking

Let us begin with a fundamental question: How do we know that politics matters when it comes to carrying out the provisions of the ESA? To answer this question, one approach is to take particular legislation and observe patterns of correlation between legislative support and measures of interest group influence on the legislators. Then, depending on the patterns, we can draw conclusions regarding what political factors matter.

This approach was taken by the economists Sayeed Mehmood and Daowei Zhang in a study of political support for various amendments to the original ESA. They argued that patterns of support would be revealed in House roll-call votes on passage of the amendments. One of the amendments was to exempt the Tellico Dam project from operation of the ESA, which we encountered in our earlier discussion of the snail darter. The other amendments were all to ease restrictions imposed by the ESA on businesses. The authors found several interesting patterns of support for easing business restrictions, including that they were more likely to be supported by southern Republicans representing rural districts with a large construction sector. They concluded that both economic and ideological factors help explain support for protection of endangered species [Mehmood and Zhang (2001)].

### II. Agency behavior

Another possibility is that political considerations influence which species are listed as endangered, or the length of time it takes for species to be listed. As we have seen, under the ESA, primary responsibility for listing decisions rests with a government agency, the Fish and Wildlife Service. The question is: How could political factors influence the listing decisions of the FWS?

There are two possible political avenues for this to occur. One is through direct pressure from interest groups. This can occur in several ways: by nominating species for listing, filing suit if the FWS fails to act quickly enough, and providing public comments on proposals for listing. A second is through their congressional representatives sitting on committees charged with agency oversight. If committees engage in effective oversight, they may be able to influence the listing process.

It turns out that both of these avenues have been used to influence FWS listing decisions, according to multiple studies. A study by the economist Amy Ando found, for example, that public pressure on the FWS can affect how quickly individual species get listed [Ando (1999)]. This is a potentially important factor, recalling our earlier discussion of preemptive harvesting of habitat by logging companies. The longer it takes species to be listed, the greater is the ability of companies to preemptively harvest in order to avoid ESA regulations.

Two other economic studies found that listing decisions can be influenced by congressional oversight of the FWS. One study found that states that enjoyed greater representation on the committee charged with budgetary oversight of the FWS had fewer listings of endangered species. Another study came to a similar conclusion, finding that more "environmentally friendly" oversight committees, as measured by the average legislator score from the League of Conservation Voters, tended to have more species listings.[6]

The bottom line is that political factors seem to play an important role in the United States in determining both how much endangered species protection occurs and how quickly it occurs. This occurs through at least two channels. One is decisions by legislators to modify the overall stringency of ESA regulation through occasional legislative amendments. The other is the effect of both direct public and legislative pressure on the FWS decisions to list endangered species. Thus, the public choice framework provides a good deal of insight into the way policy for endangered species protection is actually implemented in the United States.

## Key takeaways

- As of 2019, roughly 30,000 species are threatened with extinction. This constitutes over one quarter of all species worldwide. Furthermore, the rate of extinction appears to be accelerating.
- Both the United States and Europe have enacted major protective legislation to combat species extinction within their borders. In addition, international trade in endangered species products is regulated under CITES.
- The economic approach to endangered species protection involves the weighing of the costs and benefits of preservation. A key element of the benefits of protecting a species is its contribution to overall biodiversity.
- There is evidence of bias in endangered species protection in the United States toward disproportionate protection of large animals and birds, at the expense of fish and reptiles.
- The habitat set-asides policy may provide perverse incentives to private landowners, resulting in destruction of wildlife habitat.

- Two types of approaches for protecting endangered species are the top–down fortress approach and the bottom–up participatory approach. We are seeing increased reliance on the participatory approach in recent years, but it has experienced uneven success.
- International trade bans on ivory are predicted to have mixed effects in protecting elephants.
- There is evidence of political influence in both the creation and implementation of endangered species protection policies in the United States.

## Key concepts

- Noah's Ark problem
- Charismatic megafauna
- Fortress approach
- Participatory approach
- International ivory trade bans

## Exercises/discussion questions

17.1. Consider the Weitzman model in equation (17.1). With only four variables, it is obviously a really good example of the economist's preference for parsimonious models. Whenever we create a parsimonious model like this one, there is always the danger that we are omitting something important. Critique this model. In your opinion, is there anything important for sensible endangered species protection policy that this model is omitting? How might you modify the model to take it/them into account?

17.2. You are considering which of two wildlife habitats to protect from development, one a coastal estuary and the other a deciduous forest. You only have the funds to protect one. You want to apply the Weitzman framework to decide which one to protect. Come up with a list of specific questions that you would want to answer in order to help you decide.

17.3. Suppose you are the FWS. How might you reform the policy of setting aside habitat in order to avoid, or at least reduce, preemptive habitat destruction?

17.4. Critically evaluate the fortress approach and the participatory approach to managing wild elephant herds. What are the strengths and weaknesses of each? Can you identify conditions under which one approach might be preferred as a policy over the other? What might those conditions be? And what sorts of measures might be successful in increasing "buy-in" of local communities to the participatory approach?

17.5. Show that the desirability of an international ivory trade ban depends upon the price elasticity of demand for ivory and ivory products. Explain.

17.6. Consider Kremer and Morcom's proposed policy of governments accumulating a stockpile of ivory and releasing it onto markets if their elephant populations get too small. Why not adopt a policy releasing it onto markets if the price of ivory gets too high? In principle, this should also discourage poaching. What would be the relative merits of the two approaches?

17.7. Consider the evidence in Table 17.1 regarding which species receive the most endangered species protection in the United States. Can you provide a compelling public choice story for why these ten species received over half of all spending on endangered species protection? What evidence would you try to collect to support your story?

# Notes

1 CITES, cites.org.
2 See Weitzman (1993) for a rigorous attempt at operationalizing the notion of diversity.
3 For more on these examples of preemptive harvesting, see Brown and Shogren (1998), p. 7; Lueck and Michael (2003), p. 30.
4 Bulte and van Kooten (1996), p. 433. It is challenging to obtain precise numbers of wild elephants. The IUCN's African Elephant Specialist Group breaks down elephant populations into four categories: definitive, probable, possible, and speculative. In 1996, for example, it estimated that there were nearly 580,000 wild African elephants, of which nearly 200,000 fell into the "possible" and "speculative" categories. See Bulte and Van Kooten (1999), p. 453.
5 See, for example, Child (1993); Matzke and Nabane (1996).
6 *Importance of committee representation*: Rawls and Laband (2004); *Environmentally friendly oversight committees*: Harlee et al. (2009).

# Further readings

## I. Ivory trade

Kaempfer, William H. and Anton D. Lowenberg. "The ivory bandwagon: International transmission of interest-group politics," *Independent Review* 4(Fall 1999): 217–39.
*Argues that relatively small interest groups can have disproportionate impacts on policy through a kind of contagion effect on the positions of other interest groups. Also argues that this helps explain the establishment of the international ban on the ivory trade.*

## II. Public choice and the Endangered Species Act.

Harllee, Bonnie; Myungsup Kim; and Michael Nieswiadomy. "Political influence on historical ESA listings by state: a count data analysis," *Public Choice* 140(2009): 21–42.
*Examines the effect that congressional representation on the subcommittee that oversees the US Fish and Wildlife Service has on listings of endangered species for the state. Finds evidence that greater (less) representation by pro-environment legislators results in more (fewer) species listings.*
Lopez, Edward J. and Daniel Sutter. "Ignorance in congressional voting? Evidence from policy reversal on the Endangered Species Act," *Social Science Quarterly* 85(December 2004): 891–912.
*Examines the initial passage of the ESA in 1973 and its subsequent weakening five years later in the Tennessee Valley Authority case; investigates and rejects the hypothesis that congresspersons may have been ignorant of the full implications of the ESA when initially passed. Instead find that the subsequent congressional reversal occurred as a result of an influx of conservative Democrats in the intervening time.*

# References

Ando, Amy Whritenour. "Waiting to be protected under the Endangered Species Act: The political economy of regulatory delay," *Journal of Law and Economics* 42(April 1999): 29–60.
Breed, Allen G. "Town, woodpeckers fight over nest eggs," *NBCNews.com*, September 26, 2006.
Brown, Gardner M. and Jason F. Shogren. "Economics of the Endangered Species Act," *Journal of Economic Perspectives* 12(Summer 1998): 3–20.
Bulte, Erwin H. and G. Cornelis van Kooten. "A note on ivory trade and elephant conservation," *Environment and Development Economics* 1(1996): 433–43.
Bulte, Erwin H. and G. Cornelis van Kooten. "Economics of antipoaching enforcement and the ivory trade ban," *American Journal of Agricultural Economics* 81(May 1999): 453–66.
Child, Brian. "Zimbabwe's CAMPFIRE programme: using the high value of wildlife recreation to revolutionize natural resource management in communal areas," *Commonwealth Forestry Review* 72(1993): 284–96.

Convention on International Trade in Endangered Species of Wild Fauna and Flora [CITES]. cites.org.

CITES. "New report highlights continued threat to African elephants from poaching," May 10, 2019. https://www.cites.org/eng/news/new-report-highlights-continued-threat-to-african-elephants-from-poaching_10052019

Evans, Douglas; András Demeter; Peter Gajdoš; and Luboš Halada. "Adapting environmental conservation legislation for an enlarged European Union: Experience from the Habitats Directive," *Environmental Conservation* 40(June 2013): 97–107.

Frost, Peter and Ivan Bond. "The CAMPFIRE programme in Zimbabwe: payments for wildlife services," *Ecological Economics* 65(April 2008): 776–87.

Haller, Tobias; Marc Galvin; Patrick Meroka; Jamil Alca; and Alex Alvarez. "Who gains from community conservation? Intended and unintended costs and benefits of participative approaches in Peru and Tanzania," *Journal of Environment and Development* 17(June 2008): 118–44.

Harllee, Bonnie; Myungsup Kim; and Michael Nieswiadomy. "Political influence on historical ESA listings by state: A count data analysis," *Public Choice* 140(July 2009): 21–42.

Hilborn, Ray et al. "Effective enforcement in a conservation area," *Science New Series* 314(November 24, 2006): 1266.

International Union for Conservation of Nature (IUCN). *Red List 2020*, version 2020-2.

Kremer, Michael and Charles Morcom. "Elephants," *American Economic Review* 90(March 2000): 212–34.

Lueck, Dean and Jeffrey A. Michael. "Preemptive habitat destruction under the Endangered Species Act," *Journal of Law and Economics* 46(April 2003): 27–60.

Matzke, Gordon Edwin and Nontokozo Nabane. "Outcomes of a community controlled wildlife utilization program in a Zambezi Valley community," *Human Ecology* 24(1996): 65–85.

Mehmood, Sayeed R. and Daowei Zhang. "A roll call analysis of the Endangered Species Act Amendments," *American Journal of Agricultural Economics* 83(August 2001): 501–12.

Metrick, Andrew and Martin L. Weitzman. "Patterns of behavior in endangered species protection," *Land Economics* 72(February 1996): 1–16.

Metrick, Andrew and Martin L. Weitzman. "Conflicts and choices in biodiversity preservation," *Journal of Economic Perspectives* 12(Summer 1998): 21–34.

Nelson, Fred; Rugemeleza Nshala; and W.A. Rodgers. "The evolution and reform of Tanzanian wildlife management," *Conservation & Society* 5(2007): 232–61.

Rawls, R. Patrick and David N. Laband. "A public choice analysis of endangered species listings," *Public Choice* 121(October 2004): 263–77.

Smith, David. "Execute elephant poachers on the spot, Tanzanian minister urges," *Guardian*, October 8, 2013.

United Nations. Intergovernmental Science-Policy Platform on Biodiversity and Ecosystem Services (IPBES). *Global assessment report on biodiversity and ecosystem services*. 2019.

Weitzman, Martin L. "What to preserve? An application of diversity theory to crane conservation," *Quarterly Journal of Economics* 108(February 1993): 157–83.

Weitzman, Martin L. "The Noah's Ark problem," *Econometrica* 66(November 1998): 1279–98.

Wittemeyer, George et al. "Illegal killing for ivory drives global decline in African elephants," *PNAS* 36(September 9, 2014): 13117–21. https://www.pnas.org/content/111/36/13117

Yeoman, Barry. "Why the passenger pigeon went extinct," *Audubon*. May/June, 2014.

# 18 Conclusion
## Institutional environmental economics

We have covered a lot of material in this textbook, so let me try to make sure that the most important lessons are clear. And let us begin with a key point of departure: the way that standard economics treats many environmental issues, like air and water pollution, hazardous waste disposal, agricultural runoff, and climate change. All of these are modelled as so-called negative externalities, which commonly give rise to prescriptions for government interventions such as performance standards, quotas, emissions taxes, or carbon taxes. The idea is to achieve greater efficiency either by constraining actions, or providing private parties with financial disincentives, to keep them from overproducing various outputs. This way, resources are devoted to their highest-value uses. All laudable and worthy goals.

### The normative vs. the positive view of government

There are, however, several ways in which this approach falls short. I guess the point I want to stress the most is the distinction between *ideal* government policy and the policy that is actually likely to appear from the political process. Here, when I say ideal, I am using the economist's primary criterion: efficiency. From a purely economic point of view, for example, a carbon tax set at the appropriate level is a good example of what many economists would consider an "ideal" government policy for addressing climate change. We can point to good theoretical reasons for its adoption, and these are certainly useful and important insights. This is the *normative* view of government: what the government *should be doing* to address environmental problems.

However, this view is incomplete at best because it does not tell us what the government is *actually going to do* (the *positive* view of government). It is not at all clear that governments will, as a matter of course, behave like the models say they should in order to promote outcomes that are efficient, fair, sustainable, or whatever other criterion you may value. And you do not even need to go all the way back to the Chinese Cultural Revolution or the Stalinist policies of the mid-20th century Soviet Union, let alone 16th century Spain, to convince yourself this is true. Just consider the numerous present-day examples of corrupt or ineffective governments in various countries that lack effective rules of law. Or just consider the many examples of ineffective or inefficient environmental policies in the United States that we have discussed in this textbook.

Much more likely is that politics will dictate government policies. Sure, politics driven in part by economic considerations, but politics nonetheless. So, rather than getting policies that effectively and efficiently address environmental problems, we get policies that reflect the demands of interest groups like fossil-fuel energy producers, eastern and Midwestern coal-producing states, commercial fishing interests, and real estate developers. To name but a few.

There are a couple of important implications here. One is a matter of interpretation. The perspective that many (most?) policies derive ultimately from special interest demands should perhaps make us more cautious about recommending that the government take any action at all. Because once we start down that road, special interests may exert influence in order to swing the outcomes in their favor. As we have seen, they might be able to do this in the initial crafting of the policy, or later on by influencing how the policy is implemented over time. In either case, we may well end up with policies that serve not the general pressing needs of society but, rather, the narrow interests of certain groups or organizations.

I am, of course, not arguing that the government should *never* undertake environmental policy. Rather, I am saying that we need to take the positive view of government seriously. Studying how the government works, and what economic factors determine the likely political outcomes, serves an important interpretive function. It helps us understand why we get the policies we do and the objectives that those policies serve. It is always best to be as clear-eyed as we can about the way the world really works, and environmental policy is no exception.

I believe there is a second important implication, which has to do with the way that economists ply their trade. Regarding the environment, economists often prescribe policies based on the normative view of government. We prescribe, for example, carbon taxes because of their undeniable efficiency properties. But what often happens is that economists, in pushing our theoretically ideal policies, hit a political brick wall. That is, the policies we prescribe, like carbon taxes or cap-and-trade policies, are simply unable to gain any political traction. This does not always happen. For example, we saw in chapter ten that a perfect storm of political events made it possible to enact the market-based Clean Air Act Amendments in 1990. But for every success story like that, we can probably point to ten other cases where economically supported policies go nowhere.

None of this is to argue that we should throw away our tools of analysis or just give up. Rather, we may want to consider acknowledging political realities. This might mean that we do not necessarily go for the ideal policy but, rather, the best we can get, at least for the time being when the political headwinds are fierce. As the saying goes, sometimes the perfect can be the enemy of the good. At the very least, we should not be surprised or gnash our teeth when our favored policy prescriptions go nowhere.

A challenge in doing this, of course, is that it is often unclear what is politically feasible and what is not. We are all probably pretty good at recognizing lost causes, but the question is: How lost does it have to be before we settle for less? Obviously, there is no magic answer. However, the discussion in this book does provide clues and a bit of guidance. Though I have tried to stress that it has certain predictive limitations, the public choice approach can tell us much. For example, it can tell us when it is more and less likely that the policy process will be captured by special interests. The existence of political entrepreneurs who galvanize public opinion – say, against taxes – may place greater obstacles in the way of ideal policies like carbon taxes. Party politics and who happens to be "in charge" at the time may militate against – or in favor of – certain types of policies. All of these political factors may be worthy of consideration when considering what policies to prescribe.

## Transaction costs and environmental policy: Coase expanded

For a more systematic way to bring political factors into the picture, let us think about all this through the lens of transaction costs. In the basic world of Coase, transaction costs play a key role in determining the efficiency and equity of outcomes involving transactions. However, for studying policy, the basic Coase framework does not go far enough because it does not consider the transaction costs involved in assigning property rights in the first place. Here again is where the institutional approach provides guidance. Because the assignment of property rights occurs in a previous stage where, for example, courts make rulings and legislatures enact statutes. And this previous stage can involve significant transaction costs as well.

Consider, for example, the transaction costs associated with the operation of Congress or Parliament. All the negotiations over support for bills, the logrolling, the committee infighting, the monitoring challenges of the committee oversight process, the rent-seeking of interest groups doing their worst to try to obtain favorable policies, the negotiation costs that take place with interest groups (especially heterogeneous ones), and so forth. All of this suggests that the transaction costs of the legislative process could be enormous. Any analysis that omits this first stage is in danger of missing an important piece of the picture and, possibly, a major source of inefficiency and resource misallocation.

We first encountered this notion in chapter six, when we drew a distinction between *policymaking* transaction costs and *policy operation* transaction costs (see Figure 6.3). In the context of environmental policy, policy operation transaction costs are the transaction costs that Coase was concerned with: the costs associated with private negotiations over transactions, maintaining property rights, and so forth. Policymaking transaction costs are the costs associated with assigning property rights in the first place, which include rent-seeking costs, the negotiating costs of interest groups, the costs of enacting legislation in Congress or Parliament, and so forth. Whenever we have delved into public choice matters in each of the topic chapters, we were opening the black box of property rights assignment. In traditional economic analysis, that does not consider public choice matters, this black box is largely kept shut.

When we think about policy in this way, there may well be a tradeoff between the two types of transaction costs. Policies that have low policy operation transaction costs may have extremely high policymaking transaction costs, and the other way round. So imagine setting up a smoothly running water market, or a market for tradable emissions permits, or a local system of transferable fishing quotas. Just focusing on policy operation transaction costs may make these policies look quite good, leading to prescriptions to establish them in local watersheds, airsheds, or fisheries. However, expanding the focus to include policymaking transaction costs may well make them less appealing on overall cost grounds. This is not to argue that we should not pursue these policies, it is merely an exhortation to look at costs holistically. Doing so may paint a different picture than just focusing on the transaction costs that Coase emphasized.

As another example, consider an emissions tax. As we have seen, many economists favor this policy, on the theoretical grounds that it results in cost-efficient abatement of emissions. This conclusion is generally based on an analysis only of the policy operation stage, which typically considers the transaction costs of that stage alone, if it does at all. However, levying an emissions tax may be a political non-starter, making it impossible for a legislature to enact it in the first place. Taking into account potentially large policymaking transaction costs, we may prefer to consider an alternative policy with larger policy operation transaction costs because its political feasibility makes it more likely to be adopted.

The bottom line of this transaction cost discussion is this: in a world of scarce resources, it is important to try to minimize the costs, in terms of resources, of achieving policy goals. But transaction costs are real costs and they come in every form and at all stages of the process: everywhere from the political battles over enacting policy in the first place all the way to the operation of the policy in the real world. Ignoring any aspect of these transaction costs may lead us to prescribe policies that do not minimize overall costs. The result may be the enactment of inefficient policies.

## No panaceas

Another key takeaway of the discussion in this textbook concerns the discussion of common-pool resources that started in chapter five and extended over the last five substantive chapters of the textbook. I presume you heard loud and clear the message that tragedies of the commons are not inevitable. However, that is not the takeaway message to which I am referring. Rather, I am referring to the way we interpret the institutional arrangements themselves.

What I mean is, if you think about the standard economic model of externalities, it seems to be portraying two possible worlds: freely operating, unregulated markets, and government intervention. That is, either we have one or we have the other. In this view, you can imagine *either* a world where polluting steel firms inefficiently overproduce steel and create social deadweight loss *or* a world where a government steps in and imposes a policy – an emissions tax, performance standards, quotas, and the like – in order to correct the market failure. There is no question that this approach has provided important insights. However, it may present a false dichotomy.

One lesson of the discussion of CPRs and the SES framework is that these are not the only two possible worlds. In a third possible world, local communities self-organize to create local governance structures to manage local resources on their own. We saw this occur, for example, with local water resources, fisheries, forests, and threatened species. You can imagine it occurring with grazing commons and other local resources. In none of these cases is a central government stepping in and imposing a policy. Yet in many cases, successful management occurred that was able to avert a tragedy of the commons. We still have a long way to go in terms of really understanding the constellation of factors that will enable successful local management. But we are now keenly aware of the possibility of this third way, and much study is being devoted to trying to improve it as a practical tool for improved resource management.

The reason I labeled this subsection *No panaceas* is that it is something that Elinor Ostrom used to say. She was referring to the tendency for some scholars and policymakers to view policy in black and white terms. For example, some argue that the solution to resource allocation problems is to rely on freely operating markets. Others argue that the solution is for the government to step in and regulate. This is, of course, exactly the dichotomy suggested by the standard externalities model. When Ostrom was saying "No panaceas," she was speaking against the tendency to assume that one or the other approach is always *the solution*. This is the tendency to view one solution as a cure-all, the only suitable path forward.

It seems clear from Ostrom's writings that she worked so hard on her approach at least partly in order to provide a counterweight to these perspectives, which she viewed as simplistic, especially within the context of CPRs. No, she would say, we cannot always rely on individual private actions to address environmental problems. Nor do we need the government to step in at every opportunity. In her view, many environmental problems can be addressed through collaboration, communication, and people being invested in the health of their communities and the natural resources on which this health depends. I never heard her actually say this, but I can easily envision her pleading: Yes, there may be limits to this third way. It will not solve all of our environmental problems. But isn't this vision compelling enough to merit as much attention as we can provide to see how well it can work? Only time, and hard work by present and future scholars, will enable us to fully assess the value of this approach.

## Where to now?

In a way, the approach of this textbook reflects some recent tendencies in scholarly study of the environment, which acknowledge the complexity of many environmental issues and the potential importance of various disciplinary perspectives. This complexity was an important factor in Ostrom's conclusion that no one approach was always going to work. Many economists are now engaging in collaborations with natural scientists and colleagues in other social sciences to tackle various environmental problems. Just to take one example, the *Natural Capital Project* at Stanford University advertises itself as an interdisciplinary team of academics – including economists – and other professionals working to develop better ways to place values on natural resources and amenities, what it calls natural capital. And some economists, including Nordhaus himself, who used to be quite partial to simple models, are now coming to recognize the importance and practical usefulness of modeling the complexity of natural systems and their connection to the economy. This trend is likely to continue into the future.

The institutional approach of this textbook is also an attempt to expand the focus of economics beyond its traditional domain. Though I have not managed to describe all of the connections in this book, this general approach brings in matters of law and political science, as well as cognitive psychology and other social sciences. These other disciplines are all relevant to the extent they provide insights into the relationship between economic actors and the institutional environment in which they operate. Many economists have found this broader approach to be fruitful in providing new and useful insights into economic behavior. It has also expanded the set of action arenas in which economic behavior can be fruitfully studied, such as within interest groups and, more generally, in the political sector. Both of these have proven to be extremely useful in study of the environment and environmental policy.

In the end, any approach to the environment should be judged by how well it supports our ability to provide sensible solutions to pressing environmental problems. My main objective in writing this textbook has been to provide you with a broad and nuanced understanding of the economic approach to studying the environment. To accomplish this, I have: (a) provided you information on a wide range of environmental issues and topics, and (b) presented a wide variety of economic concepts and tools of analysis. I hope you will find all of this useful for studying whatever issues you find most urgent, interesting, or puzzling. I also hope I have conveyed my sense for why it is important to consider the institutional environment in which environmental behavior takes place and environmental policy gets enacted. But, most of all, I hope all of this has whetted your appetite to learn more so that, ultimately, you may one day be able to make important contributions of your own to addressing the pressing and increasingly urgent environmental issues of our time.

# Index

Printed and bound by CPI Group (UK) Ltd, Croydon, CR0 4YY

23/10/2024

01778256-0001